THE DRAKE EQUATION

Estimating the Prevalence of Extraterrestrial Life through the Ages

In this compelling book, leading scientists and historians explore the Drake Equation, which guides modern astrobiology's search for life beyond Earth. First used in 1961 as the organizing framework for a conference in Green Bank, West Virginia, it uses seven factors to estimate the number of extraterrestrial civilizations in our galaxy. Using the equation primarily as a heuristic device, this engaging text examines the astronomical, biological, and cultural factors that determine the abundance or rarity of life beyond Earth and provides a thematic history of the search for extraterrestrial life.

Logically structured to analyze each of the factors in turn, and offering commentary and critique of the equation as a whole, contemporary astrobiological research is placed in a historical context. Each factor is explored over two chapters, discussing the pre-conference thinking and a modern analysis, to enable postgraduates and researchers to better assess the assumptions that guide their research.

DOUGLAS A. VAKOCH is Director of Interstellar Message Composition at the SETI Institute and Professor of Clinical Psychology at the California Institute of Integral Studies. He also serves as Chair of the International Academy of Astronautics Study Group on Interstellar Message Construction and has edited numerous books in the field of astrobiology and space exploration.

MATTHEW F. DOWD is an editor at the University of Notre Dame Press and teaches periodically in the Program of Liberal Studies. He has been instrumental in organizing Notre Dame's biennial History of Astronomy conferences and in 2011 received the Adler-Mansfield Prize for contributions to the history of astronomy.

Cambridge Astrobiology

Series Editors

Bruce Jakosky, Alan Boss, Frances Westall, Daniel Prieur, and Charles Cockell

Books in the series

1. Planet Formation: Theory, Observations, and Experiments
   Edited by Hubert Klahr and Wolfgang Brandner
   ISBN 978-0-521-18074-0

2. Fitness of the Cosmos for Life: Biochemistry and Fine Tuning
   Edited by John D. Barrow, Simon Conway Morris, Stephen J. Freeland, and
   Charles L. Harper ISBN 978-1-107-40655-1

3. Planetary Systems and the Origins of Life
   Edited by Ralph Pudritz, Paul Higgs, and Jonathon Stone
   ISBN 978-1-107-41286-6

4. Exploring the Origin, Extent, and Future of Life
   Philosophical, Ethical, and Theological Perspectives
   Edited by Constance M. Bertka ISBN 978-0-521-86363-6

5. Life in Antarctic Deserts and Other Cold Dry Environments
   Edited by Peter T. Doran, W. Berry Lyons, and
   Diane M. McKnight ISBN 978-0-521-88919-3

6. Origins and Evolution of Life: An Astrobiological Perspective
   Edited by Muriel Gargaud, Purificación Lopez-Garcia, and
   Hervé Martin 978-0-521-76131-4

7. The Astrobiological Landcape: Philosophical Foundations
   of the Study of Cosmic Life Milan M Ćirković ISBN 978-0-521-19775-5

8. The Drake Equation: Estimating the Prevalence of
   Extraterrestrial Life through the Ages
   Edited by Douglas A. Vakoch and Matthew F. Dowd
   ISBN 978-1-107-07365-4

THE DRAKE EQUATION

Estimating the Prevalence of Extraterrestrial Life through the Ages

Edited by

DOUGLAS A. VAKOCH
SETI Institute

MATTHEW F. DOWD
University of Notre Dame Press

CAMBRIDGE
UNIVERSITY PRESS

CAMBRIDGE
UNIVERSITY PRESS

University Printing House, Cambridge CB2 8BS, United Kingdom

Cambridge University Press is part of the University of Cambridge.

It furthers the University's mission by disseminating knowledge in the pursuit of education, learning and research at the highest international levels of excellence.

www.cambridge.org
Information on this title: www.cambridge.org/9781107073654

First published 2015

A catalogue record for this publication is available from the British Library

Library of Congress Cataloguing in Publication data
The Drake equation : estimating the prevalence of extraterrestrial life
through the ages / edited by Douglas A. Vakoch, SETI Institute, Matthew F. Dowd,
University of Notre Dame Press.
pages cm. – (Cambridge astrobiology; 8)
ISBN 978-1-107-07365-4 (Hardback)
1. Extraterrestrial beings. 2. Life on other planets. I. Vakoch, Douglas A. II. Dowd, Matthew F.
QB54.D73 2015
576.8′39–dc23 2014050301

ISBN 978-1-107-07365-4 Hardback

To Amy Dowd, for her patience and support
To Joe Castrovinci, for his hard work and dedication

Contents

Contributors

Danielle Briot, PhD, is an astronomer at the Paris–Meudon Observatory. She first specialized in the spectroscopy of Be stars, which are hot stars with emission lines originating in a circumstellar envelope. She has also studied binary stars. She has been studying these celestial objects since soon after the discovery of the first extrasolar planets. She investigates the search for life in the universe, researching indicators of chlorophyll in Earthshine spectra. Dr. Briot is also working on the history of astrobiology, and she is especially interested in the discovery of unknown pioneers in this science. Because she thinks that the diffusion of science is important and wants to share her enthusiasm about astronomy, a part of her time is dedicated to conferences, talks, and articles in popular astronomy magazines. For over a decade, she has gone into prisons to speak about astronomy by giving courses and lectures for inmates. She is the primary author of the book *Ce que disent les étoiles*.

Garry Chick, PhD, is a professor and former head of the Department of Recreation, Park, and Tourism Management as well as a professor in the Department of Anthropology at the Pennsylvania State University. He received his doctorate in anthropology from the University of Pittsburgh in 1980, and his primary research interests are in expressive culture, including leisure and health, adult play and games, and research methods. Dr. Chick has conducted field research in Mexico, the northeastern United States, mainland China, and Taiwan. He is a past president of the Association for the Study of Play and the Society for Cross-Cultural Research. In addition to Penn State, he has taught at the University of Illinois at Urbana–Champaign, where he received fellowships to the Center for Advanced Study and the Center for the Study of Ethics and Values, the University of Wageningen in the Netherlands, and the University of Costa Rica. He was the founding editor of the journal *Play & Culture* and is a past editor of *Leisure Sciences*.

Michael J. Crowe, PhD, is the Rev. John J. Cavanaugh, C.S.C. Professor Emeritus in the Program of Liberal Studies and Graduate Program in History and Philosophy of Science at the University of Notre Dame. Among his nine books, the two most relevant to this volume are *The Extraterrestrial Life Debate 1750–1900: The Idea of a Plurality of Worlds from Kant to Lowell*, which has appeared in a three-volume Japanese translation, and his *The Extraterrestrial Life Debate, Antiquity to 1915: A Source Book*. In 2010, Crowe received the biennial LeRoy E. Doggett Prize from the American Astronomical Society for his lifetime contributions to the history of astronomy. He has developed a course titled "The Extraterrestrial Life Debate: A Historical Perspective." He is the founder of Notre Dame's Graduate Program in History and Philosophy of Science and of its Biennial History of Astronomy conferences.

Paul Davies, PhD, is Regents' Professor and Director of the Beyond Center for Fundamental Concepts in Science at Arizona State University. His research interests focus on the "big questions" of existence, ranging from the origin of the universe to the origin of life, and include the nature of time, the search for life in the universe, and foundational questions in quantum mechanics. In astrobiology, he champions the theory that Earth may host a shadow biosphere of alternative life forms. His twenty-eight popular and specialist books have been translated into more than twenty languages and are notable for presenting complex ideas in accessible terms. Among his best sellers are *The Mind of God*, *About Time*, *How to Build a Time Machine*, *The Fifth Miracle*, and *The Goldilocks Enigma*. His latest book, *The Eerie Silence*, is about the search for intelligent life in the universe. Dr. Davies has won many awards including the Templeton Prize, the Kelvin Medal from the UK Institute of Physics, and the Michael Faraday Prize from the Royal Society for promoting science to the public.

David Des Marais, PhD, is a senior space scientist at the NASA Ames Research Center. His long-term research interests are the biogeochemistry of carbon, the early evolution of the biosphere, and Mars exploration. His areas of specialization have included the stable isotope geochemistry of carbon in lunar samples, meteorites, and oceanic basalts; the biogeochemistry of microbial communities in extreme environments; and the biogeochemistry of ancient (Precambrian) carbonates and organic matter. Dr. Des Marais has published more than 180 research articles, and he serves on the editorial boards of the journals *Astrobiology* and *Geobiology*. He is a participating scientist with the Mars Exploration Rover, Mars Reconnaissance Orbiter, and Mars Science Laboratory missions. Dr. Des Marais was elected fellow of the Geochemical Society, the European Association of Geochemistry, the American Geophysical Union, and the International Society for the Study of the Origins of Life. He has received NASA awards for science and leadership. The Geochemical Society recently awarded him the Alfred Treibs Medal for career achievement in Organic Geochemistry.

David DeVorkin, PhD, is a senior curator in the Space History Division at the Smithsonian Institution's National Air and Space Museum. He holds a PhD in the history of astronomy from the University of Leicester and a Master of Philosophy in astronomy from Yale University. In 2008 he was awarded the LeRoy E. Doggett Prize of the American Astronomical Society for his work on the origins and development of modern astrophysics and the origins of the space sciences in the twentieth century. Dr. DeVorkin has published more than one hundred scholarly and popular articles, and he is author or editor of nine books including *The Hubble Space Telescope: New Views of the Universe*; *Beyond Earth: Mapping the Universe*; *Henry Norris Russell: Dean of American Astronomers*; and *The American Astronomical Society's First Century*. He curated the major museum exhibitions *Stars*; *V-2: The World's First Ballistic Missile System*; and *Explore the Universe*.

Steven J. Dick, PhD, is the Baruch S. Blumberg NASA/Library of Congress Chair in Astrobiology. He has been the Charles A. Lindbergh Chair in Aerospace History at the National Air and Space Museum, and from 2003 to 2009 he served as the NASA Chief Historian and Director of the NASA History Office. Among his books are *Plurality of Worlds: The Origins of the Extraterrestrial Life Debate from Democritus to Kant*; *The Biological Universe: The Twentieth Century Extraterrestrial Life Debate and the Limits of Science*; *Life on Other Worlds*; *The Living Universe: NASA and the Development of Astrobiology*; *Many Worlds: The New Universe, Extraterrestrial Life, and the Theological Implications;* and *Discovery and Classification in Astronomy: Controversy and Consensus*. In 2006, Dr. Dick received the LeRoy E. Doggett Prize from the American Astronomical Society for a career that has significantly influenced the field of the history of astronomy. In 2009, minor planet 6544 Stevendick was named in his honor.

Matthew F. Dowd, PhD, received his doctorate in the history and philosophy of science from the University of Notre Dame in 2004. His dissertation examined the astronomical and computistical works of Robert Grosseteste, an English scholar of the twelfth and thirteenth centuries. He has since worked as an editor at the University of Notre Dame Press, and periodically teaches courses for Notre Dame's undergraduate Program of Liberal Studies. He has co-taught with Michael Crowe a course titled "The Extraterrestrial Life Debate: A Historical

Perspective." He has been instrumental in organizing Notre Dame's Biennial History of Astronomy conferences, and in 2011 received the Adler–Mansfield Prize for contributions to the history of astronomy for his work on the conferences.

Frank Drake, PhD, is Professor Emeritus of Astronomy and Astrophysics at the University of California, Santa Cruz, where he was previously Dean of the Natural Sciences Division. He is also Chairman Emeritus of the Board of Trustees of the SETI Institute, as well as former Director of the Carl Sagan Center for the Study of Life in the Universe at the SETI Institute. In 1960, Dr. Drake conducted Project OZMA, the first organized search for radio signals from extraterrestrial intelligence. The following year he devised the widely known Drake Equation, giving an estimate of the number of communicative extraterrestrial civilizations that we might find in our galaxy. In 1974, he constructed and transmitted the Arecibo interstellar message. Three additional messages were sent to outer space using the techniques and methods he developed: the Pioneer 10 and 11 plaques and the Voyager interstellar recording. He is author of *Is Anyone Out There?* (with Dava Sobel).

David Dunér, PhD, is Professor of History of Science and Ideas at Lund University, Sweden, and researcher at the Centre of Cognitive Semiotics at Lund University. His research concerns the development of science, medicine, mathematics, and technology during the scientific revolution and onward. His latest book is *The Natural Philosophy of Emanuel Swedenborg: A Study in the Conceptual Metaphors of the Mechanistic World-View*. Dr. Dunér studies the history of astrobiology, and he was guest editor of the special issue "The History and Philosophy of Astrobiology" of the journal *Astrobiology*. His astrobiological research includes "astrocognition," the study of cognition in space, detailed in his chapter "Astrocognition: Prolegomena to a Future Cognitive History of Exploration," which appeared in *Humans in Outer Space – Interdisciplinary Perspectives*. Dr. Dunér's work on the semiotic and cognitive foundations of interstellar communication appeared in the volumes *Communication with Extraterrestrial Intelligence* and *Extraterrestrial Altruism*.

Patrick François, PhD, is an astrophysicist at the Paris–Meudon Observatory. He specializes in the chemical composition of very old stars, witnesses of the early evolution of the galaxy. He also studies globular clusters and dwarf spheroidal galaxies. He has worked on theoretical aspects of models of galactic chemical evolution. Dr. François was among the first to confirm the existence of a planet orbiting the star 51 Pegasi. He worked several years at the European Southern Observatory (ESO) in Chile as support astronomer and team leader, and he has been involved in the development of new instruments for the ESO, in particular high-resolution spectrographs. He is currently involved in a large spectroscopic survey of stars using the *Gaia* spacecraft, launched by the European Space Agency in 2013.

Chris Impey, PhD, is a university distinguished professor and deputy head of the Department of Astronomy at the University of Arizona. His research is on observational cosmology, gravitational lensing, and the evolution of galaxies. He has over 170 refereed publications and 60 conference proceedings, and his work has been supported by $20 million in grants from NASA and the NSF. Dr. Impey is past Vice President of the American Astronomical Society and has been NSF Distinguished Teaching Scholar, a Phi Beta Kappa Visiting Scholar, and the Carnegie Council on Teaching's Arizona Professor of the Year. He has written thirty popular articles on cosmology and astrobiology, two introductory textbooks, and three popular books: *The Living Cosmos*, *How It Ends*, and *How It Began*. He was a panel co-chair for the Astronomy Decadal Survey of the National Academy of Sciences. In 2009, he was elected a Fellow of the American Association for the Advancement of Science.

Lori Marino, PhD, is a neuroscientist and expert in animal behavior and intelligence who was on the faculty at Emory University for nineteen years. She is internationally known for her work comparing the evolution of dolphin and whale brains with primate brains and has worked with the astrobiology and SETI community for two decades. Presently, she is the Executive Director of the Kimmela Center for Animal Advocacy focused on applying scientific data and expertise to animal advocacy issues such as marine mammal captivity, nonhuman primate rights, and elephant protection. Dr. Marino has authored more than one hundred peer-reviewed papers, book chapters, and magazine articles on marine mammal biology and cognition, comparative neuroanatomy, self-awareness in other animals, human–nonhuman animal relationships, and animal welfare and the captivity industry. Dr. Marino was recently featured as a National Geographic Innovator and appears in several films and television programs on the above topics, including the influential documentary *Blackfish*, about killer whale captivity. She currently teaches online courses in animal behavior and evolution.

Florence Raulin Cerceau, PhD, is Associate Professor in the Department of Men, Nature, and Societies at the National Museum of Natural History (MNHN) of Paris, France. From 1988 to 1997, she was in charge of the preparation of several exhibitions at the MNHN. In particular, she collaborated on the preparation of the *Grande Galerie de l'Evolution*, which opened in 1994. Since 1997, she has been a researcher at the Centre Alexandre Koyré (UMR 8560 – EHESS/CNRS/MNHN), a Center for the History of Science and Techniques partner of the MNHN. Her research is focused on the history and epistemology of exobiology and astrobiology. Her current interests deal with the concept of planetary habitability and the first attempts to contact planets beginning in the nineteenth century. Among her books are *A l'écoute des planètes, Les origines de la vie: histoire des idées,* and *D'autres planètes habitées dans l'univers?*

Jean Schneider, PhD, is Senior Researcher in the Laboratory for the Universe and Theory (LUTh) at Paris Observatory. He joined Paris Observatory after five years of high-energy physics research at Orsay University (France) and CERN (Switzerland). For fifteen years, Dr. Schneider focused on relativistic astrophysics, and he was the first to measure the correct time delay in the lensed double quasar 0957+561. In 1988, he turned his attention toward exoplanets and exobiology. He was the promoter of the search for transiting exoplanets with the CoRoT spacecraft, which discovered more than thirty-five planets and made the first measurement of a super-Earth, CoRoT-7 b. He was one of the main promoters of the European Space Agency (ESA) Darwin space mission project and principal investigator of the ESA SEE (Super-Earth Explorer) proposal. Dr. Schneider was the first to propose the detection of exomoons by the timing of exoplanet transits and, along with several other exoplanet studies, was the organizer of the first spectroscopic detection of terrestrial vegetation in the Moon's Earthshine. Having a substantial academic background in philosophy, he also devotes part of his research to the concept of time, to language theory, to the definition of life, and to the foundations of scientific concepts through a new discipline called epistemo-analysis.

Seth Shostak, PhD, is the Senior Astronomer and Director of the Center for SETI Research at the SETI Institute, and host of the weekly radio program *Big Picture Science*. He earned his undergraduate degree in physics from Princeton University, and his PhD in astronomy from the California Institute of Technology. His books include *Sharing the Universe: Perspectives on Extraterrestrial Life*, *Cosmic Company: The Search for Life in the Universe*, and *Confessions of an Alien Hunter: A Scientist's Search for Extraterrestrial Intelligence.* Dr. Shostak is a Fellow of the Committee for Skeptical Inquiry, and he has served as chair of the SETI Committee of the International Academy of Astronautics. In 2004, he won the Astronomical Society of the Pacific's Klumpke-Roberts Award in recognition of his contributions to the public's understanding and appreciation of astronomy.

Stéphane Tirard, PhD, is Professor in History of Science and Director of the Centre François Viète d'Epistémologie et d'Histoire des Sciences et des Techniques (EA 1161) at the University of Nantes. His research interests include the epistemology and history of theories of the origins and limits of life. He is particularly interested in the epistemology of interdisciplinary fields. Tirard served as the field editor in epistemology for the *Encyclopedia of Astrobiology*, and he has published *Histoire de la vie latente*.

Douglas A. Vakoch, PhD, is Director of Interstellar Message Composition at the SETI Institute, as well as Professor in the Department of Clinical Psychology at the California Institute of Integral Studies. He serves as chair of both the International Academy of Astronautics (IAA) Study Group on Interstellar Message Construction and the IAA Study Group on Active SETI: Scientific, Technical, Societal, and Legal Dimensions. He has edited several books including *Astrobiology, History, and Society: Life Beyond Earth and the Impact of Discovery*; *Communication with Extraterrestrial Intelligence (CETI)*; *Extraterrestrial Altruism: Evolution and Ethics in the Cosmos*; *Archaeology, Anthropology, and Interstellar Communication*; *Psychology of Space Exploration: Contemporary Research in Historical Perspective*; and *Feminist Ecocriticism: Environment, Women, and Literature.*

Foreword

Humble and simple though it may be, the equation attempts to quantify a phenomenon of prime interest to both the scientific world and the general public: the abundance of observable technical civilizations in space. It is obvious that learning of those civilizations, many of them much older than our own, will enrich us greatly in facts of biology, sociology, philosophy, and science in general. We will learn how other civilizations have conducted their affairs over times much longer than we have existed. We will perhaps learn of the problems they faced and how the problems were dealt with. We will learn the limitations, opportunities, and possible dangers of technologies much more advanced than our own. We will learn the ways of life of creatures and places much different from those on Earth. It will be the ultimate exciting journey to exotic foreign countries.

To achieve these results, we must detect and study those distant worlds. No doubt the search will be lengthy, expensive, and resource-intense, including the time of talented people and sophisticated search instruments of great power and cost. Already, for more than fifty years, we have embarked on such searches without success. This is not surprising, however, since even the most optimistic analysis of the amount of searching required for success is far beyond what we have done to date.

To formulate future searches, optimize them, and make a case for the resources needed for success, we need to create, as best we can, an estimate of how much searching will be required to have a good chance of success. To do this, we need to estimate the number of places, which we call N, from which evidence of intelligent life might be coming, and compare that to the totality of galactic stars, which number approximately 200 billion.

The equation was invented to provide an estimate of N. It is a simple arithmetical product of seven variables. It works to quantify the production of galactic planets, their habitability, the possible origin of life on a planet, the evolution of intelligence, the evolution of technology we might detect, and the length of time the technology is detectable. The equation is described in detail elsewhere in this book.

The equation works as a recipe to make detectable civilizations. Take seven ingredients in the proper amounts, mix them together, let simmer for about three billion years, and you get N advanced civilizations! Indeed, the equation is basically very simple: just seven variables, all to the first power. No exponents, no exponentials, no trig functions. This is all to the delight of students with math anxiety when they are presented with the equation in elementary astronomy courses, as is common these days. It is even simpler than Kepler's laws. There is also a sigh of relief from many others, as well as students, when they see it for the first time. So simple, yet it is dealing with something of the most profound importance. It gives the answer, when solved, to one of the oldest and most important questions, scientific or philosophical, which exists for humanity: What is the abundance of intelligent life in the universe?

Of course, it is not all that simple. What are the values of the variables? When the equation was first formulated in 1961, we had crude knowledge of only one variable: the rate of star formation. This hardly justified a book. But much has happened in the fifty-three years since, particularly the blossoming of astrobiology, which is, after all, the study of everything involved in the equation. Fifty-three years ago very few people worked on the variables; today, the number of workers is not really known but is surely in the many thousands. Now it is time to stand back and look at where we are. That is what this book is about.

In those fifty-three years, we have seen the discovery of thousands of other planetary systems – none were known in 1961. Some of these planetary systems belong to red-dwarf stars, stars that we didn't think, back then, could harbor habitable planets. Planets in the habitable zone of the red dwarfs have been found, increasing the number of habitable objects in the galaxy by about ten times. We have achieved a growing understanding of the habitability of those planets. We have even found more than four other objects in the solar system itself with large liquid lakes, possibly suitable for life, on them. This has expanded our deduced limits on the "habitable zone," where life as we know it may survive. That qualifies many more objects as potential abodes of life than we thought fifty-three years ago.

There have been numerous experiments simulating the chemical situations on young planets; these have identified a variety of chemical pathways that can produce the basic molecules of life. It almost has gone unnoticed, but in none of these experiments has a "showstopper" surprised us. No complex or rare catalyst or unnatural molecule like a freon is required to produce life's basic building blocks, just molecules using very common elements are required. Studies of the brains of mammals have shown, again, that there seem to be no showstoppers when it comes to evolving intelligence. All of this makes us optimistic that the development of living things and high intelligence elsewhere in space should occur, and possibly very prolifically.

You will see that some of the writers of the chapters here are limited in their ability to quantify the factor in the equation they are writing about. This is as it should be. The last four variables in the equation, in particular, are ones where our only information, right now, is from Earth history. The writers are reluctant, as ethical scientists, to draw any firm conclusions from a sample of one. So there is still much to be learned before we will be able to arrive at any firm quantitative results from the equation. We wish, in particular, to make a discovery of at least a second, independently produced, system of life. For now, we can only speculate and use our intuition to arrive at possibly plausible answers.

The "limitation of the sample of one" applies particularly strongly to the variable L. Not only do we have only ourselves as a source of knowledge about L, but we have seen in ourselves the perceived limits on L to be changing rapidly, even on the time scale of human lives. The variable L represents the length of time a civilization uses a technology we can detect. We have only our technological history to rely on for this information. What do we see? In just fifty years, our ability to detect various technologies has greatly improved, and there is reasonable room for improvement. This can increase the operative value of L if we are typical. At the same time, improvements in technology have decreased the strength of some of the strongest signs of our existence, such as television broadcasts and military radar systems. This acts to reduce L if we are typical. Overall, we just don't know what "typical" is for civilizations; it is a great unknown, and has a huge impact on any calculation of N. We are in a catch-22 situation: We won't know how much effort it will take to succeed in our searches until we have succeeded.

History is repeating itself as we explore the cosmos for other creatures. We have become like Columbus, setting sail to discover a rich new world without knowing how far we must sail before we will reach that land. Nevertheless, we are confident that there are glorious new worlds to be discovered out there among the stars.

Frank Drake

Preface

The Drake Equation: Estimating the Prevalence of Extraterrestrial Life through the Ages provides a thematic history of the search for extraterrestrial life to the present day, using the terms of the Drake Equation as organizing categories. This equation was originally proposed by the astronomer Frank Drake, who conducted the first modern experiment in the search for extraterrestrial intelligence (SETI). He used the equation as the organizing theme for a conference on interstellar communication held in 1961 at the National Radio Astronomy Observatory in Green Bank, West Virginia. By multiplying the following seven terms, Drake provided an estimate of the number of extraterrestrial civilizations currently transmitting in our galaxy:

- R^* Rate of formation of stars suitable for the development of intelligent life
- f_p Fraction of stars with planetary systems
- n_e Number of planets, per solar system, with an environment suitable for life
- f_l Fraction of suitable planets on which life actually appears
- f_i Fraction of life-bearing planets on which intelligent life emerges
- f_c Fraction of civilizations that develop a technology that releases detectable signs of their existence into space
- L Length of time such civilizations release detectable signals into space

From the outset, the Drake Equation has served as a means for quantitatively estimating the number of extraterrestrial civilizations in our galaxy that are capable of making contact at interstellar distances. Some continue to use the equation in that manner, and an examination of efforts to estimate the specific values for various terms will be discussed in this book. More people, however, use the equation as a heuristic device to consider the factors relevant to the search for evidence of life beyond Earth without a strong emphasis on quantifying these terms, and this book will similarly explore the equation in this manner.

Fifty years after it was first introduced, the Drake Equation continues to influence scientists involved in astrobiology – the study of the origin, evolution, distribution, and future of life in the universe. While scholarly works, textbooks, and popular writings in astrobiology often include sections on the Drake Equation, until now there has been no stand-alone volume on the topic.

We have two reasons for writing this book. First, we wish to provide a comprehensive review and analysis of each term of the equation. This will provide a resource for scientists and graduate students actively conducting research in the many disciplines related to astrobiology. Second, the book places contemporary astrobiological research in historical context by including two chapters for each term of the Drake Equation: one covering the concept before the 1961 Green Bank conference, and the other ranging from 1961 to the present. To promote dialogue between practicing scientists and historians, authors cross-reference one another's chapters liberally, highlighting the complex interplay between the terms. By understanding the historical context of contemporary science, we hope that scientists will better recognize and appreciate the suppositions that guide their own research, often in unspoken ways.

Acknowledgments

To the authors of *The Drake Equation*, we express our appreciation of the innovation, depth, and sensitivity of the work they share here. They deserve special thanks for thoughtfully engaging one another's ideas, as reflected in the numerous cross-references between chapters throughout the volume, highlighting the links between the terms of the equation. It is especially meaningful for us to include a chapter by Mike Crowe, who first introduced both editors, separately, to the history of discussions about the prevalence of extraterrestrial life.

We are grateful to Joe Castrovinci for so capably copyediting each chapter in their early stages, enhancing the consistency across chapters while still encouraging authors to keep their own voices. His organizational ability combined with his expertise in technical writing allowed us to move ahead smoothly and on schedule.

Doug Vakoch would especially like to thank his many colleagues from the SETI Institute, who have shared with him their insights into the Drake Equation over the past fifteen years. He especially thanks Molly Bentley, David Black, James Brewster, Steve Brockbank, Edna DeVore, Frank Drake, Andrew Fraknoi, Dane Glasgow, Gerry Harp, Ly Ly, David Morrison, Michelle Murray, Chris Munson, Chris Neller, Karen Randall, Jon Richards, Pierre Schwob, Seth Shostak, and Jill Tarter. More recently, Doug warmly acknowledges the administration, faculty, staff, and students of the California Institute of Integral Studies (CIIS), especially for support from Joseph Subbiondo, Judie Wexler, and Mera Atlis. In addition, he thanks Harry Letaw and Joyce Letaw for their ongoing support of this work.

For encouraging us to develop this project and then shepherding the book through the editorial process at Cambridge University Press, we are indebted to Vince Higgs. Also at Cambridge University Press, Sara Werden has our gratitude for helping to move the book swiftly and efficiently into production.

We wish to thank our families, including our parents, Alvin Vakoch and Stella Vakoch, and Jim Dowd and Betty Dowd, for their never-ending love and support. Doug is especially grateful to his wife, Julie Bayless, in more ways than he can or will share here. Matt is similarly grateful to his wife, Amy, and apologizes to his sons, Ben and Mike, for a missed cross-country season.

Douglas A. Vakoch
Matthew F. Dowd

Introduction

The Drake Equation in context

Steven J. Dick

Abstract

The Drake Equation, a method for estimating the number of communicative civilizations in the Milky Way galaxy, was a product of its time in several important ways. After a period of several decades during which the idea of life on other planets had reached a low point due to rise of the "rare collision" hypothesis for planet formation, by the 1950s the nebular hypothesis was once again in favor, whereby planets would form as a common byproduct of stellar evolution. The Miller–Urey experiments in the early 1950s produced complex organic molecules under simulated primitive-Earth conditions, indicating life might easily originate given the proper conditions. And while little was known about the gap between primitive life and intelligent life, and a sophisticated understanding of intelligence was lacking, the Lowellian Mars still lingered in the cultural background and, along with contemporary astronomical advancements, stimulated the scientific imagination to consider aliens. The original emphasis on "radio communicative" reflected the new era of radio astronomy, exemplified by the radio telescopes under construction at the newly founded National Radio Astronomy Observatory (NRAO) in Green Bank, West Virginia, where Frank Drake was working at the time when the equation originated as the meeting agenda for an informal conference there in 1961. Drake's ability to undertake such a controversial subject, including the first radio search for extraterrestrial intelligence in 1960, was aided by senior scientists Lloyd Berkner and Otto Struve. Assessments of the probabilities of extraterrestrial life and intelligence had been sporadically undertaken in the course of the twentieth century, but most particularly by former Harvard Observatory Director Harlow Shapley in his book *Of Stars and Men* (1958); Drake had recently graduated from the Harvard astronomy program, and had cited the book. Here we look at the origins and development of the

The Drake Equation, ed. Douglas A. Vakoch and Matthew Dowd. Published by Cambridge University Press.

equation over time, including significant variations in the equation; examine positive and negative views of its epistemological status and utility ranging from scientists to popular authors such as Michael Crichton; and attempt to tease out the scientific and metaphysical assumptions behind the equation. We conclude by discussing the future of the equation, and the cultural hopes and fears it embodies.

Origins of the equation

The Drake Equation was born during an informal conference on "Extraterrestrial Intelligent Life" held on November 1–2, 1961, at the nascent NRAO in Green Bank, West Virginia. The meeting, sponsored by the Space Science Board of the National Academy of Sciences, was held in the wake of the excitement generated by Project Ozma, the first search for interstellar communications, conducted at the NRAO by Frank Drake, a young astronomer on its staff (Figure I.1). The two-hundred-hour search, with the observatory's 85 foot Tatel radio telescope in April, 1960 (Figure I.2), targeted only two nearby Sun-like stars, Tau Ceti and Epsilon Eridani, around the 21 centimeter line of neutral hydrogen (Drake and Sobel 1992, chapter 2). Although it failed to detect any extraterrestrial civilizations, the project captured the imaginations of scientists and public alike. The Ozma search (though independently conceived by Drake) followed the landmark publication by Giuseppe Cocconi and Philip Morrison that argued on theoretical grounds that such a search should be undertaken (Cocconi and Morrison 1959). The Green Bank meeting was therefore the last in a troika of events from 1959 to 1961 that launched the modern SETI era (Dick 1996, 1998).

The standard story, even from Drake himself, is that press coverage of Project Ozma triggered the interest of the National Academy (Drake 1992, 14–15; Drake and Sobel 1992, 46). But National Academy records demonstrate that the immediate cause was actually a lecture Drake gave on the subject at the Philosophical Society of Washington on March 10, 1961 (Drake 1961a; Pearman 1961). In the audience was biologist J. P. T. Pearman of the National Academy's Space Science Board staff, who that night after the lecture discussed with Drake the possibility of such a meeting (Pearman 1961). By March 13, Drake replied to Pearman with a letter stating that NRAO Director Otto Struve not only approved such a meeting but also offered to hold it at NRAO. The observatory had living accommodations for about thirty people, Drake noted, and "the isolation of Green Bank would also help solve the problem of keeping the symposium quiet and scientific" (Drake 1961b).

The National Academy's records indicate Pearman immediately set to work, handling much of the logistics for the meeting. But the organization of the scientific content fell largely to Drake. Thinking in the days before the meeting about how to

Figure I.1 Frank Drake at the National Radio Astronomy Observatory, 1962, where he had conducted Project Ozma two years before. Drake, recently graduated from Cornell and Harvard, had been interested in extraterrestrial life from an early age, and had been influenced during his Cornell years by a lecture on planetary systems given by Otto Struve.
Credit: NRAO/AUI/NSF

proceed, Drake decided to arrange the discussions of extraterrestrial intelligence around an equation that concisely represented the relevant factors. Thus appeared for the first time the formulation that would be used repeatedly in the following decades in attempts to determine the likelihood of radio-communicative civilizations in our galaxy – and thus the likelihood of success in any such search.

The original form in which Drake wrote the equation was $N = R^* f_p n_e f_l f_i f_c L$, where each symbol on the right side of the equation represents a factor bearing on the number of radio-communicating civilizations in the galaxy (N) (Figure I.3). The first three factors were astronomical, estimating respectively the rate of star formation in the galaxy, the fraction of stars with planets, and the number of planets per star with environments suitable for life. The fourth and fifth factors

Figure I.2 Drake at the 85 foot Tatel telescope during his visit to give the October 1999 Jansky Lecture. The Tatel telescope was completed in early 1959 at the nascent NRAO.
Credit: NRAO/AUI/NSF

were biological: the fraction of suitable planets on which life developed and the fraction of those life-bearing planets on which intelligence evolved. The last two factors were social: the fraction of civilizations that were radio-communicative over interstellar distances and the lifetime of radio-communicative civilizations. The uncertainties, already shaky enough for the astronomical factors, nevertheless increased as one progressed from the astronomical to the biological to the social. Taken together, they represented cosmic evolution writ large.

Although Drake was the first to put these factors in simple equation form, he was not the first to ask the question in terms of probabilities. Assessments of the probabilities of extraterrestrial life and intelligence had been sporadically undertaken in the course of twentieth-century discussions of the subject. On the eve of

Figure I.3 The Drake Equation, inscribed on a plaque in the conference room at the National Radio Astronomy Observatory in Green Bank, West Virginia, where Drake first formulated the equation in 1961. N represents the number of technological communicating civilizations in the Milky Way galaxy, and the right side of the equation embodies various parameters of star and planet formation, the likelihood of the origin and evolution of life and intelligence, and the lifetimes of technical civilizations.
Photo courtesy of author

the events of 1959–61, former Harvard Observatory Director Harlow Shapley had calculated the number of intelligent civilizations in the universe based on probabilities, but had not discussed interstellar communication (Shapley 1958, 73–74). Drake had recently graduated from the Harvard astronomy program, and had cited Shapley's calculations prior to the Green Bank meeting (Drake 1959). Probabilities had also been used by radio astronomer Ronald Bracewell in another early discussion of the number of advanced communities in the Milky Way (Bracewell 1960, 670). Bracewell, however, had couched his discussion in graphical rather than equation form. And astronomer Sebastian von Hoerner had used probabilities to conclude that one in three million stars might have a technical civilization, but that the longevity of a technical civilization (a concept he credited to Bracewell) might be very limited (von Hoerner 1961).

When Drake began the Green Bank meeting by writing his equation on the board, he could not have known that he was establishing a paradigm for SETI discussions that would last into the twenty-first century. But by considering in turn astrophysical, biological, and social factors he did just that, and Green Bank was only the first of many occasions where experts would discuss the factors that Drake proposed. In the wake of the Green Bank meeting, discussions centered on the likelihood of communicative extraterrestrial civilizations using radio technology. The calculations of N varied wildly, over a range not seen before in the history of science (Dick 1996, 441; 1998, 217). One could take this as an indication of a very unsettled protoscience, though one that held promise for the future.

In the task of calculating the number of radio-communicative civilizations, the compelling nature of an equation – even one whose parameters were not well known – was not to be denied, since an equation is a symbol of science and lends authority to any scientific discussion. The meteoritic career of the Drake Equation, rather than one of the other probabilistic assessments, is evidence of such authority. Only a month after the Green Bank meeting in November 1961, Philip Morrison used a similar equation in a NASA lecture (Morrison 1962). The equation first saw print not in an article by Drake but in Pearman's account of the Green Bank conference published in a 1963 volume of collected articles on the subject entitled *Interstellar Communication* (Cameron 1963a; Pearman 1963). In the same volume its editor, the astrophysicist A. G. W. Cameron, used a similar equation (Cameron 1963b, 1963c). Sagan was also among the first to publish the equation (Sagan 1963), and Drake himself used it in a paper presented at a JPL symposium on exobiology in February 1963 (Drake 1965). Although not known at first as the Drake Equation, after a period of uncertainty when it was called the Sagan Equation or the Green Bank Equation, the originator was given due credit (Drake 1992). Perhaps the decisive events in the spread of the Drake Equation were Walter Sullivan's popularized account of it in *We Are Not Alone* (1964) and Sagan's incorporation of it into his translation and expansion of Russian astrophysicist Joseph Shklovskii's book *Intelligent Life in the Universe* (Shklovskii and Sagan 1966), which became the Bible of the SETI movement. These books assured the rapid diffusion of the Drake Equation to the public and interested scientists alike.

Although not immediately used in the Soviet Union, the Drake Equation, with its emphasis on radio communication, focused attention on the electromagnetic-radio search paradigm. Already by 1966 this concept and all of the assumptions that went with it were sufficiently entrenched that physicist Freeman Dyson labeled it the "orthodox view" of interstellar communication, characterized not by interstellar travel but by "a slow and benign exchange of messages, a contact carrying only information and wisdom around the galaxy, not conflict and turmoil" (Dyson 1966). As anyone who read science fiction knew, this was not the only possible view of the universe. But it was a practical method, a logical extension of the new field of radio astronomy, and one that at least some of its practitioners were keen to carry out. For these reasons, the discussion of rationale and strategy within the radio search paradigm continued its upward climb.

By 1971, ten years after its origin, the Drake Equation was the centerpiece for the first international SETI meeting, held at the Byurakan Astrophysical Observatory in Yerevan, the Soviet Union (Sagan 1973). This time the organizers of the meeting, sponsored by the Academies of Sciences of both the United States and the Soviet Union, included not only Drake, but also Carl Sagan and Philip Morrison of the United States, as well as Victor Ambartsumian, Nikolai Kardashev, Joseph Shklovskii, and V. S. Troitskii of the Soviet Union. Instead of the eleven participants at the Green Bank meeting in 1961, twenty-eight Soviets, fifteen Americans, and four scientists from other nations participated. They concluded that perhaps a million technical civilizations existed in the galaxy. SETI, though still a small endeavor by science standards, was growing, and the Drake Equation was its central icon.

The equation in context

The equation was a product of its time, triggered by the ability of radio telescopes to search for artificial signals from nearby stars (Drake and Sobel 1992, chapter 2). Drake has given us an inside look at the eleven participants as they gathered at the Green Bank meeting (Drake and Sobel 1992, chapter 3): Drake himself was the expert young radio astronomer. His boss, Otto Struve (Figure I.4), and Struve's former student Su-Shu Huang, were the experts on planetary systems. Other participants were recruited for their expertise in a particular factor in the Drake Equation. Collectively, they represented most but not all of the factors in the equation. Notably, no social science or humanities experts were present to discuss the number of civilizations or their lifetimes, in part a reflection of the gulf between the two cultures of science and the humanities in the early 1960s.

At the time of the Green Bank meeting, the idea of extraterrestrial life was gaining momentum. After a period of several decades, during which the idea of life on other planets had reached a low point due to rise of the "rare collision" hypothesis for planet formation, by 1960 the nebular hypothesis was once again in favor and held that planets would be a common by-product of stellar evolution. At the Green Bank meeting, Struve was enthusiastic about the number of planetary systems, based primarily on his work on stellar rotation, and was supported by Huang, who had concluded from his own research on habitable zones around stars that the number of planets in the galaxy suitable for life was indeed very large (Dick 1996).

Figure I.4 Otto Struve, director of the NRAO, July 1, 1959 through December 1, 1961. Neither Project Ozma nor the Green Bank meeting on Interstellar Communication could have been undertaken without his enthusiastic acceptance.
Credit: NRAO/AUI/NSF

In the wake of NASA's founding in 1958, planetary science was also on the upswing, with the real possibility of sending spacecraft to study planetary surfaces and atmospheres. Indeed, that is precisely what happened, with the search for life on Mars often in the forefront as a driver of space science. Although no one represented NASA at the meeting, the young planetary scientist Carl Sagan, already involved in many planetary projects, was in attendance and well aware of NASA's planetary efforts. Joshua Lederberg, who had just coined the word "exobiology," had some input to the meeting, but could not attend (Lederberg 1961).

The origin of life was also a hot topic at the time. The 1953 Miller–Urey experiments in simulating life under primitive-Earth conditions indicated life might easily originate given proper stimulus. Melvin Calvin, an expert on chemical evolution, argued at the meeting that the origin of life was a common and even inevitable step in planetary evolution, and his already formidable credentials were given another boost when he received notification during the meeting that he would be awarded the Nobel Prize for his work on the chemical pathways of photosynthesis. And while little was known about the gap between primitive life and intelligent life, or even the definition of intelligence, the

Lowellian Mars of artificially constructed canals still lingered in the background, stimulating imaginative scientists to think about aliens.

It was a very large step from the origins of life to intelligence, and the concept of "intelligence" was neither well defined nor understood, which is still the case today. Perhaps not surprisingly in this environment, the organizers looked for a participant doing practical research in the field. John Lilly, who had just come out with his controversial book *Man and Dolphin*, met that criterion and argued at Green Bank that dolphins were an intelligent species with a complex language, and that we might even be able to communicate with them. Dolphins thus became a kind of symbol for interspecies communication.

The equation's emphasis on "radio-communicative" reflects the new era of radio astronomy, exemplified by the radio telescopes being built at the newly founded NRAO. This early history and subsequent events are elaborated in Dick (1996, 1998), and there is no need to repeat it here. Summarizing the results of their discussions, the members of the conference concluded that, depending on the average lifetime for a civilization, the number of communicative civilizations in the galaxy might range from less than one thousand to one billion. Opting for the more optimistic figure (likely an unfounded bias based on their interest in the subject), most of the members felt the higher number was likely closer to the truth.

Hidden assumptions

Even as it grew in popularity, the Drake Equation embodied many hidden assumptions, perhaps responsible for both confusion and its enduring legacy. Nowhere is this truer than it its first and last factors, R* and L, which are the only parameters with dimensions (stars forming per year and number of years). It is often forgotten that Drake's formulation was an eminently practical exercise, driven by Project Ozma and the desire to estimate the chances of its success by estimating the number of communicative civilizations existing *now*. This explains why the first parameter in the equation was not simply the number of stars existing today in the galaxy, whose formation began some eleven or twelve billion years ago. Nor was it even the number of stars existing 4.5 billion years ago, since they were all in different stages of development; if those stars had spawned civilizations, they would all be in different stages of development. Rather, Drake was interested in civilizations that were communicating now and at about the same stage of development as ours. He therefore used as the first parameter of the equation a rate of star formation rather than a number of stars. And he used L because it was the bottleneck that restricted technological civilizations to those communicating *now*.

The rate of star formation in our galaxy was the best-known quantity in the equation, and by the evidence of the time, it was calculated in a straightforward, "quick and dirty" way, not taking into account current theories of star formation as discussed in Chapter 2. In Pearman's account of the meeting, the calculation went as follows:

> If stars of solar type only are considered, a rough estimate of R* is given by the total number of such stars in the galaxy divided by their average lifetime. Thus $R^* = 10^{10}/10^{10} = 1$ per year. This is perhaps a conservative estimate and less restrictive considerations permitting the inclusion of some Population II stars would give values as high as 10 stars per year. (Pearman 1963, 289)

In other words, estimating ten billion solar-type stars in the galaxy, each with a lifetime of about ten billion years, yields one star forming per year. Including Population II stars (still Sun-like stars but older than our Sun), one could raise this estimate to ten per year, thus the often-used estimate in the Drake Equation of one to ten stars forming in our galaxy per year.

Needless to say, this assumes a uniform rate of star formation over the lifetime of the galaxy, which we know today not to be the case. The same can be said for the calculation sometimes used that employs the number of solar-type stars in the galaxy divided by its age. Strictly speaking, R* today is defined not as the rate of star formation over the lifetime of the galaxy but as the rate of star formation 4.5 billion years ago when our Sun and its planets were formed. At least that is the way Drake defines it. Responding to an inquiry about his current usage of R*, Drake wrote,

> I prefer it because it more accurately quantifies the process by which current intelligent technology-using life came about. There are two versions of the equation which occur in various textbooks, etc. One uses number of stars/age of galaxy. The other uses R*. The first conceals a somewhat important aspect of the whole picture, since it implies that the relevant star formation rate is the mean rate during the existence of the galaxy. But that is not the one which applies to the calculation of how many technology civilizations are out there to be found *now*. That number is governed by not the mean rate of star formation, but the rate of star formation which existed at the time stars of about the same age as the Sun were formed, namely about 4.5 billion years ago. (Drake 2014a, emphasis in original)

This formulation assumes that extraterrestrial technological civilizations develop at about the same rate as on Earth, a very large assumption indeed. Drake fully recognizes the assumption, but finds it necessary considering our ignorance:

> What we really need to know is the statistics of star formation over a substantial period 4.5 billion years ago, since the process of producing an intelligent species will take some range of time intervals. We won't know that until SETI succeeds. However, the rate of star formation 4.5 billion years ago is the best estimate we can use in our current state of

knowledge ... the rate of star formation started high early on, and has been gradually decreasing over the history of the galaxy. So, to be as rigorous as possible, I always use the form of the equation with R*. (Drake 2014a)

How different was the rate of star formation over the history of the galaxy? In particular, how different was it 4.5 billion years ago compared to now? The first two chapters in this volume address the history of astronomers' ideas about star formation.

But what about planets that could have formed much earlier in the history of the galaxy, say, as much as eight billion years ago, according to current theories (Dick 2003; Larson and Bromm 2001; Lemonick 2014)? Drake's answer is that they would not be at the same stage of civilization as ours. Drake emphasizes that his use of R* corresponding to about 4.5 billion years ago "is based on a quiet, unwritten rule, we use in SETI. 'Only assume phenomena you know exist.' Of course this is limiting, because history tells us that there must be powerful inventions yet to be made. But without such a rule, wild speculation can run rampant, and where do you draw the line?" (Drake 2014b). One such example is that civilizations that old might even be postbiological (Dick 2003). Moreover Drake points out that older civilizations might no longer be detectable by radio:

Our civilization, in an effort to save resources and money (probably fair to believe all civilizations practice economies), is moving to communication techniques which release minimal energy into space, therefore releasing minimal energy to serve as a sign of our existence, and wasting minimal energy. Prime examples are cable TV and direct-to-home TV from satellites. Therefore, it appears that our detectability may last only a few hundred years, unless, of course, much more sensitive search systems such as using a solar gravitational lens are developed. (Drake 2014b)

This brings us full circle to L, which we need to remember is not the lifetime of a technological civilization, but the lifetime of a *communicating* technological civilization. Despite the limitation mentioned above, there remains the possibility of civilizations continuing to send out beacons, possibly altruistically and for the benefit of others. Drake also recognizes this possibility:

Of course, we should always have in the back of our head the thought that maybe there are possibly a small fraction of civilizations which are altruistic and maintain a bright, easily detectable, signal for very long times to enrich the knowledge of other civilizations. This would change the value of N a lot. If just one percent of civilizations maintained a 'contact' beacon for a billion years (not crazy!), then L would be ten million years! A byproduct of this scenario is that the right strategy is to search in the directions where you will test the maximum number of stars for signals, which is in directions close to the galactic plane. (Drake 2014b)

The extent of extraterrestrial altruism might seem unknowable, but an entire volume of essays has been written on just this subject (Vakoch 2014).

The same kinds of hidden assumptions are present in the dimensionless factors sandwiched between R* and L. Referring to the third factor in the equation, the number of planets that can potentially support life, Drake points out:

> [A]s another example of a complicated parameter, the ecosphere is a loose concept because in the simplest form usually given, the ecosphere is bounded by the boiling and freezing points of water. But we know of ways by which it extends to much larger distances. A deep atmosphere, and/or lots of CO_2, or having [an] ice layer covering an ocean, all extend it out to much greater distances. A surface like that of Mars or Earth extends it way out because life can exist at suitable depths beneath the surface, since the temperature of a solid surface always rises with depth. In fact, just about everything planets can have – solid body, atmosphere, ocean – extends the ecosphere outwards. Think Enceladus. (Drake 2014a)

One more example of a hidden assumption involves L again. "Is L the total time a technical civilization exists?" Drake asks. He continues,

> Using that definition gives you how many such civilizations there are. This L might be a billion years. But if you want to know how many *detectable* civilizations there are, then L is the length of time civilizations manifest themselves in some way a plausible detection system (another thing which has a wide range of possibilities) could detect. This could be only a couple of hundred years, if we are an average example, and our radio transmissions are the detectable signs of our existence. If we go to nothing but cable TV, and direct-to-home TV from satellites, our L will be measured in hundreds of years or less. On the other hand, if civilizations really can build useful telescopes using their star as a gravitational lens, then L possibly becomes a billion years! That is a big uncertainty and exciting possibility! (Drake 2014a, emphasis in original)

The Drake Equation is again quite conservative in this sense: "when it comes to L, we are stuck with using ourselves as a model, which is all we have to go on until we discover another civilization" (Drake 2014b).

To many, the Drake Equation parameters between R* and L immediately make sense as fractional factors winnowing the possibilities of communicating civilizations. But the meaning of the product of R* and L is not so intuitive. What does the rate of star formation have to do with the lifetime of a radio-communicative civilization? Given all we have said above about R* and L, the bottom line is that if one radio-communicative civilization is forming per year, and if L is 100, then any observer can see such civilizations in the radio domain for only 100 years. So, on the conservative view that we are only talking about radio-communicative civilizations, and on the admittedly shaky assumption that those civilizations are developing at the same rate as we are, one needs only to look at 100 years around the star-forming time domain 4.5 billion years ago because all the other civilizations would have blinked out. Many more technological civilizations *could* exist; we just cannot see them because they are no longer communicative, at least in the radio spectrum. To put it another way, L acts as a kind of gateway that rejects all years during which a civilization does not communicate or have detectable radiation in the radio spectrum, no matter how many civilizations are forming. Drake likes the analogy of a Christmas tree, where each light blinks only once at various times (Drake 2014c). This is the source of the often-repeated phrase in connection with the Drake Equation (and a source of license plates among SETI pioneers), that "N EQLS L," or N approximates L.

In the context of the lifetimes of civilizations, one must remember that the early 1960s were the height of the Cold War. Estimates for L were remarkably optimistic for a time when civilization on Earth might have been wiped out by nuclear war at any moment. The variable L is at one and the same time the potential bottleneck for the success of all SETI searches and conservative in its own way. As radio astronomer Seth Shostak points out, L for a radio-communicative civilization might be shorter than L in the optical region of the spectrum or for communication modes based on other technology, or for a civilization not communicating at all. If we cannot detect a civilization, L is interesting for sociological reasons, but not for SETI reasons (Shostak 2009). Moreover, L should not be based on the lifetime of any particular civilization on Earth, such as Chinese, Greek, or Roman (Denning 2011). If Earth is any guide, the civilization in question will not be global; all that matters is that it be detectable.

The parameters between R* and L, while straightforward as members of the Drake Equation, are increasingly unknown as one moves to the right. One begins to see how the equation does an excellent job of stimulating thought and discussion, and will likely continue to do so into the foreseeable future.

Criticisms and variations

Not everyone has praised the Drake Equation. In fact, it has been highly criticized by people ranging from fiction writers to real scientists. Representative of the former is Michael Crichton, the best-selling author of books and movies such as *The Andromeda Strain, Jurassic Park*, *Sphere,* and *Prey*, which feature the failures of technology in society. In 2003, in the context of denying global warming arguments, Crichton used the Drake Equation as another example of bad science:

> This serious-looking equation gave SETI a serious footing as a legitimate intellectual inquiry. The problem, of course, is that none of the terms can be known, and most cannot even be estimated. The only way to work the equation is to fill in with guesses. And guesses – just so we're clear – are merely expressions of prejudice. Nor can there be "informed guesses." If you need to state how many planets with life choose to communicate, there is simply no way to make an informed guess. It's simply prejudice. The Drake Equation can have any value from "billions and billions" to zero. An expression that can mean anything means nothing. Speaking precisely, the Drake Equation is literally

meaningless, and has nothing to do with science. I take the hard view that science involves the creation of testable hypotheses. The Drake Equation cannot be tested and therefore SETI is not science. SETI is unquestionably a religion. (Crichton 2003)

SETI proponents would argue that not only *could* the results of the Drake Equation be tested, they *had* been tested with Project Ozma and could be tested further with improved telescopes searching various targets at a range of frequencies. But the point about shaky values for individual parameters is indisputable. In this respect, Crichton was not the first, nor the last, to take the equation too seriously. Most of the creators and users of the equation realized, and often explicitly stated, its limitations; Drake himself was amazed at its popularity, and SETI pioneer Bernard Oliver referred to the equation in his Project Cyclops report as "a way of compressing a large amount of ignorance into a small space." (Oliver 1971, 26). No one claimed the Drake Equation had the status of a scientific law such as $F = Ma$ or $E = mc^2$. Although criticism is always welcome in science, Crichton's declaration seems suspiciously harsh, indicating he may have had some ideological agenda, not unusual in the context of climate change. His outburst brought responses from people ranging from bloggers to scientists. One journalist, William M. Briggs, argued that since we exist, it is not improbable that extraterrestrials exist, unless we were created in a single, unique event, which is a religious explanation, not science (Briggs 2008). Scientists assuredly cannot assume what they are trying to prove, but they are allowed to make probability estimates. The equation must be seen primarily as a useful heuristic, a way of contemplating the problem, not as a law of nature. As such it cannot be seen as strong justification for SETI searches, but rather as suggestive that such searches *could* be successful.

Viewed in this way, as an organized method for stimulating discussion, the equation has been a smashing success. This is evident not only in its appearance in numerous textbooks, lectures, and TV presentations but also in the number of variations it has generated (which, it must be pointed out, may also take the equation too seriously). Already in 1971, J. G. Kreifeldt noted the equation was defective from a temporal point of view, in the sense that it did not allow for the time dependence of its terms. He went on to present a dynamic formulation that took into account different star generation rates, civilization lifetimes, and so on, resulting in an expression for the number of communicative civilizations in the galaxy as a function of time, including a variance for this estimate (Kreifeldt 1971). Wallenhorst (1981) elaborated on this temporal deficiency, and concluded that as a result N might only be one hundred rather than one million. In a similar vein, Cirkovic (2004) argued that the galaxy was not habitable during its entire lifetime, since time is required for heavy elements to form the terrestrial planets. One result of this consideration is that civilizations are more concentrated in a given period of the galaxy's history.

A second type of variation on the Drake Equation involves taking interstellar colonization into account, in the wake of claims made by Michael Hart (1975) that colonization of the galaxy would take place over relatively short time scales, thus leading to the "Fermi Paradox." The latter, which brought a crisis to SETI community thinking in the 1970s, states that if there are so many civilizations, then they would have been here already, so "where are they?" In 1980, Walters, Hoover, and Kotra suggested adding a new parameter, "C," to the Drake Equation, which takes into account the fraction of civilizations that wish to colonize, the fraction of stars with planets suitable for colonization, and the ability to reach those stars. If no civilizations wish to colonize, C is 1, reverting to the original Drake Equation. Based on a variety of considerations, the authors concluded that C would be less than ten, not the devastating impact on SETI that Hart had suggested. In 1983, astrophysicist and science fiction writer David Brin took a similar tack with the concept of a "contact cross section" to explain what he called "The Great Silence" (Brin 1983). While these variations may or may not give a more accurate picture of what is really going on, the more complex forms of the equation are not likely to appear in popular lectures. Its simplicity remains one of its enduring features, and is in part responsible for its longevity and continued utility (Drake 2013).

A third line of reasoning relevant to the Drake Equation was begun by Gonzalez, Brownlee, and Ward (2001), who argue that the Galactic Habitable Zone (GHZ) further narrows the possibilities for life. The GHZ, analogous to the circumstellar habitable zone often used in estimates of the fraction of planets where the conditions arise necessary for life, is the region in the galaxy where planets can retain liquid water on their surfaces and provide a long-term habitat for complex life. The central concept here is "metallicity," the existence of heavy elements in a stellar nursery in amounts high enough to build a rocky terrestrial planet, a condition that does not occur everywhere in the galaxy. These arguments fed into Ward and Brownlee's best-selling book *Rare Earth*, the title of which indicates their conclusion (Ward and Brownlee 2000). In the process they also developed their own version of the Drake Equation, dubbed the "Rare Earth Equation," which takes into account not only the GHZ but also other factors such as mass extinction events (Ward and Brownlee 2000, 270–75). Lineweaver, Fenner, and Gibson (2004) have further elaborated this concept, taking into account not only metallicity but also an environment free of life-extinguishing supernova and other factors.

The Drake Equation has also spawned discussion in more exotic directions, including modes of communication, difficulties of communication, epistemological considerations (Hetesi and Regaly 2006), the messaging (METI) factor (Zaitsev 2005), and a "statistical Drake Equation," that assumes some standard deviation for each of the seven parameters (Maccone 2010). Again, any of the ideas behind these elaborations could have been undertaken without the equation, and do not stand or fall with it, but were in fact stimulated by it.

Does all this mean the original Drake Equation is obsolete? That was the argument of the Canadian futurist, science writer, and ethicist George Dvorsky, whose article "The Drake Equation is Obsolete" pulled no punches (Dvorsky 2007). He argues it is arbitrary, does not account for cosmological changes over time, and that the radio window on Earth is closing and so accounts for only a narrow class of civilizations radiating in the radio spectrum.

Future of the equation

The Drake Equation continues to inspire discussion, as evident in a special issue devoted to it in the *International Journal of Astrobiology* (2013). Yet others have concluded that the equation was being left behind by events (Burchell 2006). In an article cleverly titled "W(h)ither the Drake Equation?" Burchell concludes that with the rise of astrobiology, in which SETI is only one small intellectual part (and no federally funded programmatic part), the Drake Equation has become less important. The vast bulk of research in astrobiology today applies to microbes, and many consider not only that microbes may be the first extraterrestrial life we discover but also that microbes, not intelligence, may rule the universe. This may well be true. But it is also true that the existence of extraterrestrial intelligence remains a major scientific question, one whose funding has suffered mainly due to congressional politics in the United States. SETI may rise again and even become an integral part of astrobiology (Dick 2013), and when it does, the Drake Equation will remain as relevant as ever.

Some of the criticisms of the Drake Equation are constructive and well taken. But existential threats to its existence are exaggerated. We can see the continued utility of this type of equation in the fact that it has also inspired similar equations. Most recently, MIT astrophysicist Sara Seager adapted it to the current effort to search for biosignatures in exoplanet atmospheres. The "Seager Equation," written as $N = N^{*} F_Q F_{HZ} F_O F_L F_S$, estimates N, the number of planets with detectable biosignature gases, where N^{*} is the number of stars within the sample, F_Q the fraction of quiet stars, F_{HZ} the fraction with rocky planets in the habitable zone, F_O the fraction of observable systems, F_L the fraction with life, and F_S the fraction with detectable spectroscopic signatures. The new equation (Seager 2013) is sometimes referred to as a "revised Drake Equation," although it is not primarily addressed to the search for radio-communicative civilizations. But like the Drake Equation, it is driven by a practical consideration: if a system is built to detect biosignatures, what are the chances of success? The Transiting Exoplanet Survey

Satellite (TESS) and the James Webb Space Telescope (JWST) are systems under development to address this problem, among others, and the Seager Equation was developed for this purpose. Such biosignatures might range from simple microbial biosignatures to the most complex, technological ones. The Seager Equation is subject to many of the same uncertainties as the Drake Equation, but it is seen to be useful in any case.

To summarize, we may state the following: (1) It is important to specify what assumptions are made when putting numbers into the Drake Equation. Part of the enduring utility of the equation is that the parameters may be defined slightly differently, if one wants to know the number of civilizations, or the number of radio-communicative civilizations, or the number of civilizations communicative in some other region of the spectrum, and so on. All such calculations are possible as long as one defines how each term is being used. Even with such definitions, hidden assumptions abound, which some see as a weakness, but which may also be a strength in that they generate more discussion. (2) As is often stated, most of the parameters themselves are wildly uncertain, even at the beginning of the twenty-first century. Nevertheless, they hold the promise of improved estimates over time, as demonstrated by the current daily improvement in our knowledge of the fraction of stars with planets, based on both ground-based observations and the Kepler spacecraft results. Still, the equation should not be taken too seriously; its epistemological status is as a heuristic device rather than a law. (3) The equation itself is quite conservative in many ways, calculating only the number of radio-communicative civilizations, a number that may be much smaller than the total number of civilizations. And (4) the equation was a product of its time, but its utility is evident in its longevity as well as in the birth of new equations of the same kind such as the Seager Equation, which is also a product of its time.

When all is said and done, of course, the equation remains only a guideline. This, after all, is what Drake intended it to be when he first wrote the equation on the board at the Green Bank meeting in 1961 in an attempt to estimate the chances of success for any SETI search using radio telescopes. As in all areas of science, in the end there is no substitution for observations, no matter what the theory or expectations are. Even given its inherent limitations, the future of the original Drake Equation remains bright, precisely because it is simple but begs for more rigorous elaboration taking into account hidden assumptions, increasingly accurate parameters based on observation, and our subjective hopes and fears. Drake's image of the Christmas tree is a vivid and haunting one, each light representing a civilization blinking on, for a period short or long depending on its lifetime, then blinking off again, perhaps forever, whether due to a change in technology, self-destruction, or some unfathomed reason. That, too, is a reflection of the age during which the equation was constructed, with the threat of nuclear war blinking out one light on the tree of civilizations.

References

Bracewell, Ronald. 1960. "Communications from Superior Galactic Communities." *Nature* 186: 670–71. Repr. in Cameron 1963a, 243–48.

Briggs, William M. 2008. "Michael Crichton and SETI." November 7. Available online at http://wmbriggs.com/blog/?p=249/.

Brin, David. 1983. "The 'Great Silence': The Controversy Concerning Extraterrestrial Intelligent Life." *Quarterly Journal of the Royal Astronomical Society* 24: 283–309.

Burchell, Mark J. 2006. "W(h)ither the Drake Equation." *International Journal of Astrobiology* 5: 243–50.

Cameron, A. G. W. 1963a. *Interstellar Communication*. New York: W. A. Benjamin.

Cameron, A. G. W. 1963b. "Future Research on Interstellar Communication." In Cameron 1963a, 309–15.

Cameron, A. G. W. 1963c. "Communicating with Intelligent Life on Other Worlds." *Sky and Telescope* 23: 258–61.

Cirkovic, Milan. 2004. "On the Temporal Aspect of the Drake Equation and SETI." *Astrobiology* 4: 225–31.

Cocconi, G., and P. Morrison. 1959. "Searching for Interstellar Communications." *Nature* 184: 844–46.

Crichton, Michael. 2003. "Aliens Cause Global Warming." Available online at www.cfa.harvard.edu/~scranmer/SPD/crichton.html.

Denning, Kathryn. 2011. "L on Earth." In Vakoch 2011, 74–83.

Dick, Steven J. 1996. *The Biological Universe: The Twentieth Century Extraterrestrial Life Debate and the Limits of Science*. Cambridge University Press.

Dick, Steven J. 1998. *Life on Other Worlds*. Cambridge University Press.

Dick, Steven J. 2003. "Cultural Evolution, the Postbiological Universe and SETI." *International Journal of Astrobiology* 2: 65–74. Repr. as "Bringing Culture to Cosmos: The Postbiological Universe," In Dick and Lupisella 2009, 463–88.

Dick, Steven J. 2013. Congressional Testimony Before U.S. House Committee on Science. Available online at https://science.house.gov/hearing/full-committee-hearing-astrobiology-search-biosignatures-our-solar-system-and-beyond/.

Dick, Steven J., and Mark Lupisella. 2009. *Cosmos and Culture: Cultural Evolution in a Cosmic Context*. Washington, DC: NASA, SP-2009–4802.

Drake, Frank. 1959. "How Can We Detect Radio Transmissions from Distant Planetary Systems?" *Sky and Telescope* 19:140. Repr. in Cameron 1963a, 165–75.

Drake, Frank. 1961a. *Philosophical Society of Washington Lecture in Smithsonian Institution Archives*. Washington, DC.

Drake, Frank. 1961b. Drake to Pearman, March 13, 1961. NAS-NRC Archives: Space Science Board Records Group: ADM: C&B: SSB: Conferences: Extraterrestrial Intelligent Life: Green Bank (West Virginia): 1961–62.

Drake, Frank. 1965. "The Radio Search for Intelligent Extraterrestrial Life." In *Current Aspects of Exobiology*, ed. G. Mamikunian and M. H. Briggs, 323–45. Oxford University Press.

Drake, Frank. 1992. Oral History Interview by Steven Dick, May 29 and May 30, 1992. NASA HQ History Office.

Drake, Frank. 2013. "Reflections on the Equation." Special Issue on the Drake Equation. *International Journal of Astrobiology* 12(3): 173–76.

Drake, Frank. 2014a. Drake to Steven J. Dick, private communication, March 6.
Drake, Frank. 2014b. Drake to Steven J. Dick, private communication, March 10.
Drake, Frank. 2014c. Drake to Steven J. Dick, private communication, March 17.
Drake, Frank, and Dava Sobel. 1992. *Is Anyone Out There?* New York: Delacorte Press.
Dvorsky, George. 2007. "The Drake Equation Is Obsolete." Available online at www.sentientdevelopments.com/2007/05/drake-equation-is-obsolete.html.
Dyson, F. 1966. "The Search for Extraterrestrial Technology." In *Perspectives in Modern Physics*, ed. R. E. Marshak, 641–55. New York: Interscience Publishers.
Goldsmith, Donald. 1980. *The Quest for Extraterrestrial Life*. Mill Valley, CA: University Science Books.
Gonzalez, Guillermo, Donald Brownlee, and Peter Ward. 2001. "The Galactic Habitable Zone: Galactic Chemical Evolution." *Icarus* 152: 185–200.
Hart, Michael. 1975. "An Explanation for the Absence of Extraterrestrial Life on Earth." *Quarterly Journal of the Royal Astronomical Society* 16: 128–35. Repr. in Goldsmith 1980. 228–31.
Hetesi, Zsolt, and Zsolt Regaly. 2006. "A New Interpretation of the Drake Equation." *Journal of the British Interplanetary Society* 59: 11–14.
International Journal of Astrobiology. 2013. Special Issue on the Drake Equation. vol. 12, no. 3.
Kreifeldt, J. G. 1971. "A Formulation for the Number of Communicative Civilizations in the Galaxy." *Icarus* 14: 419–30.
Larson, R., and V. Bromm. 2001. "The First Stars in the Universe." *Scientific American* 285: 64–71.
Lederberg, Joshua. 1961. Lederberg to Pearman, April 11, 1961. NAS-NRC Archives: SSB Records Group: ADM: C&B: SSB: Conferences: Extraterrestrial Intelligent Life: Green Bank (West Virginia): 1961–62.
Lemonick, Michael D. 2014. "The First Starlight." *Scientific American* 310(April): 38–45.
Lineweaver, C. H., Y. Fenner, and B. K. Gibson. 2004. "The Galactic Habitable Zone and the Age Distribution of Complex Life in the Milky Way." *Science* 303: 59–62.
Maccone, Claudio. 2010. "The Statistical Drake Equation." *Acta Astronautica* 67, 1366–83.
Morrison, Philip. 1962. "Interstellar Communication." *Bulletin of the Philosophical Society of Washington* 16: 58. Repr. in Cameron 1963a, 249–71.
Oliver, B. M. 1971. *Project Cyclops: A Design Study of a System for Detecting Extraterrestrial Intelligent Life*. Moffett Field, CA: NASA Ames.
Pearman, J. P. T. 1961. Pearman to O. G. Villard, Jr., Dept. of Engineering, Stanford University, NAS-NRC Archives: SSB Records Group: ADM: C&B: SSB: Conferences: Extraterrestrial Intelligent Life: Green Bank (West Virginia): 1961–62.
Pearman, J. P. T. 1963. "Extraterrestrial Intelligent Life and Interstellar Communication: An Informal Discussion." In Cameron, 1963a, 287–93.
Sagan, Carl. 1963. "Direct Contact among Galactic Civilizations by Relativistic Interstellar Spaceflight." *Planetary and Space Science* 11: 485–98. Repr. in Goldsmith 1980, 205–13.
Sagan, Carl, ed. 1973. *Communication with Extraterrestrial Intelligence (CETI)*. Cambridge, MA: MIT Press.
Seager, Sara. 2013. "An Equation to Estimate the Probability of Identifying an Inhabited World within the Next Decade." Available online at www.cfa.harvard.edu/events/2013/postkepler/Exoplanets_in_the_Post_Kepler_Era/Program_files/Seager.pdf.
Shapley, Harlow. 1958. *Of Stars and Men*. Boston: Beacon Press.

Shklovskii, I. S., and Sagan, C. 1966. *Intelligent Life in the Universe*. San Francisco: Holden-Day.

Shostak, Seth. 2009. "The Value of 'L' and the Cosmic Bottleneck." In Dick and Lupisella 2009, 399–414.

Sullivan, Walter. 1964. *We Are Not Alone*. New York: McGraw-Hill.

Vakoch, Douglas. 2011. *Civilizations Beyond Earth: Extraterrestrial Life and Society*. New York: Berghahn Books.

Vakoch, Douglas. 2014. *Extraterrestrial Altruism: Evolution and Ethics in the Cosmos*. Berlin: Springer.

von Hoerner, Sebastian. 1961. "The Search for Signals from Other Civilizations." *Science* 134: 1839–43. Repr. in Cameron 1963a, 272–75.

Wallenhorst. S. G. 1981. "The Drake Equation Reexamined." *Quarterly Journal of the Royal Astronomical Society* 22: 380–87.

Walters, C., R. A. Hoover, and R. K. Kotra. 1980. "Interstellar Colonization: A New Parameter for the Drake Equation?" *Icarus* 41: 193–97.

Ward, Peter, and Donald Brownlee. 2000. *Rare Earth: Why Complex Life is Uncommon in the Universe*. New York. Copernicus Books.

Zaitsev, A. 2005. "Messaging to Extra-Terrestrial Intelligence." Available online at http://arxiv.org/abs/physics/0610031/.

Notes

I am thankful to the ever helpful archivists at the National Academy of Sciences, the National Radio Astronomy Observatory, the Smithsonian Institution, and the Library of Congress (Carl Sagan papers) for locating relevant documents at their institutions. I am grateful to David H. DeVorkin for comments on the paper, and to Ken Kellermann for consultation on the origin of the Drake Equation. And final thanks to Frank Drake for correspondence on the subject.

1

Rate of formation of stars suitable for the development of intelligent life, R*, pre-1961

David DeVorkin

Abstract

The rate of star formation, past and present, in the galaxy is an important astronomical question that lies at the heart of our modern understanding of the nature of stars, their distribution in space and time, and the history of the galaxy itself. Here we look at history, to 1961, concentrating on the question of how stars form and what is the rate of formation of stars in the galaxy. It is related to what astronomers call the luminosity function for stars, or the frequency distribution of luminosities for all stars in the galaxy. Our recounting will encompass the correlation era of stellar statistics, when astronomers began to conceptualize evolutionary models for the galaxy: that the luminosity function changes with time, and that the rate of change is determined not only by the rate of star formation but also by the nature and rate of the evolution of stars once they have been formed. We will show that, throughout much of the first half of the twentieth century, these rates were debated and revised many times, first by the recognition that there exists a main sequence of stars, which was interpreted as the distribution of stars existing in some form of quasi-stability. Later came the recognition of giant stars and instability, then the mass–luminosity law for stars, then an awareness of the energy sources of the stars and the subsequent revision of the evolutionary place and nature of giant stars, and finally, the fact that populations of stars existed within the galaxy that were generationally distinct. Based on this review, we will then argue that Drake estimated R* (regarded by most as the number of stars forming per year in the galaxy) in a manner consistent with the state of knowledge at the time.

The Drake Equation, ed. Douglas A. Vakoch and Matthew Dowd. Published by Cambridge University Press.

Estimating R*

In a 2003 reminiscence about how he formulated the eponymous Drake Equation, which attempts to estimate the number of radio-detectable civilizations (N) in the galaxy, Drake identified the first factor as being "proportional to the rate of star formation, which we write as R*, because the more stars you make, the more civilizations there will be, eventually. That's an easy one" (Drake 2003). In a 1970 review of the problem of interstellar communication, Carl Sagan summarized how the factor influenced N: "The more frequently stars are formed, the more likely it is that the galaxy will contain inhabited planets." Assuming that there are 10^{11} stars in our galaxy, and its age is some 10^{10} years, Sagan concluded that the factor was ten stars per year and considered this a lower limit (Sagan 1974, 13). Thinking this way, it appears to have been the easiest factor to discern – certainly the least controversial. But a detailed examination reveals historical complexities that, though probably of less concern to those who today pragmatically employ the Drake Equation to guess the number of communicating civilizations in the galaxy, provide an opportunity for historians of twentieth-century astronomy to explore the state of knowledge in past years of how stars form.

In his introduction to this book, Steven J. Dick points out that in more recent years, Drake clarified his definition of R* from the number of stars forming per year over the lifetime of the galaxy, to "the rate of star formation 4.5 billion years ago when our Sun and its planets were formed." That Drake made this change in the meaning of R* is not surprising. From what we know today, it makes sense to estimate the rate in the past, at a time sufficiently distant to give a star enough time to form planets and produce civilizations that might well exist today that express intelligence through an organized technological capability and desire to communicate over interstellar distances. The question is, would it have been sensible in 1961, given Drake's goals? We will explore this question after we make a quick reconnaissance of evidence for evolutionary change, especially the formation of stars, from antiquity to the eve of the Drake Equation.

Descriptive evidence for evolutionary change

The universe of classical antiquity was static, save for the orderly motions of the Sun, Moon, and planets. No change was allowed. But over the many intervening centuries, first with discoveries made during the Renaissance and then by sensibilities fostered during the Enlightenment, the concept of time emerged as a dimension through which change was possible. To paraphrase A. O. Lovejoy, the "inventory" of nature (the classification of forms presumed static throughout all time) became the "programme" of nature, "which is being carried out gradually and exceedingly slowly in the cosmic history" (Lovejoy 1960, 244).

In the age of Newton, gravity became the way stars form from nebulae. As early as 1692, Newton envisioned the process in a letter to the Reverend Dr. Bentley (Turnbull 1961, 234), and Laplace built the solar system on the idea. But William Herschel put the idea on a firmer observational footing (Holden 1881, 202–4; Hoskin 1959, 30; 1967, 80). By the 1830s, Herschel's description of the many forms of nebulae, and his classification that suggested an evolutionary course, were not universally accepted. But they introduced the concept of change, or process, into astronomy. The work of two generations of Herschels caused Hermann von Helmholtz, G. B. Airy, and others to remark from time to time that the various forms of observed nebulae stimulated "the idea of change" in the heavens, and that the old universe, which remained static with time, had been replaced by one where processes of structural change were everywhere evident (Helmholtz 1856, 503). Even by mid-century, when the observations of Lord Rosse suggested that all nebulae were in fact unresolved clusters of stars, the existence of true nebulae was still believed by some, though the concept had certainly been weakened. The problem was that, without nebulous matter, out of what do stars form? During the 1850s, meteoritic theories became temporarily popular, not only as the source of heat for the stars but also as the medium out of which they formed. Though meteoritic theories survived for some time, notably as championed by J. Norman Lockyer, by the 1860s, spectroscopic studies of nebulae by William Huggins and others demonstrated that true nebulae existed. They also showed that stars formed from varieties of nebulae through the action of gravity and then were thought to either heat or cool, depending on the degree to which astronomers believed these supposedly largely gaseous bodies behaved according to the newly articulated ideal gas laws. The direction of stellar evolution, if it was indeed revealed somehow through the various classes of stellar spectra, remained a topic of debate through the first half of the twentieth century (DeVorkin 1978; Meadows 1972).

The rise of the conservation laws, especially that of energy, brought these questions into sharper focus: What was the source of stellar light and heat? Were nebulae initially hot, with stars forming as hot bodies? Or were nebulae cold, with the source of heat for stars coming from another yet-unknown mechanism such as combustion or electrical action? Before mid-century, gravity was thought merely to be the cause of the process of contraction, and it was not considered to be an agent that produced stellar heat until the rise of kinetic gas theory. And even then, by the turn of the century, time scales for stars could not match uniformitarian estimates of the age of the Earth (Brush 1996; Burchfield 1990; Hufbauer 1991). Another nagging question that remained after the introduction of the spectroscope was: Did all nebulae produce stars, or were some nebulae in fact giant clusters of stars at distances too vast to resolve?

These questions vexed astronomers throughout the first half of the twentieth century, but by 1955, all had been, if not resolved, at least put on a firm-enough basis to create a working consensus. Historians have extensively documented both the solution to the stellar energy problem and the history of the recognition that galaxies exist and are moving away from one another (Hufbauer 1981; Smith 1982). Less attention has been given to tracing theories of star formation.

Star formation in the past

Given this general framework, what was the state of knowledge, circa 1960, about past star formation rates? We ask this question to set the stage for appreciating the thread we develop later concerning both star formation and galaxy evolution. In 1957, in a popular *Leaflet* essay for the Astronomical Society of the Pacific, Su-Shu Huang at Berkeley put it bluntly. He identified the problem of star formation as being two separate questions: star formation at present and star formation "at an early time when the galaxy itself was being shaped." He limited his discussion to the present because the second "is obviously more speculative ... because we really do not know the physical and dynamical conditions of our galaxy in its early stages of evolution." Further, "we do not have observations to guide our thinking" (Huang 1957, 1). Huang, a student of Subrahmanyan Chandrasekhar at Yerkes in the late 1940s who also worked with Otto Struve and then joined him at Berkeley in the 1950s, blended mathematical theory and computational methods with observational data, devoting his career to problems in stellar astrophysics along lines Struve had established – namely, the evolutionary effects of stellar rotation, turbulence in stellar atmospheres, and binary star formation and evolution. By 1959, influenced by Struve, he showed interest in "The Problem of Life in the Universe and the Mode of Star Formation," contributing essays to *American Scientist* and to another *Leaflet* in that year (Huang 1959). Even though it was implied in the title, Huang provided no commentary on the state of knowledge of star formation. But he did reveal the deep influence Struve had on him, who evidently talked constantly and passionately in the halls and warrens of Berkeley's Leuschner Observatory about the lingering time-scale problems, the sizes of habitable zones, and the present impossibility of detecting planets around other stars that could be friendly to life, only three of which were known to exist within five parsecs (the Sun, ε Eridani, and τ Ceti). He also adopted Struve's three lines of indirect evidence for the occurrence of planetary systems: the abrupt drop in axial rotation of stars near F5, Harold Urey's arguments for the origin of meteorites in "pre-stellar" nuclei, and the ubiquity of binary systems (Huang 1959). Huang's disinclination to speculate on primordial star-formation rates was acquired from Struve, who lectured on the subject throughout the 1950s. At the outset of his 1950 monograph, *Stellar Evolution*, he illustrated the "hypothetical nature of the evolutionary process" in a manner reminiscent of William Herschel:

In trying to discover the origin and the evolution of the stars the astronomer is confronted with only a snapshot of the galaxy as it is now. (Struve 1950, ix)

Struve's monograph derived from his Vanuxem lectures at Princeton in 1949, a series devoted to exploring evolution in the "material universe" as manifest across the physical and biological sciences. Though he framed his discussion as "hypothetical," he was optimistic that recent advances in astronomical observation would no doubt further constrain speculation. For the first time in history, he claimed, astronomers had at least gained the "few fleeting glimpses of the truly magnificent and awe-inspiring panorama of creation":

We are no longer limited to our own ability to reason out what went before and what will come next, but we can actually see, with our own eyes, how stars shed matter at their equators and produce gaseous rings and envelopes around them, thereby losing mass and rotational momentum; or how newly formed stars gather up dust from the clouds of diffuse interstellar material in which they are embedded, and thereby gain mass, and probably rotational momentum. (Struve 1950, x)

In 1950, with Caltech's 200-inch and McDonald's 84-inch telescopes fully operational, and with a new 120-inch one on the way for Lick, all amplified by exciting new electronic detectors and new federal patrons, Struve could think positively about observational capability in the future. But his positive statement also carried with it clear indecision. One can find at least two scenarios in his introductory remarks above that reflect contemporary views of how stars might evolve.

Struve's text also provides a data point for other issues at the time relating to the interface between stellar evolution and the evolution of the galaxy. He fully acknowledged the implications of Walter Baade's identification and evolutionary interpretation of the two stellar populations, which finally settled the long-known curiosity of why mean velocities increased with advancing spectral type among main sequence stars. But he shied away from declaring that the 1944 work would soon become the major watershed for understanding the evolution of the galaxy, at least by 1953.

In the 1940s, during wartime blackouts and using red-sensitive photographic emulsions on the 100-inch Hooker reflector, Baade was able to obtain high-resolution images of the nucleus of M31 as well as its companion galaxies M32 and NGC 205, successfully resolving stars there. He found that when plotted in a Hertzsprung–Russell (H–R) Diagram these stars were distributed differently from stars in the solar neighborhood. Their distribution resembled stars that were found to be moving at high velocities relative to the Sun and the spiral arm stars in the Milky Way. In so doing, he established two "populations" of stars distinguished by kinematics, location, and, eventually, composition (Baade 1944).

Struve was also not ready to discuss explicitly how the galaxy looked in past times, though he tentatively acknowledged the theoretical studies of Anders Reiz, Martin Schwarzschild, and Robert Richardson, who were just then building star models with convective cores and radiative envelopes that, with hydrogen exhaustion, caused them to become transformed, by the Bethe-Weizsäcker process, from main sequence stars to the realm of the giants (Struve 1950, 149). Struve presented this possibility along with other mechanisms for evolution, including through rotational breakup and a view held persistently by some astronomers that evolution occurred by accretion of mass as the star passed through a dense interstellar cloud. But he did not express how profoundly this changed the evolutionary status of giants, which in his lifetime had started out as stars in the process of formation and therefore indicative of early life, but were now seen as evolved stars, far from their origins (DeVorkin 2006).

Struve felt equally uncomfortable relating how stars form, even though he had rich observational evidence linking stars associated with nebulae of various types. He built upon recent work by Lyman Spitzer, Fred Whipple, and Bart Bok who were extending, and linking to observation, theoretical studies by James Jeans and others in the 1930s on how self-gravitating bodies could form out of a diffuse medium as a function of the temperature and density of that medium. Although Spitzer and others had great difficulties showing how stars of single solar mass could form if the general medium had the properties it did at the present time, he could show how massive clouds could collapse into a multitude of nuclei forming clusters of stars, a class of object commonly observed in the galaxy. Even so, at the time of Struve's lectures, there were lingering suggestions by highly respected theorists ranging from von Weizsäcker to Hoyle and Lyttleton that star formation and evolution was also an accretion process, and not just a process of gravitational collapse (Struve 1950, 105). Struve expressed confidence that somehow these mechanisms would be reconciled. After all, there was clear evidence from observations by Bok of the commonality of dark clouds of gas and dust that were more or less spherical, with observed diameters as small as 0.06 parsecs and masses on the order of the Sun. Even here, though, the globules could be nuclei of material gathering mass with time, or they could be self-gravitating, collapsing into stars. Struve cited Fred Whipple's recent efforts using Mount Wilson radial-velocity data to show that a nonrotating cloud that was sufficiently turbulent could possibly collapse before it suffered much accretion. Nevertheless, so powerful was the descriptive evidence that Struve wanted to conclude that all "indications are that interstellar matter can and does condense into protostars which later form real stars" but oddly added that "we do not definitely know whether the newly created stars appear at the upper end or at the lower end of the Main Sequence in the H-R diagram" (Struve 1950, 106).

A statement like this neatly illustrates Struve's strong dependence on observational evidence. The H–R Diagram was a powerful tool for describing the course of stellar evolution, and Struve was especially adept at applying it. He well knew that hot early-type stars were intimately associated with gas and dust clouds, and they were formed recently according to the implications of the mass–luminosity relationship. But he also knew he had no definitive observations for the formation of stars with less mass than the Sun or if those stars had started farther up or lower down the main sequence. Indeed, he alluded to theoretical evidence by the brilliant Soviet theorist V. A. Ambartsumian that stars did not move much up or down the main sequence, as previously thought. Where they were observed at present on the main sequence was close to where they arrived originally. Still, Struve concluded: "It is at present somewhat disconcerting that we cannot tell definitely whether a dwarf is or is not formed directly out of the interstellar matter" (Struve 1950, 106).

The sidereal problem

Struve's hesitancy belies the legacy of what was at the turn of the century variously called the "sidereal problem" or "sidereal question": the constitution, structure, and dynamical behavior of the sidereal system, or the "Sidereal Universe" (Rolston 1907; Plaskett 1915; Paul 1993). The great photographic star-mapping projects of the time, the Astrographic Chart and the Carte du Ciel, promised positional information and proper motions, but there had been no reconnaissance of the distribution of the spectroscopic character of the stars, neither the stellar types nor their radial motions, with magnitude and distance. This was the lifework of Jacobus C. Kapteyn, who established statistical astronomy as a research program, and in 1904–5 announced that stellar motions were not random but consisted of two major "streams" (Paul 1993, 89). Building on the descriptive efforts of R. A. Proctor and B. A. Gould, and the statistical efforts of W. H. S. Monck and H. Kobold in the 1890s, Kapteyn had by the first and second decades of the twentieth century established the existence of star streaming, which soon took on an evolutionary interpretation. He did this by encouraging a coordinated international effort to secure astrophysical information, mainly radial velocities, spectral types, and magnitudes, on stars within selected regions of the sky. By 1901, he had derived a refined value for the solar motion from proper motion data and had created a statistical method for measuring the distances to groups of stars by means of their parallactic components of proper motion due to solar motion – the technique of secular parallaxes. In three more years, he announced his theory of star streaming. In an early discussion of his two streams, he acknowledged that Kobold had anticipated streaming in the 1890s, but it was not until Kapteyn separated the effect of solar motion that a clear picture emerged. In 1905, Kapteyn did not apply the existence of streams to evolution, or evolution to the interpretation of his streams, but by 1910, others made this connection, including W. W. Campbell, A. S. Eddington, and George Ellery Hale (DeVorkin 1978, 243–44).

The evolutionary significance of Kapteyn's streams linked the origins of the stars to the structural and dynamical history of the galaxy, or the "sidereal problem." Astronomers had been correlating the various spectral classes with other observed properties of the stars for some time. The mere association of B-type stars with nebulae and the plane of the Milky Way was reason enough for classifiers like Fleming, Maury, and Cannon at Harvard to place these stars before the A types in the Draper classification (DeVorkin 1978, 237). The wealthy private astronomer Frank McClean, who among other accomplishments discovered oxygen lines in what were believed to be the hottest, or the so-called Helium stars, also remarked on their close association to the Milky Way. McClean had further noted in the late 1890s that these stars were also the most closely associated with the nebulae, and both they and the nebulae were the youngest of the classes. He also stated that, overall, "as the stellar types of spectra become more advanced they are found to be more evenly distributed . . . stars of the solar type . . . [which] started their lives as helium stars . . . before the condensation of the galaxy" (McClean 1896, 428; see also McClean 1897).

These descriptive observations of evolutionary order, based on rough statistics, were soon highly constrained by Kapteyn's star streams, but all of them eventually supported an evolutionary scenario. Basically, what Kapteyn found was that stars moved preferentially "in two distinct and diametrically opposite directions" (Paul 1993, 88–89). All stars, Kapteyn claimed, belonged to one or the other stream. In the first expressions of his discovery between 1902 and 1905, he called for more observations, specifically for radial velocities that might well reveal other correlations, such as with spectral class and galactic latitude.

W. W. Campbell of the Lick Observatory had already initiated a vast radial-velocity program with Lick's 36 inch refractor, and in fact was also finding correlations that revealed systematic trends with spectral class. After 1910, he speculated on the meaning of stream motion, by now offered not only by Kapteyn but also in variations by Karl Schwarzschild and Arthur Stanley Eddington. Campbell listed various phenomena associated with the increase of stellar velocity and interpreted them as correlating "with increasing effective stellar ages." His was not the only voice, but became the most authoritative as the producer of the data. He also remarked repeatedly on the rarity of B stars and emphasized their close relationship to the Milky Way, their association with irregular nebulae (through similar velocities), and their apparent clustering tendency – all of which suggested stars of young age and recent formation.

In his 1914 book *Stellar Movements and the Structure of the Universe*, Eddington also felt that the increase of velocity with advancing spectral type was significant for the earlier classes, but that the data at hand for the later classes was not yet conclusive. He did, however, believe in the fundamental reality of this discovery and interpreted it along evolutionary lines: "How then is it that the M stars show practically no galactic concentration, whereas the A stars are strongly condensed?" (Eddington 1914, 168). Type B, on account of their low individual speeds and their evident youth, remained "strongly condensed in the plane. In succeeding stages the stars have had time to stray farther from the galactic plane. . . . In the latest type, M, the stars have become almost uniformly scattered" (169).

By the 1930s, star streaming was shown to be caused by the differential rotation of the Milky Way, and in the 1940s, it was reset wholly within the context of Baade's populations. So its evolutionary overtones lasted for decades, embodied in an associated concept Kapteyn established, which held that there existed a "luminosity function" for the distribution of stars through space. Tracing the development of this concept brings us back to Struve.

The luminosity function

In 1958, Walter Baade was invited to Harvard to lecture on the "Evolution of Stars and Galaxies," and soon after his death, Cecilia Payne-Gaposchkin edited the transcripts into a monograph published by Harvard in 1963 (Baade 1963). At the outset, he looked back upon Kapteyn's legacy, describing the luminosity function and how Kapteyn originally wanted to use it as a probe of the structure and extent of the galaxy, his solution to the "sidereal problem," or as astronomers referred to it later, as the "Kapteyn Universe." To this, Baade bluntly stated: "Today nobody would use the luminosity function to probe the structure of the Galaxy." But, he quickly added, "Currently the luminosity function interests us for a very different reason: it gives valuable information about the history of star formation in the Galaxy." (Baade 1963, 2–3).

Kapteyn had been developing this concept since the 1880s, well aware that others were finding correlations between spectral type with proper motions and with galactic latitude. By 1901, he had gathered and correlated statistics on enough stars of observed magnitude and proper motion to begin estimating their mean parallaxes as a function of apparent magnitude. From these first steps, he set down arguments for the determination of what he called "Absolute luminosity or absolute magnitude" and from his statistics was able to derive estimates for the density of stars in space as a function of magnitude and from this a "luminosity curve" (Kapteyn 1902, 12). Among its other uses, Kapteyn believed that directional variations in the luminosity function might reveal systematic effects. He was most interested in detecting a galactic latitude effect for "galactic and extra-galactic regions" and through this, with far more extensive observations than were available at the time, "we will be led to a better understanding of the real structure of the galactic system" (Kapteyn 1902, 20).

Kapteyn also looked for systematic differences in star density and luminosity in the gross spectral classes, limiting attention to Vogel's Type I (Draper A, F, and G) and II (Draper K) stars. He separated them into magnitude and galactic latitude groups, searching for evidence that the space density of stars increases or decreases with distance from the Sun. As E. Robert Paul has argued, this was the essence of Kapteyn's agenda, one he shared with Hugo von Seeliger and others, to elucidate the "sidereal question," which was nothing less than, as he expressed it in 1920, "a law of nature, a law which plays a dominant part in the most diverse natural phenomena" (Kapteyn and van Rhijn 1920; quoted in Paul 1993, 100–1).

Kapteyn keenly knew that his density distributions in the Milky Way would not be reliable unless he accurately knew the absorption of light in the galaxy. It became P. J. van Rhijn's mission to separate luminosity values from possible systematic effects due to interstellar absorption. By the 1930s, van Rhijn, who had been Kapteyn's student and was now his intellectual successor, was ready to provide refined coefficients for both visual and photographic absorption. He derived photographic absorption from Lick and Harvard data on radial velocities and absolute magnitudes of eight open clusters, following techniques also developed by Robert Trumpler. Armed now with the knowledge that the galaxy is in differential rotation, as described by Jan Oort and Bertil Lindblad, van Rhijn derived mean distances employing the Oort formula to find a photographic absorption of 1.00 magnitude per kiloparsec and visual absorption of 0.5 magnitude per kiloparsec (van Rhijn 1929; 1936).

With these results, van Rhijn turned to the luminosity function, developing expressions for the density distribution in the Milky Way from longitudes 0 to 200 degrees using star count numbers as a function of apparent photographic magnitudes and the distribution of absolute magnitudes in the solar neighborhood (van Rhijn 1925; 1929; 1936). In the first solution of the integral relating the density, star counts, and distribution of absolute magnitudes, he neglected to account for the effect of absorption. But then, in 1936, he applied a more sophisticated statistical procedure developed by F. H. Seares that included the effect of absorption, solving the resulting integral relation numerically to derive densities.

Van Rhijn derived the luminosity by two somewhat independent methods. The first, due originally to Kapteyn, used probable parallaxes derived from mean parallax data based upon a summation over the proper motion range. This yielded the total number of stars in each parallax range and their apparent magnitude interval. Of course, knowing the apparent magnitude and parallax yields absolute magnitude, and from that comes the total number of stars for each absolute magnitude interval.

His second method of deriving the luminosity function by trigonometric parallaxes was more direct but limited. Using Allegheny, McCormick, Mount Wilson, Greenwich, and Yerkes data (but carefully excluding binary effects), he counted the number of stars between determined limits of parallax and apparent magnitude. Here he also accounted for effects due to selection and errors of observation in the parallaxes. The primary selection effect was the choice of stars for parallax study, which were typically of high proper motion.

As a check on his methods, van Rhijn re-derived mean parallaxes from his derived density distribution and found that the mean parallaxes computed by means of the adopted densities agreed pretty well with the values derived from parallactic motion. Even so, he cautioned that his derivation was preliminary because he had assumed that the variation of absorption in latitude was independent of longitude and that the decrease in density was constant anywhere in the galaxy in a direction perpendicular to the plane. Through the rest of the 1930s, van Rhijn's followers, including Willem J. Luyten, worked to refine the luminosity function.

Van Rhijn's application of Robert Trumpler's open cluster studies, dating from the late 1920s, brought to bear another descriptive correlative tool that had been articulated first by Ejnar Hertzsprung using mean proper motion data and then independently by Henry Norris Russell using parallax data, known today as the H–R diagram (DeVorkin 1978; 2000). Indeed, since its discovery in the first decades of the century, what was called at first the Russell diagram and then the H–R diagram has become a powerful tool for exploring the stellar universe, specifically theories of how stars are born and live their lives. Equating the intrinsic brightness of stars against their observed colors (interpreted as temperature), they found that the vast majority of stars occupy a diagonal band called the main sequence, from bright blue stars at the upper left to dim red stars at the lower right. But there were a few stars, both blue and red, that were all intrinsically bright. Astronomers soon adopted these two sequences as two classes of stars, distinguished by size: the main sequence contained the "dwarfs" and the bright ones were the "giants."

Moreover, the H–R diagram demonstrated that the vast majority of stars were associated in a definite sequence that to both Russell and Hertzsprung had evolutionary significance. Russell, more explicitly, viewed it as evidence that stars began their lives as red giants, and under gravitational contraction heated to the point where densities in their interiors became too great to allow further contraction in an ideal state, thus causing the body to cool. The main sequence therefore became the graphical locus where stars lost their ideal gas state and cooled to extinction.

Russell's theory remained popular until the mid-1920s. But it was seriously questioned after James Jeans, following a suggestion by F. A. Lindemann, showed that contracting gas spheres remained in the perfect gas state even when on the main sequence. Eddington's masterful rationalization known as the mass–luminosity law was then the final blow to Russell's "giant to dwarf" theory. Not only was the rate of evolution of a star highly sensitive to its mass, but its very formation was now also dependent upon the mass ranges for collapsing clouds of interstellar material and how those masses were distributed according to what was seen in the sky – namely, clusters of stars.

In the late 1920s through the 1930s, Robert Trumpler at Lick had explored the appearance of open clusters on the H–R diagram, taking great pains to reduce observational errors mainly through more sensitive means of determining membership and rejecting field stars. His meticulous efforts – taking spectra, deriving color indices, and measuring radial velocities – resulted in the first systematic evidence for general interstellar reddening and absorption. But his efforts also revealed another characteristic among clusters: The distribution of stars for each cluster along the main sequence seemed to be loosely correlated with the physical degree of condensation, or the structure of the cluster, in space. Tightly concentrated clusters contained blue hot stars and clusters more distributed in space contained more common types of the middle spectral classes. Trumpler wondered what this meant as early as 1925, speculating that either stars in a cluster are not coeval or they were all of the same age, indicating that they ran "through their evolutionary course with unequal speed" (Trumpler 1925, 315). Trumpler suggested evolutionary scenarios for his clusters based upon Russell's scheme, which of course arrived stillborn. On the other hand, he did establish a new way to look at cluster diagrams, which Gerard Kuiper followed up on later in the 1930s.

Trumpler's classification intrigued Kuiper, who had spent some time at Lick but was, by the mid-1930s, at the Yerkes Observatory. Kuiper was also aware of Bengt Strömgren's efforts to use the H–R diagram to plot the evolutionary course of stars according to their fractional hydrogen content. Strömgren showed heuristically that if a star consumed hydrogen during its lifetime and was always fully mixed, it would expand with time, moving from the region of the main sequence on the H–R diagram up and to the right into the realm of the giants. By the late 1930s, with both Strömgren and S. Chandrasekhar at hand at Yerkes, and with Struve's encouragement, Kuiper re-examined some fourteen Trumpler clusters, superimposing them on a single H–R diagram. What was striking about this exercise was that the lower main sequences coincided, but the upper main sequences diverged for higher luminosities, curving up and to the right just as Strömgren's evolutionary scenario suggested (DeVorkin 2006, 435–36). Struve's staff then started wondering what Kapteyn's luminosity function was really telling them about stellar formation and evolution.

Kuiper's analysis remained inconclusive until the postwar years, when in the wake of the solution to the stellar energy problem by Bethe and von Weizsäcker, George Gamow campaigned to establish giant suns as evolved main sequence stars. His freewheeling style made the field somewhat controversial throughout the 1940s, but resulted in solid work at the observational–theoretical interface by a number of teams, including Chandrasekhar and his students at Chicago, as well as Schwarzschild and students and colleagues at Princeton and Mount Wilson-Palomar. In a series of collaborations with Baade, William Baum, Allan Sandage, B. Oke, and later Fred Hoyle, Schwarzschild managed to establish giants as evolved main sequence stars.

At first, as we've seen in Struve's review, there was cautious acceptance of the idea that giants are evolved stars, but solid acceptance came quickly by the mid 1950s, where the new evolutionary order was being generally incorporated into a wide range of problems (DeVorkin 2006). In a highly cited paper, Ed Salpeter showed in 1955 that the observed luminosity function for main sequence stars in the solar neighborhood could now be used along with these new theories of stellar evolution to produce what he called an initial luminosity function (Salpeter 1955, 161). Salpeter employed the Schoenberg-Chandrasekhar limit, which required that stars move off the Main Sequence after 10 percent of the hydrogen in the core was consumed. The initial luminosity function then became a measure of the rate of star creation as a function of mass.

So why did Drake use a simple ratio at first?

Salpeter's work was a watershed in modeling the history of the galaxy in terms of the luminosity function, but it was far from complete. In fact, he had assumed that stars were being created at a uniform rate in the solar neighborhood for the last five billion years. He, and virtually everyone else, remained wary of what was still, to be sure, a murky field: the evolution of the galaxy and its impact on the rate of star formation. In the late 1950s, Martin Schwarzschild cautioned his students to be wary of any sort of initial "birth-rate function." Estimates for young Population I stars (bright blue stars confined to the plane of the galaxy), where a large fraction of the available mass preferentially creates the brighter stars according to observation, did not mean the same preferences applied to the older Population II stars (halo stars punctuated by the red giants), since the formation of that generation "is a major unsolved problem at present and a serious stumbling block for the theory of the over-all galactic evolution" (Schwarzschild 1958, 290).

Maarten Schmidt made this quite clear at the Berkeley International Astronomical Union meetings in 1961. Surveying the question of the evolution of the stellar content of galaxies, he began by admitting that no firm conclusions could be drawn, and "our considerations will be necessarily of a somewhat speculative nature" (Schmidt 1962, 170). Starting within a framework based upon the Hubble classification for galaxies and employing its evolutionary implications, Schmidt made the "fairly safe" assumption that "the rate of star formation in a galaxy as a function of time does not increase." Given that, and the changing fraction of stars in various states of development due to the effects of the mass–luminosity law, he concluded that galaxies could be expected to redden over time. Throughout his review he employed enough qualifiers to make almost everything seem tentative. And in describing the behavior and growth of amorphous regions in galaxies of various types, following Hubble's scenario, he quipped, "One gets the impression that a disease that prevents the formation of bright stars spreads from the center of a system as it evolves and, eventually, affects the whole system" (Schmidt 1962, 171). He made many suggestions, calling for improved quantitative estimates of the change in the luminosity of a galaxy over time, as well as changes in gas content and in the abundance of the heavy elements in the interstellar medium. But he then observed that, in fact, "the only way to find out about the past formation rate is through the end products of stellar evolution: white dwarfs, helium, and heavy elements." Even so, he knew that such analyses were fraught with many assumptions. If the local interstellar medium was constant over the lifetime of the galaxy, then such analyses would possibly yield useful insights. If, however, the medium has not been constant, but has been somehow replenished or altered in any way, "the situation is too complicated and practically nothing can be derived about star formation" (Schmidt 1962, 174–75).

Thus in the early 1960s, a framework had been provided by Salpeter, but there was considerable debate and doubt over the nature of galactic evolution and hence the history of star formation. For example, in 1961 Hayashi was only just beginning to reform ideas about pre–main sequence stellar evolution, especially for solar mass stars and below, showing that contracting masses were fully convective, and this, though impressive, reminded astronomers of how much about the past was still unknown.

In the inimitable words of Virginia Trimble, who was speaking of first light and reionization in 2006, it is always safer to be a uniformitarian than a catastrophist (Trimble 2006). She knew only too well, from the brilliant work of Beatrice Tinsley in a highly cited paper derived from her 1967 University of Texas PhD thesis, "Evolution of Galaxies and its Significance for Cosmology," that the whole picture of galaxy evolution had been thrown into deep debate. Tinsley showed, for the first time, that the chemical history of the galaxy since the formation of Population I stars should indeed be significant and moreover that evolutionary corrections – far larger than those that had been suggested by Sandage in 1961 – were required. She showed that these changes were significant enough to be observable in distant galaxies (Faber 1981; Tinsley 1968).

Given his singular goal and the state of knowledge at the time, Frank Drake had every reason to keep it simple. On the eve of the symbolic launch of SETI, which was tied to the 500th anniversary of Columbus's arrival in the New World, John Noble Wilford reported for the *New York Times* that NASA was funding a ten-year program, which, if successful, "would surely rank as one of the transcendental discoveries." R*, for the *Times*, was simply "the number of new stars formed in our galaxy each year" (Wilford 1992, C1). Given the fanfare that continues to this day, there is every reason to believe that Drake achieved his goal. Typing "Drake Equation" into Google on April 18, 2014, returned 174,000 hits in 0.44 seconds, but only 332 hits from a full word search in the comprehensive search engine for astronomy, the Astrophysics Data System (ADS N.d.).

References

Astrophysics Data System (ADS). N.d. Available online at http://adsabs.harvard.edu/.

Baade, Walter. 1944. "The Resolution of Messier 32, NGC 205, and the Central Region of the Andromeda Nebula." *Contributions from the Mount Wilson Observatory* 696: 1–10. Washington, DC: Carnegie Institution of Washington.

Baade, Walter. 1963. *Evolution of Stars and Galaxies*. Cecilia Payne-Gaposchkin, ed. Cambridge, MA: Harvard University Press.

Brush, Stephen G. 1996. *Transmuted Past: The Age of the Earth and the Evolution of the Elements from Lyell to Patterson*. New York: Cambridge University Press.

Burchfield, Joe D. 1990. *Lord Kelvin and the Age of the Earth*. University of Chicago Press.

DeVorkin, D. 1978. *An Astronomical Symbiosis: Stellar Evolution and Spectral Classification, 1860–1910*. Unpubl. diss., U. Leicester. Available online at https://lra.le.ac.uk/handle/2381/10909/.

DeVorkin, D. 2000. *Henry Norris Russell: Dean of American Astronomers*. Princeton University Press.

DeVorkin, D. 2006. "The Changing Place of Red Giant Stars in the Evolutionary Process." *Journal for the History of Astronomy* 37: 429–69.

Drake, Frank. 2003. "The Drake Equation Revisited: Part I." Available online at www.astrobio.net/index.php?option=com_retrospection&task=detail&id=610.

Eddington, A. S. 1914. *Stellar Movements and the Structure of the Universe*. London: Macmillan.

Faber, Sandra. 1981. "Beatrice Tinsley." *Physics Today* 34(9): 110.

Helmholtz, H. von. 1856. "LXIV. On the Interaction of Natural Forces." *Philosophical Magazine*, 4th series, 11: 489–518.

Holden, E. S. 1881. *Sir William Herschel*. New York: C. Scribner's Sons.

Hoskin, M. 1959. *William Herschel and the Construction of the Heavens*. London: Sched and Ward. Reprinted 1963, London: Oldbourne.

Hoskin, M. 1967. "Apparatus and Ideas in Mid-Nineteenth-Century Cosmology." *Vistas in Astronomy* 9: 79–85.

Huang, Su-Shu. 1957. "Star Formation." *Astronomical Society of the Pacific Leaflet* 7 (340), September.

Huang, Su-Shu. 1959. "The Problem of Life in the Universe and the Mode of Star Formation." *Publications of the Astronomical Society of the Pacific* 71: 421–24.

Hufbauer, Karl. 1981. "Astronomers Take Up the Stellar-Energy Problem, 1917–1920." *Historical Studies in the Physical Sciences* 11: 277–303.

Hufbauer, Karl. 1991. *Exploring the Sun: Solar Science since Galileo*. Baltimore: Johns Hopkins University Press.

Kapteyn [Kapteijn], J. C. 1902. "No. II. On the Luminosity of the Fixed Stars." *Publications of the Kapteyn Astronomical Laboratory Groningen* 11: 3–32. Formerly *Publications of the Astronomical Laboratory at Groningen*.

Kapteyn, J. C., and P. J. van Rhijn. 1920. "No. 188. On the Distribution of the Stars in Space Especially in the High Galactic Latitudes." *Contributions from the Mount Wilson Observatory* 188: 1–16. Washington: Carnegie Institution of Washington.

Lovejoy, A. O. 1960. *The Great Chain of Being: A Study of the History of an Idea*. New York: Harper.

McClean, Frank. 1896. "Photographs of the Spectra of Twenty-Three Characteristic Helium Stars; also Photographs of the Spectra of Six Stars of the Third Magnitude, Showing the Transitions from Type to Type." *Monthly Notices of the Royal Astronomical Society* 56: 428–29.

McClean, Frank. 1897. "On Comparative Photographic Spectra of Stars to the 3–1/2 Magnitude." *Proceedings of the Royal Society of London* 61: 213–16.

Meadows, A. J. 1972. *Science and Controversy: A Biography of Sir Norman Lockyer*. London: Macmillan. Second ed., Palgrave Macmillan, 2008.

Paul, E. Robert. 1993. *The Milky Way Galaxy and Statistical Cosmology, 1890–1924*. New York: Cambridge University Press.

Plaskett, J. S. 1915. "The Sidereal Universe." *Journal of the Royal Astronomical Society of Canada* 9: 37–56.

Rolston, W. E. 1907. "A Proposed International Attack on the Sidereal Problem." *Nature* 75: 427–28.

Sagan, Carl. 1974. "An Introduction to the Problem of Interstellar Communication." In *Interstellar Communication: Scientific Perspectives*, ed. C. Ponnamperuma and A. G. W. Cameron, 1–25. Boston: Houghton-Mifflin.

Salpeter, E, 1955. "The Luminosity Function and Stellar Evolution," *Astrophysical Journal* 121: 161–67.

Schmidt, M. 1962. "Evolution of Stellar Content of Galaxies." In *Problems of Extra-Galactic Research, Proceedings from IAU Symposium no. 15*, ed. G. McVittie, 170–76. New York: Macmillan.

Schwarzschild, Martin. 1958. *The Structure and Evolution of the Stars*. Princeton University Press.

Smith, Robert W. 1982. *The Expanding Universe: Astronomy's "Great Debate," 1900–1931*. Cambridge University Press.

Struve, Otto. 1950. *Stellar Evolution: An Exploration from the Observatory*. Princeton University Press.

Tinsley, Beatrice. 1968. "Evolution of the Stars and Gas in Galaxies." *Astrophysical Journal* 151: 547–65.

Trimble, Virginia. 2006. "'The Light of Other Days,' First Light, and Reionization: Theoretical Study and Experimental Detection of First Luminous Sources in the Universe." *New Astronomy Reviews* 50: 13–18.

Trumpler, R. J. 1925. "Spectral Types in Open Clusters." *Publications of the Astronomical Society of the Pacific* 37: 307–18.

Turnbull, H. W. 1961. *The Correspondence of Sir Isaac Newton*, vol. 3. Cambridge University Press.

van Rhijn, P. J. 1925. "The Number of Stars between Definite Limits of Proper Motion, Visual Magnitude and Galactic Latitude for Spectral Class." *Publications of the Kapteyn Astronomical Laboratory Groningen* 36: 1–16.

van Rhijn, P. J. 1929. "Distribution of Stars According to Apparent Magnitude, Galactic Latitude and Galactic Longitude." *Publications of the Kapteyn Astronomical Laboratory Groningen* 43: 1–104.

van Rhijn, P. J. 1936. "The Absorption of Light in Interstellar Galactic Space and the Galactic Density Distribution." *Publications of the Kapteyn Astronomical Laboratory Groningen* 47: 1–34.

Wilford, John N. 1992. "Astronomers Open New Search for Alien Life: A Message, Like Columbus, Would Mean a New World." *New York Times*, Oct. 6, 1992: C1.

2

Rate of formation of stars suitable for the development of intelligent life, R*, 1961 to the present

Patrick François and Danielle Briot

Abstract

The first term in the Drake Equation is R*, the number of newly formed stars in the galaxy per year. The estimate given in 1961 was ten stars per year. Over the past fifty years, new instruments and methods have allowed us to better understand how stars begin their lives and how efficiently gas can create new ones.

Powerful instruments specifically adapted to the study of star formation include both space facilities – the Galaxy Evolution Explorer (GALEX), the Spitzer Space Telescope (SST), the Herschel Space Observatory (HSO), and the Hubble Space Telescope (HST) – and a host of ground-based optical, infrared, submillimeter, and radio telescopes. These instruments have described in unprecedented detail the key phases and physical processes that lead to the formation of individual stars.

In-depth case studies of individual star-forming regions have yielded an understanding of the central physical processes that determine how molecular clouds contract and fragment into clumps and cores and, finally, clusters and individual stars. The determination of the global star formation rate (SFR) for the Milky Way is rigorously based on measurements of the global parameters of several local star-forming regions. In general, any total flux measure that is related to the SFR of a galaxy (including the Milky Way) is completely dominated by high-mass stars, since these are responsible for virtually all of the luminosity of a galaxy.

The detailed picture of how gas is transformed into stars requires not only knowledge of the SFR but also the distribution of mass of stars at their birth, a function called initial mass function (IMF). Theoretical simulations have explored how large molecular clouds fragment into stars under very different physical conditions. These works have permitted us to identify the most important physical parameters and have led to analytical formulations of the SFR and the IMF.

The Drake Equation, ed. Douglas A. Vakoch and Matthew Dowd. Published by Cambridge University Press.

In particular, they give estimates of a factor that is particularly important for the Drake Equation: the fraction of stars that are binary.

Most estimates of the SFR of the Milky Way have relied on global observables. Such studies generally rely on indirect tracers of massive (O- and early-B-type) stars to determine a massive SFR. This value is then extrapolated to lower masses to derive a global SFR for our galaxy. For example, an analysis from the late 1970s led to a value of five solar masses per year by making use of the fact that the integrated flux density from an HII region is a direct measure of the number of ionizing photons required to maintain that HII region, and is therefore an indirect measure of the number of O- and early B-type stars. In 2006, an estimate of four solar masses per year was derived from observations using the European Space Agency's International Gamma-Ray Astrophysics Laboratory (INTEGRAL) mission, which measured the gamma rays emitted by radioactive aluminum as a proxy for the massive star population of the Milky Way. Another study from 2006 gave a value of 2.7 solar masses per year, by using the total 100-micron flux of our galaxy. Along the line of these examples, this chapter will review in detail the evolution of estimates of R*, which is now closer to five solar masses per year than the ten assumed in 1961.

Introduction

Although it seems natural to have the number of stars in our galaxy as a single term in the Drake Equation, the first term, R*, instead represents the SFR of the galaxy. The latter is used because the last term of the equation, L, is the length of time civilizations release detectable signals into space. If we want to obtain the number of communicating civilizations, N, the last term L should instead be the term L divided by the lifespan of the Milky Way. Taking the number of stars in our galaxy and dividing it by its age, we have a fair approximation of the SFR.

In 1968, Beatrice Tinsley discovered that the diversity of galaxies reflects a diversity in star-formation histories. The discovery in 1996 of a population of star-forming Lyman-break galaxies (Steidel et al. 1996), and the prediction based on quasi-stellar object (QSO), or quasar, absorption-line evolution (Fall, Charlot, and Pei 1996) of a significant evolution in the cosmic SFR, indicate a complex history for the SFR (Lilly et al. 1996). The famous Madau plot published the same year (Madau et al. 1996) clearly demonstrates that the SFR density in galaxies, in general, was ten times larger in the past with a peak value some 5 to 8 billion years ago.

When we assign a value to R* in the Drake Equation, we use a mean R* averaged over the lifetime of our galaxy. However, the situation is far from that simple. R* changes not only with time but also with place. This chapter will describe how the determination of R* has evolved over the years. It will also discuss the methods and the instruments used to evaluate the SFR in different astrophysical environments. In addition, we will look at the theoretical aspects of the conversion of gas into stars.

The first estimates

The rate of supernova events in the Milky Way galaxy contains information on the present death of stars. Combining this result with the mass range of stars leading to supernovae and the IMF, the present supernova rate can be used to compute the present rate of star formation. Such computations have been performed by Truran and Cameron, who in 1971 obtained a rate of about two solar masses per year in the entire galaxy.

In 1978, Smith, Bierman, and Mezger collected data relevant to giant HII regions of the Milky Way. These regions host recently formed O stars, which are massive and very luminous, and the intense radiation they emit ionize the dense surrounding gas, which re-emits the radiation due mostly to the recombination of excited hydrogen levels. The authors counted the Lyman alpha continuum photons that are absorbed by both the gas and the dust in giant radio HII regions in the galaxy and have estimated the average lifetime of these regions. From these quantities and some theoretical assumptions, such as the IMF and the relation between SFR and gas fraction, they derived that the present rate of star formation is five solar masses per year.

In 1979, Miller and Scalo examined observational and theoretical considerations related to the history of the stellar birthrate and the IMF in the solar neighborhood. Using the present-day luminosity function, which is simply the distribution of all stars in the solar neighborhood as a function of their luminosity, and using relations between this function, the IMF, and the SFR, they deduced a SFR in the solar neighborhood ranging from three to ten solar masses per year.

But what is the current situation? What are the latest estimates of the SFR in our galaxy? We will review these below. Before doing so, however, we shall present the most important theoretical aspects that have helped astrophysicists better understand the mechanisms that convert gas into stars.

Key physical processes

Clouds

The work performed on star formation is schematically divided into two main branches, the first of which looks at the star formation processes on large scales (galactic or cosmic) whereas the second explores the star formation processes in

Figure 2.1 Composite image of M51. Image data from the Hubble's Advanced Camera for Surveys was reprocessed to produce this alternative portrait of the well-known interacting galaxy pair. The bright spots along the spiral structure are zones of star formation, which are used to determine the parameter R*.
Credit: NASA, Hubble Heritage Team, (STScI/AURA), ESA, S. Beckwith (STScI). Additional Processing: Robert Gendler

individual regions. The main physical processes responsible for the fragmentation of molecular clouds into clumps that ultimately form stars can be studied only at a very high spatial resolution, and hence are limited to the Milky Way.

The interstellar medium (ISM) is a very complex system that covers a large range of physical conditions (e.g., temperature, density, and ionization state). The ISM of the Milky Way consists of two kinds: (1) hot ionized gas with low density and temperatures in excess of 100,000 degrees Kelvin, and (2) warm ionized or neutral gas and cold neutral gas. The interesting part, as far as star formation is concerned, is this cold neutral gas. This cold neutral medium (CNM) has a density $n > 10\ cm^{-3}$ and temperatures on the order of 70 degrees Kelvin. Clouds formed by CNM can be detected thanks to the hyperfine transition of HI detectable by radio telescopes. At even colder temperatures (on the order of 10 degrees Kelvin) and densities higher than $30\ cm^{-3}$, we find molecular clouds that are the coldest parts of the ISM. They are detected due to the presence of the lower rotational transitions of carbon monoxide.

Figure 2.2 Barnard 68, molecular cloud. This photograph shows an unusual sky field in the Milky Way band. It is centered on one of the classical dark globules, known as Barnard 68 (B68) after the American astronomer, Edward E. Barnard (1857–1923), who included it in a list of such objects published in 1919. It appears as a compact, opaque, and rather sharply defined object against a rich, background star field.
Credit: ESO

A fundamental property of molecular clouds is that they are not isolated structures like big cold spheres wandering between the stars. They are merely dense condensations in a large volume of atomic gas. The molecular content of an interstellar cloud may vary rapidly with time because it is very reactive to local conditions such as a change in the radiation flux from a nearby star. It has been shown that the SFR correlates better with the total gas content of a cloud than just its molecular part. Another characteristic of molecular clouds is that they are unstable: Their stability is typically on the order of 10^7 years, which is rather short on the timescale of the life of the Milky Way (Larson 1981). An important property of molecular clouds is that they are irregular structures with complex shapes, which indicates that they are not in equilibrium.

Molecular clouds: Formation and fragmentation

Molecular clouds are constantly formed and destroyed as a consequence of instabilities. The rate at which gas is accreted into molecular clouds is related to the SFR. Assuming star formation in our Milky Way of three solar masses per year and given that only 2 percent of a typical molecular cloud is converted into stars (Evans and Lada 1991), 150 solar masses of gas per year has to be converted into molecular clouds, the source of star formation. Assuming a total amount of gas of 5×10^9 solar masses for the entire Milky Way, the average time to create a molecular cloud is on the order of 30 million years. More detailed calculations made for the solar neighborhood led to a value of 50 million years (Larson 1992). Several mechanisms have been proposed for the formation of molecular clouds: accretion of smaller clouds by random collisions and coalescence, instabilities due to gravity, or accumulation triggered by shocks from supernovae. There are several hints indicating that gravitational instabilities are the primary source of molecular cloud formation. Indeed, star formation is observed in regions of the galaxy where the surface density of the gas exceeds a given threshold with a value close to the one predicted by theoretical works (Kennicutt 1990).

Collapse and star formation occur in the densest part of the molecular cloud while the cloud is still accreting gas. The classical picture is that star formation involves the collapse of a portion of the cloud under gravity and the fragmentation of the cloud into small, gravitationally bound clumps. A molecular cloud will contain many of these clumps with the Jeans mass (Jeans 1902), the minimum for gravitationally bound structures. The collapse of the densest part of the cloud will not take the shape of a large sphere but instead exhibit flattened and filamentary structures. The origin of this filamentary structure is found not only in the cloud's initial asymmetries but also in the effect of the magnetic field or turbulence. Numerical studies show that the collapse of an elongated cylinder with a small amount of rotation contracts first toward its axis to form a thin filament, which then fragments along its length into several dense clumps on the order of the Jeans mass. These clumps split into smaller fragments, which can produce stars as small as 0.1 solar mass (Bonnell and Bastien 1992).

Formation of stars and star clusters: The initial mass function

Observations of star clusters show that a given cluster is not formed out of stars with the same mass. Instead, it appears that the mass distribution follows a power law of the logarithm of the mass of the star (Salpeter 1955). More recent determinations over a wider range of mass favor a flattening of this power toward low masses (Chabrier 2003; Kroupa, Tout, and Gilmore 1993). Although variations of the IMF with environment are regularly put forward, a constant IMF seems to be a sound hypothesis (Bastian, Covey, and Meyer 2010).

On theoretical grounds, the origin of the IMF, when linked to the problem of how star clusters form, is a long-standing problem in astrophysics. There have been many attempts to solve this problem, but the major breakthrough has been to understand that the solution to these two problems was based on gas thermodynamics. Although cloud properties vary over a large range of galactic environments, there is no evidence of a corresponding variation of the IMF. These observations led to the conclusion that the shape of the IMF strongly depends on the input physics: that is, it is sensitive to departures from the simple equations of isothermal spheres (Krumholz 2011; Spaans and Silk 2000). Krumholz, Klein, and McKee (2012) performed the first simulation that reproduces the observed IMF in a cluster large enough to contain massive stars (like the Orion Nebula) and where the peak of the mass function is determined by a fully self-consistent calculation of gas thermodynamics. These simulations also reproduce an SFR close to the observed values. Their radiation hydrodynamic simulations have shown interesting features. Radiation feedback suppresses the formation of brown dwarfs, reproducing the observed turndown of the IMF. Simulations including radiation feedback are able to suppress fragmentation in very dense regions, allowing the formation of massive stars under certain conditions. Finally, these simulations produce an IMF that does not vary with the properties of the star-forming cloud in a low-mass, low-density environment, or with the gas metallicity.

Cloud destruction and recycling

Star formation has a strong impact on molecular clouds, leading to their destruction. The main mechanism is simply ionization by the massive stars that have been formed. If stars form a normal mass function, a mere 4 percent of the mass of the cloud turning into stars is sufficient to produce enough ionizing photons to completely ionize the cloud. Stellar winds are a second important mechanism that contributes to the destruction of the molecular cloud. The gas ionized and evaporated from molecular clouds is dispersed as strong outflows escape from ionized cavities. The cycle of cloud formation and destruction is a major component not only of the evolution of the clouds but also of the evolution of galaxies as a whole.

Modern view of the star formation rate

Direct method: Star counts

The most obvious way to estimate the SFR is to count the number of stars in a given age range. If we can get some information on their masses and ages, we can derive a fair estimate of the SFR. Dividing the sample in age ranges will give its temporal variation. The ideal situation would be to have a full set of masses and ages to derive the complete star formation history. However, our observations cover only part of the masses and a limited age range. For example, in the star count for a young massive star cluster, we will have a wide range of masses and a limited age range. Using a method called main-sequence fitting, we can obtain the age of the cluster and derive a mean SFR for that age. Similar techniques can be applied for star clusters of different ages or even for entire galaxies. However, extragalactic studies are often limited to the most luminous stars, O-type or Wolf-Rayet stars. The advent of ten-meter-class telescopes and powerful instruments and satellite missions have dramatically improved our understanding thanks to the study of resolved stellar populations.

In 2010, a new direct method has been used to estimate SFR. Young stellar objects (YSO) are formed in the core of giant molecular clouds composed of gas and dust. Until now, these clusters were hidden in the molecular cloud and could not be observed. Infrared observations have the capacity to see through these clouds. Using the data from a large infrared survey called GLIMPSE (Galactic Legacy Infrared Mid-Plane Survey Extraordinaire) conducted with the NASA's Spitzer infrared telescope, a team of astronomers from the Center for Astrophysics and the Space Science Institute derived the galactic SFR by comparing the number of YSOs with a refined stellar-population synthesis model (Robitaille and Whitney 2010). Their study concluded that the method they used is one of the most accurate ways to determine the galactic SFR, as it makes use of the whole mass range rather than only massive stars or indirect tracers of massive stars. The authors found that a total of 0.68–1.45 solar mass per year was able to reproduce the observed number of observed YSOs.

Indirect methods

Most estimates of the SFR of the Milky Way have relied on global observables. Such studies generally rely on indirect tracers of massive (O- and early-B-type)

Figure 2.3 Swirling dust clouds and bright newborn stars dominate the view in this image of the Lagoon nebula from NASA's Spitzer Space Telescope. This infrared telescope is able to reveal the hot gas and the massive stars that are embedded in the molecular cloud.
Credit: Spizter Space Telescope

stars to determine the SFR of massive stars and extrapolate their values toward lower mass stars to derive a global SFR for the galaxy.

Ultraviolet emissions of stars

The UV radiation observed in galaxies is mainly due to the photospheric emission of young stars and is one of the most direct tracers of the star formation rate at the present time. As this wavelength range is peaking for stars of several masses for the integrated light of a complete population of stars with a standard mass distribution, this method can be used to probe star formation over the past 10 to 200 million years. The NASA space mission GALEX, launched in 2003, has performed the first all-sky imaging and spectroscopic surveys in the far UV (1350 to 2750 Angstroms). Its first goal was to

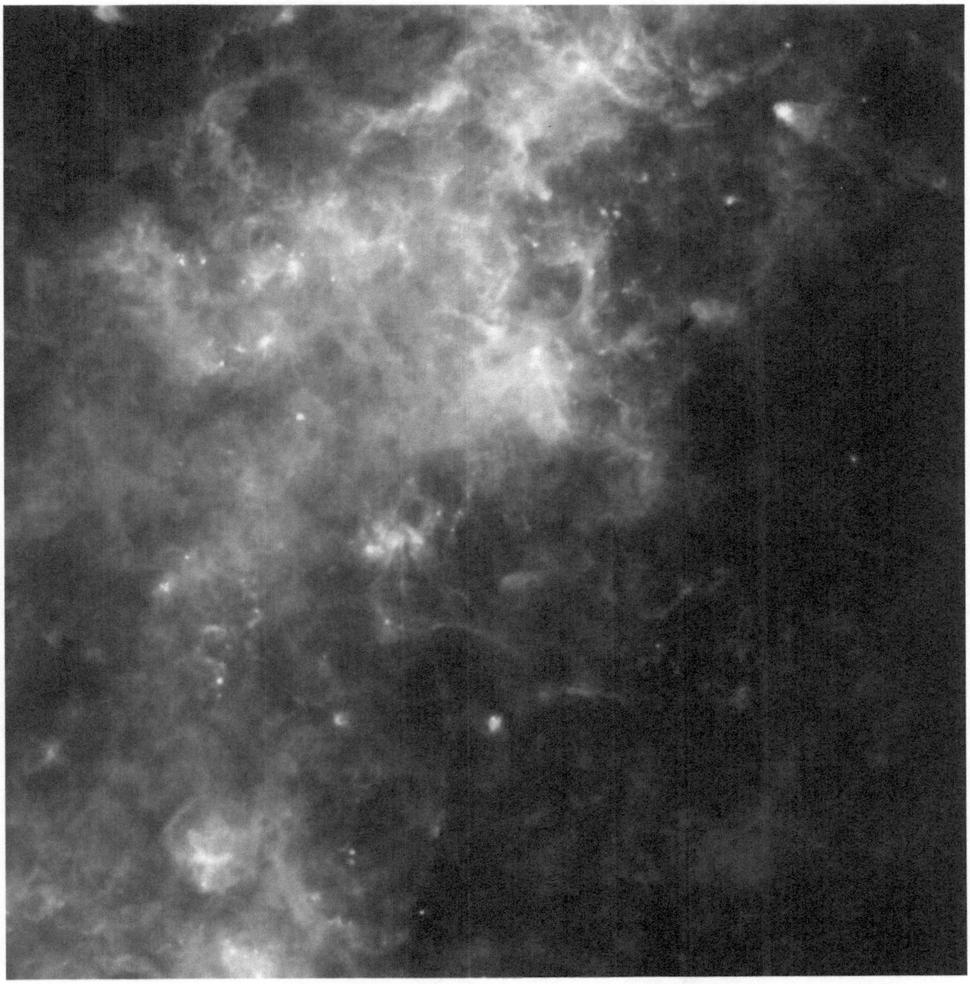

Figure 2.4 Dust cloud revealed by Herschel Space Telescope. Even the darkest patches of sky can shine brightly to Herschel. Although classical optical observations would show only a dark region in the sky, this image reveals itself to be a place of intense star formation with filaments and condensations of dust cocooning newly forming stars. These observations are used to count the number of stars formed in a given molecular cloud.
Credit: ESA and the SPIRE & PACS consortium

provide a calibration of UV and galaxy SFR taking into account extinction, starburst history, IMF, and metallicity. This calibration can then be used to derive the SFR for a wide range of star-formation environments. In particular, the GALEX survey has permitted estimates of the impact of interstellar dust absorption on UV fluxes.

Emission-line tracers

In contrast to the previous section, which was based on the continuous flux emitted by young stars, the SFR indicators presented here are based on specific emission lines and hence are associated with spectroscopic observations. The photons emitted by the massive stars ionize the surrounding gas, which re-emits photons at specific wavelengths in the optical or the near-infrared range. Although many different lines can be used as tracers, we will focus only on hydrogen lines, namely H alpha and Lyman alpha, the most-used tracers for star formation.

The H alpha emission line is a powerful indicator of local universe star formation, but, as redshift increases, the wavelength of these lines is shifted toward infrared. For large redshift, measurements are made with the Lyman alpha (121.6 nm) emission line. The strength of the line is substantially greater than H alpha, which makes it a powerful way to trace star formation at very large distances. However, many factors, among them dust, affect the transmission of Lyman alpha photons, to a level that can reach a factor of 100. As a consequence, it is particularly difficult to calibrate the strength of the Lyman alpha line that we measure with our instruments and make a correct estimate of the intensity of star formation in a given distant region. Lyman alpha emission line surveys are more often used to identify star-forming regions in a large sample of distant galaxies.

Some studies use a combination of the emission lines [NeII] 12.8-μm and [NeIII] 15.6-μm (Ho and Keto 2007) with the Spitzer Space Telescope (SST) or the Infrared Space Observatory (ISO), while others use the IR telescope of the Herschel Space Observatory (HSO) to observe the [CII] 158-μm. All studies try to calibrate their new indices with known indices.

The problem with the infrared tracers is that they are strongly affected by dust attenuation. Far-IR maps and luminosities of nearby galaxies, however, can be used to estimate this extinction.

Infrared emission from dust

Dust absorbs the photons emitted by stars and re-emits them in the infrared. Measuring the re-emitted radiation allows one to estimate the number of stars responsible for this emission and derive the star formation intensity of a given region. This method is made possible by space observatories like the SST and the HSO. Because these two telescopes are sensitive to infrared radiation, they are able to detect molecular-emission bands (in the range of 5 to 20 microns) and the emission from the thermal continuum from dust grains to longer wavelengths.

Studies of several star-forming regions have revealed that the conversion of IR luminosities to SFRs was changing as a function of the wavelength at which the observations were made. Ideally, a multi-wavelength calibration would be desirable, but complete wavelength coverage is still missing.

Multi-wavelength tracers

Multi-wavelength observations of galaxies allow us to calibrate SFR indices by combining the information carried out with different instruments that have sensitivity to given wavelengths. The most-used method makes use of UV or far-UV observations combined with IR data to create a dust-corrected SFR index. Thanks to the space telescopes GALEX, Spitzer, and Herschel, this method, limited before to the peculiar class of luminous starburst galaxies, has been extended to normal star-forming galaxies.

A similar combination of indices has been applied to compute dust-corrected emission-line luminosities of galaxies. Combinations of the observed flux in H alpha with the flux at 25 microns or the radio flux at 1.4 gigahertz provides a correction originating from the dust to the flux received in H alpha for a given star-forming region.

Gamma ray observations

Instead of observing young stars to derive the SFR, another approach is to observe stars at the end of their life and correlate these measurements with the SFR. Massive stars end their life in the form of extremely energetic events called supernovae. During this phase, a lot of energetic photons are emitted very quickly. By contrast, photons emitted by the radioactive decays of emitted nuclei can be used to trace the occurrence and the intensity of such events.

In particular, the gamma rays from radioactive by-products of the nucleosynthesis ejecta provide a direct measurement of cosmic nucleosynthesis and therefore estimate the frequency of supernova events in our galaxy. Mg^{26} decays from Al^{26} and emits gamma ray photons at 1,808.65 kiloelectron volt. Core-collapse supernovae are known to produce this rare isotope, which has a half-life of 0.75 million years.

In 2002, ESA launched the INTEGRAL, a space telescope for observing gamma rays. One paper (Diehl et al. 2006) used INTEGRAL to study the inner part of the galaxy, because the Milky Way is relatively transparent to such gamma rays. The authors found that the Al^{26} emission is dominated by the ejecta of massive stars. They also found a very high level of Al^{26} gamma ray emission from the Cygnus region, indicating that a substantial fraction of galactic Al^{26} could originate in localized star-forming regions. They estimated that the Al^{26} gamma-ray fluxes corresponded to a mass of 2.8 solar mass of Al^{26}. Given the amount of Al^{26} ejected by a supernova, they concluded that the frequency of core-collapse supernovae was around two per century, corresponding to an SFR of four solar masses per year.

Star formation rate and star formation history

The Milky Way appears to be a typical spiral disk galaxy, and it gives us the chance to acquire much more detailed information than from other galaxies. The star formation histories of the main stellar components (halo, bulge, thick, and thin disk) can be used to constrain the formation and evolution of such galaxies. For example, the detailed study of stars of different ages in the galaxy has shown that the chemical composition of these stars is different in the halo and in the disk (François 1986).

Among these components, the Sun was formed in the thin disk about 4.5 billion years ago. Stars like the Sun that we see and which might host planetary systems and planets with communicating civilizations reflect star formation history several billion years ago. This explains why it would be interesting to see if the SFR in the Milky Way disk has evolved over the life of the galactic disk.

Theoretical works have shown that it has in fact varied over the life of the Milky Way (Matteucci and François 1989). Cignoni and co-authors (2006) used data from the space telescope Hipparcos and analysed the color–magnitude diagram of the Hipparcos sample to derive star formation history in our solar neighborhood back 12 billion years. They found that recent local star formation history in the galactic disk increases from past to present with some irregularities. The mean value increases steeply from about 7 billion years ago to 2 billion years ago and shows that the SFR varies by a factor of four during the last 5 billion years.

Conclusion

From a value of $R^* = 10$, deduced by a simple approximation computed as the ratio of the number of stars in the Milky Way divided by the age of the galaxy, we have achieved an extraordinary level of refinement in the study of the star formation history. This has led to estimates of the value ranging from 2 to 16, depending on the method used and the age at which we consider R^*. All our new instruments and the data they have collected and all the theoretical work developed to compute the SFR have been used to solve some of the fundamental problems in modern astrophysics. How do galaxies form? Why do we find such large differences in galaxy shape mass stellar and gas content? What are the processes that are responsible for the conversion of gas into stars? How do planetary systems form?

References

Bastian, Nate, Kevin R. Covey, and Michael R. Meyer. 2010. "A Universal Stellar Initial Mass Function? A Critical Look at Variations." *Annual Review of Astronomy and Astrophysics* 48: 339–89.

Bonnell, Ian, and Pierre Bastien. 1992. "Fragmentation of Elongated Cylindrical Clouds. V-Dependence of Mass Ratios on Initial Conditions." *The Astrophysical Journal* 401: 654–66.

Chabrier, Gilles. 2003. "Galactic Stellar and Substellar Initial Mass Function." *The Publications of the Astronomical Society of the Pacific* 115: 763–95.

Cignoni, Michele, S. Degl'Innocenti, P. G. Prada Moroni, and S. N. Shore. 2006. "Recovering the Star Formation Rate in the Solar Neighborhood." *Astronomy and Astrophysics* 459: 783–96.

Diehl, Roland, Hubert Halloin, Karsten Kretschmer, Giselher G. Lichti, Volker Schönfelder, et al. 2006. "Radioactive ^{26}Al from Massive Stars in the Galaxy." *Nature* 439: 45–47.

Evans, Neal J., and Elisabeth A. Lada. 1991. "Star Formation in Three Nearby Molecular Cloud Complexes." In *Fragmentation of Molecular Clouds and Star Formation*, ed. E. Falgarone, F. Boulanger, and G. Duvert, 293–315. International Astronomical Union Symposium 147. Dordrecht: Kluwer.

Fall, S. Michael, Stéphane Charlot, and Yichuan C. Pei. 1996. "Cosmic Emissivity and Background Intensity from Damped Lyman-Alpha Galaxies." *The Astrophysical Journal* 464: L43–46.

François, Patrick. 1986. "Chemical Evolution of the Galaxy – A Comparison of the Abundances of Light Metals in Disk and Halo Dwarfs." *Astronomy and Astrophysics* 160: 264–76.

Ho, Luis C., and Eric Keto. 2007. "The Mid-Infrared Fine-Structure Lines of Neon as an Indicator of Star Formation Rate in Galaxies." *The Astrophysical Journal* 658: 314–18.

Jeans, James H. 1902. "The Stability of a Spherical Nebula." *Philosophical Transactions of the Royal Society A* 199: 1–53.

Kennicutt, Robert C., Jr. 1990. "Large Scale Star Formation and the Interstellar Medium." In *The Interstellar Medium in Galaxies*, ed. H. A. Thronson and J. M. Shull, 405–35. Proceedings of the 2nd Teton Conference, Grand Teton National Park, WY. Dordrecht: Kluwer.

Kroupa, Pavel, Christopher A. Tout, and Gerard Gilmore. 1993. "The Distribution of Low-Mass Stars in the Galactic Disc." *Monthly Notices of the Royal Astronomical Society* 262: 545–87.

Krumholz, Mark R. 2011. "On the Origin of Stellar Masses." *The Astrophysical Journal* 743: 110–16.

Krumholz, Mark R., Richard I. Klein, and Christopher F. McKee. 2012. "Radiation-Hydrodynamic Simulations of the Formation of Orion-Like Star Clusters. II. The Initial Mass Function from Winds, Turbulence, and Radiation." *The Astrophysical Journal* 754: 71–88.

Larson, Richard B. 1981. "Turbulence and Star Formation in Molecular Clouds." *Monthly Notices of the Royal Astronomical Society* 194: 809–26.

1992. "Galaxy Formation and Evolution." In *Star Formation in Stellar Systems*, ed. G. Tenorio-Tagle, M. Prieto, and F. Sanchez, 125–28. Proceedings of the III Canary Islands Winter School of Astrophysics. Cambridge University Press.

Lilly, Simon J., Olivier Claude Le Fevre, Francois Hammer, and David Crampton. 1996. "The Canada-France Redshift Survey: The Luminosity Density and Star Formation History of the Universe to Z Approximately 1." *The Astrophysical Journal* 460: L1–4.

Madau, Piero, Henry C. Ferguson, Mark E. Dickinson, Mauro Giavalisco, Charles C. Steidel, and Andrew Fruchter. 1996. "High-Redshift Galaxies in the Hubble Deep Field: Colour Selection and Star Formation History to Z~4." *Monthly Notices of the Royal Astronomical Society* 283: 1388–1404.

Matteucci, Francesca, and Patrick François. 1989. "Galactic Chemical Evolution: Abundance Gradients of Individual Elements." *Monthly Notices of the Royal Astronomical Society* 239: 885–904.

Miller, Glen E., and John M. Scalo. 1979. "The Initial Mass Function and Stellar Birthrate in the Solar Neighborhood." *The Astrophysical Journal Supplement Series* 41: 513–47.

Robitaille, Thomas P., and Barbara A. Whitney. 2010. "The Present-Day Star Formation Rate of the Milky Way Determined from Spitzer-Detected Young Stellar Objects." *The Astrophysical Journal Letters* 710: L11–15.

Salpeter, E. 1955. "The Luminosity Function and Stellar Evolution." *The Astrophysical Journal* 121: 161–67.

Smith, Lisa F., Peter Biermann, and Peter G. Mezger. 1978. "Star Formation Rates in the Galaxy." *Astronomy and Astrophysics* 66: 65–76.

Spaans, Marco, and Joseph Silk. 2000. "The Polytropic Equation of State of Interstellar Gas Clouds." *The Astrophysical Journal* 538: 115–20.

Steidel, Charles C., Mauro Giavalisco, Max Pettini, Mark Dickinson, and Kurt L. Adelberger. 1996. "Spectroscopic Confirmation of a Population of Normal Star-Forming Galaxies at Redshifts $Z > 3$." *The Astrophysical Journal Letters* 462: L17–21.

Tinsley, Beatrice M. 1968. "Evolution of the Stars and Gas in Galaxies." *The Astrophysical Journal* 151: 547–65.

3

Fraction of stars with planetary systems, f_p, pre-1961

Matthew F. Dowd

Abstract

The term f_p – indicating the fraction of suitable stars with planetary systems – is anachronistic to much of the time period under discussion. This chapter will survey a number of theories about planetary systems in Western thought prior to 1961 and will consider what values the term would have been assigned based on the natural philosophy or science of periods before 1961. In antiquity, two strands of thought, the Aristotelian and the atomistic, would have provided very different values for f_p. The former proposed a very clear value: zero. There are no planetary systems other than our own. The value that the latter would have proposed, however, presents a more complex situation, as atomists believed there are infinite other inhabited worlds, though none could be seen from our own. The Aristotelian view held sway in the West until the Copernican Revolution sparked a dramatic shift in cosmological ideas. Ultimately, the notion that our solar system was merely one of many planetary systems won out. Such a view was promoted, alongside the idea of plentiful extraterrestrial life, by two popularizers from the late seventeenth century: Christiaan Huygens and Bernard le Bovier de Fontenelle. The view that planetary systems were plentiful persisted through the nineteenth century, and so f_p must be understood to have a value of close to 1. Numerous authors posited life even in other places, such as comets and the Sun; planetary systems, then, were not the only abode for extraterrestrial life during this period of optimism about the plurality of worlds, suggesting that f_p was an inadequate criterion for the location of life. In the second half of the nineteenth century, William Whewell posited that little was scientifically known about other planetary systems and made various arguments

The Drake Equation, ed. Douglas A. Vakoch and Matthew Dowd. Published by Cambridge University Press.

against their commonality. This was reinforced in the first half of the twentieth century with new theories on the origins of planetary systems, which suggested that f_p would need to be assigned a much lower value.

The second term of the Drake Equation, f_p, assumes that life outside the Earth will arise on planets. Only recently has observable evidence shown that other star systems have planets. In the long era prior to 1961, Western cosmological systems did not always assume the existence of other planetary systems. Only after the Copernican Revolution did Western natural philosophers and scientists understand that the stars in the sky potentially represented planetary systems. But even during that shorter period in which the modern cosmological worldview took form, a variety of theories suggested, in turn, that planets might not be the only source of life and then that planetary systems might not be so common. The historical survey of this chapter will highlight some of the theories in which applying a number to the term f_p would be problematic.[1]

The ancient and medieval periods

The idea of extraterrestrial life has been present in Western thought since ancient Greek philosophers considered the extent of the cosmos (Dick 1982, 6). Two systems of thought that were especially prominent in the extraterrestrial life debate were the atomists and the Aristotelians (Crowe 2008, 3–13). When it comes to the term f_p, they produced very different results.

The cosmological worldview of Aristotle (384–322 BC) envisions a finite, spherical universe. At its center is the terrestrial region. Herein are found the four elements of earth, water, air, and fire, which make up the physical world in which generation and corruption occur. That is, it is the realm in which we live and in which we see constant change, including life and death. This spherical region, which includes the Earth on which we reside and the air around us, stretching out to the Moon's orbit, is surrounded by a much larger spherical region bounded on the exterior by the outer edge of the cosmos, in which lie the stars. In the vast region between the Moon and the stars, a small number of planets, including the Sun, move around the Earth. This celestial region is eternal, and the only sort of change found therein is that of locomotion, or change in place, which occurs in circular patterns, the only shape that repeats itself and hence the only shape appropriate for an eternal realm. The whole celestial region rotates around the Earth each day, which brings about the phenomenon of day and night.[2]

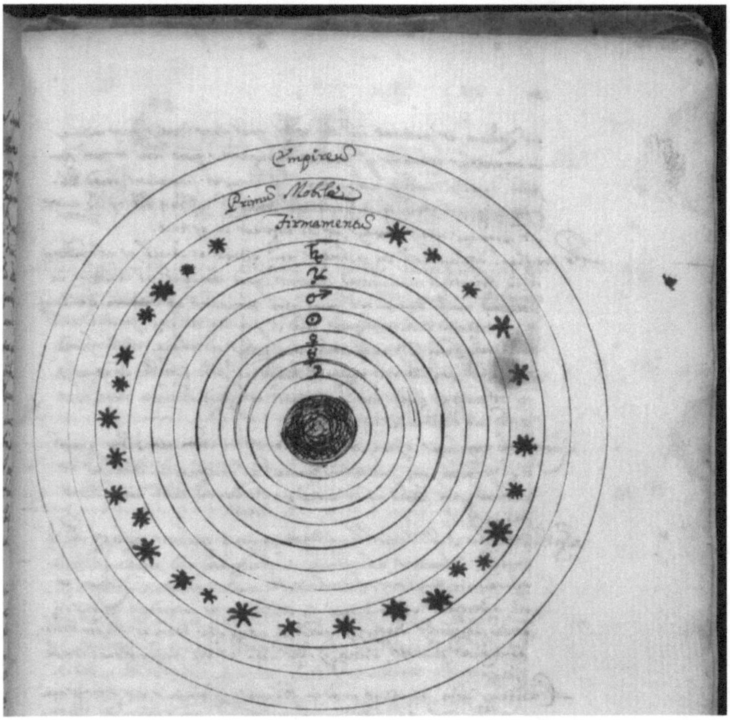

Figure 3.1 Diagram of Aristotelian system of concentric planetary spheres from a late-seventeenth-century volume, Sphaera-16xx-037r.
Credit: Image courtesy of History of Science Collections, University of Oklahoma Libraries

This spherical universe is all that exists: "the body which moves in a circle is not endless or infinite, but has its limit" (*On the Heavens*, 273a5–6, in Aristotle 1984, 454). Furthermore, "there is . . . no place or void or time outside the heaven . . . and outside the heaven . . . body neither exists nor can come to exist" (*On the Heavens*, 279a12–16, in Aristotle 1984, 462–63). For that reason, the term f_p is either inappropriate or effectively takes the value of zero. There are no systems of planets other than our own, though even that statement is misleading, because in the Aristotelian world, life does not exist on a planet. Instead, it exists on the Earth, a body that does not move, the term planet being reserved for a handful of objects that move around the Earth. Moreover, the terrestrial region is the only place in the cosmos in which change occurs, and hence is the only place where life can be generated and die. There is thus no place outside of the terrestrial realm in which life can exist.

The atomist worldview, on the other hand, posits plentiful life outside of the Earth, but again the term f_p cannot be applied legitimately to this cosmological system. For the atomists, the Earth and all we see in the sky constitute one world, or cosmos. But this is simply one finite system within the infinite bounds of space. As Epicurus (341–270 BC) states it in his letter to Herodotus:

> There are infinite worlds both like and unlike this world of ours. For the atoms being infinite in number . . . are borne on far out into space. For those atoms, which are of such a nature that a world could be created out of them or made by them, have not been used up either on one world or on a limited number of worlds, nor again on all the worlds which are alike, or on those which are different from these. (Epicurus 1926, 25)

Outside of our own system, an infinite number of other worlds exist, all having come into being through the chance collision of matter. Because space is infinite, all possibilities exist somewhere, and thus other worlds must exist that contain life:

> We must believe that these worlds were neither created all of necessity with one configuration nor yet with every kind of shape. Furthermore, we must believe that in all worlds there are living creatures and plants and other things we see in this world. (Epicurus 1926, 47)

As should be evident, however, the term f_p does not apply to such a system. When we look into the sky and see stars, we are not looking at other solar systems or other worlds like our own. Instead, we are seeing the limits of our own bounded system. On the other side of the stars, so to speak, lie other worlds, many of which will contain life merely because infinite space presents infinite possibilities. But these creatures that undoubtedly exist in the vast expanse of infinite space are not creatures we can observe nor with which we can expect to interact. So the term f_p, applied to other systems circling stars we can see, would effectively be zero, because such systems do not exist among the stars we observe. On the other hand, f_p could effectively be given a value of infinity, because an infinite number of other systems inevitably contain planets in an arrangement similar to our own; such mathematical playfulness, however, more accurately reflects the anachronism of applying the Drake Equation to a cosmological system foreign to our modern one.

The medieval period in Western Europe maintained an essentially Aristotelian cosmos. Though Aristotle's original works were largely lost to Latin Europe in the early medieval period, the Aristotelian image of a finite universe, with the Earth at its center and orbited by planets, including the Sun and Moon, was present among encyclopedic works of the early Middle Ages, such as the *Commentary on the Dream of Scipio* by Macrobius (fl. ca. AD 400) and *The Marriage of Philology and Mercury* of Martianus Capella (fl. fifth century AD).

Macrobius's *Commentary* delves into a variety of topics raised by Cicero's recounting of a dream experienced by Scipio Africanus, in which Scipio was raised up into the sky so that he could see the Earth far below him while the rest of the universe encircled him. The text is filled with examples of the Earth as the center of the universe, about which all else moves. Particularly interesting is a discussion of rain, in which Macrobius points out that, if heavy matter did not fall toward the center of the Earth, then rain that fell upon us would, at the edges of the spherical earth, fall in parallel motion past the earth and back into the celestial region, a phenomenon that obviously does not happen (Macrobius 1952, 182–84). This is merely one "proof" that the Aristotelian system is true, and that the only place where heavy matter accumulates is at the center of the universe where the Earth lies.

Martianus Capella's *The Marriage of Philology and Mercury* is divided into a number of books, and each of the seven traditional liberal arts – grammar, dialectic (logic), rhetoric, geometry, arithmetic, astronomy, and harmony (music) – has its own book in which a personified representation of the art expounds upon its content. Astronomy describes the cosmos as follows:

> The universe is formed in the shape of a globe composed entirely of four elements.[3] The heavens, swirling in a ceaseless and rotary motion, set the earth apart in a stationary position in the middle and at the bottom . . . [Above the terrestrial region] is a fifth agglomeration of corporeal matter, in which the shining heavenly bodies have their courses. (Martianus Capella 1977, 318–19)

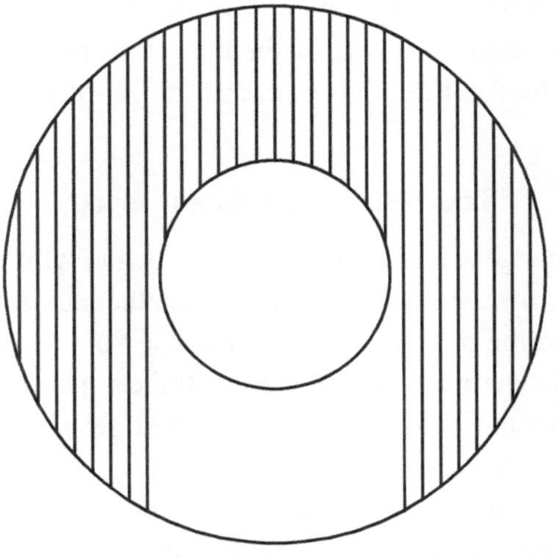

Figure 3.2 A diagram representing Macrobius's argument for the sphericity of the Earth based on the behavior of rain.
Reproduction of medieval examples of the diagram can be found in Murdoch (1984, image 248, 282–83)
Credit: Matthew F. Dowd

Clearly this describes an Aristotelian, finite, spherical universe, just as Macrobius had done.

The late Middle Ages did, however, see some speculative departures from Aristotelian cosmology. As Edward Grant has shown, even within the scholastic method that relied heavily upon Aristotelian science, various portions of the cosmological view of Aristotle were questioned and challenged, especially ideas that departed from Christian theological claims, such as a temporal creation of the universe or the notion that God could not have created the universe differently than he did (Grant 1994). Indeed, the scholastics pondered precisely the question of a plurality of worlds (Grant 1994, 150–68), allowing that, in the words of Nicole Oresme (1320–82), "God can and could in his omnipotence make another world besides this one or several like and unlike it" (Grant 1994, 166) but denying that God ever did so.

One source of significant speculation about extraterrestrial life was Nicholas of Cusa (1401–61). In his treatise *On Learned Ignorance*, Nicholas notes that we know little about the regions outside of our Earth. If there are inhabitants there, we would not know what they are like: "Were we to suppose that . . . the whole region of the other inhabited stars stands in some relation of comparison, unknown to us, to the whole region of the earth . . . not even then with this supposition could we find a relation of comparison between those inhabitants of the other stars, of whatever nature they be, and the natives of this world" (quoted in Crowe 2008, 31). Nicholas does not reject the Aristotelian cosmological schema per se, but he does make tantalizing suggestions about life elsewhere. There is no clear statement in Nicholas that the stars are the center of other planetary systems, and thus f_p would remain an anachronistic term; mathematically, its value is zero, but at the same time Nicholas does not reject extraterrestrial life of some sort.

The initial Copernican period

The move away from a geocentric, finite universe began with the 1543 publication of *De revolutionibus orbium coelestium*. In this mathematical treatise, Copernicus (1473–1543) swapped the place of the Earth and the Sun, so that the Earth became a planet. Whereas Copernicus maintained a finite, spherical universe akin to Aristotle's, heliocentrists of the late-sixteenth and seventeenth centuries broke open the bounds of the cosmos.

One famous early example of a cosmology that envisioned an extended cosmos was that of Giordano Bruno (1548–1600). Not a mathematical astronomer, Bruno's natural philosophy posited a universe that stretched out into space, filled with stars that had planets orbiting about them. This picture of a vast cosmos is present throughout a number of his works, especially *La cena de la ceneri* (*The Ash Wednesday Supper*; 1584), *De l' infinito universo e mondi* (*On the Infinite*

Universe and Worlds; 1584), *De la causa, principio et uno* (*On Cause, Origin, and Unity*; 1584), and *De immenso et innumerabilibus* (*On the Immense and Innumerable*; 1591). (For further discussion of Bruno's ideas, see Dick 1982, 63–69.) Bruno's universe did not bear much similarity to our modern one except for the assumed extension in space beyond a finite sphere; nonetheless, we can consider f_p effectively to have a value of 1 for Bruno because he assumed life to be everywhere in the universe.

Another famed natural philosopher, René Descartes (1596–1650), posited an extended universe. In his *Principles of Philosophy,* Descartes described a universe of vortices surrounding each star. This universe was constituted by a plenum of ethereal matter that swirled around the stars. When that matter accumulated to form larger bodies such as planets, they were caught up in the vortex and circled the star. Some bodies, such as comets, could pass from vortex to vortex as they passed through the plenum. Again, such a universe would provide an effective value for f_p of 1.

Not all early Copernicans believed that swapping the positions of the Earth and Sun necessarily implied an extended universe. Johannes Kepler (1571–1630), for example, argued that the finite, spherical system that Copernicus posited reflected the Trinitarian nature of God (Kozhamthadam 1994, 16–18, 29–34), thereby necessarily rejecting an extended cosmos. Other early Copernicans, such as Thomas Digges (1546–95) or Philips Landsbergen (1561–1632), argued that the stars constituted a shell around our planetary system, but that they were distant, massive objects, some as large as the whole orbit of the Earth around the Sun. Such a claim was necessary to account for the lack of parallax established by Tycho Brahe (1546–1601) and based on the spurious identification of a visually apparent diameter for stars when viewed through early telescopes.[4] In cosmoi such as these, similar to the Aristotelian worldview with the position of the Sun and Earth swapped, the variable f_p is not a relevant one, as there are no other planetary systems; alternatively, one could say that f_p is effectively zero, provided one ignores the misapplication of the term to a system in which it makes no sense.

Eventually, the notion of a universe extended in space would become the norm, and the Newtonian system of physics would become the dominant explanation for physical behaviors of planetary systems. Isaac Newton (1642–1727) himself had little to say about extraterrestrials (see Crowe 1986, 22–25), but his system was compatible with multiple planetary systems, and perhaps even reinforced it by uniting the celestial and terrestrial regions under a single system of forces acting on physical bodies. James Ferguson (1710–76), an influential popularizer of science, published in 1756 a book titled *Astronomy Explained upon Sir Isaac Newton's Principles*. Regarding planetary systems, he writes,

> Instead then of one Sun and one World only in the Universe ... Science discovers to us such an inconceivable number of Suns, Systems, and Worlds, dispersed through boundless

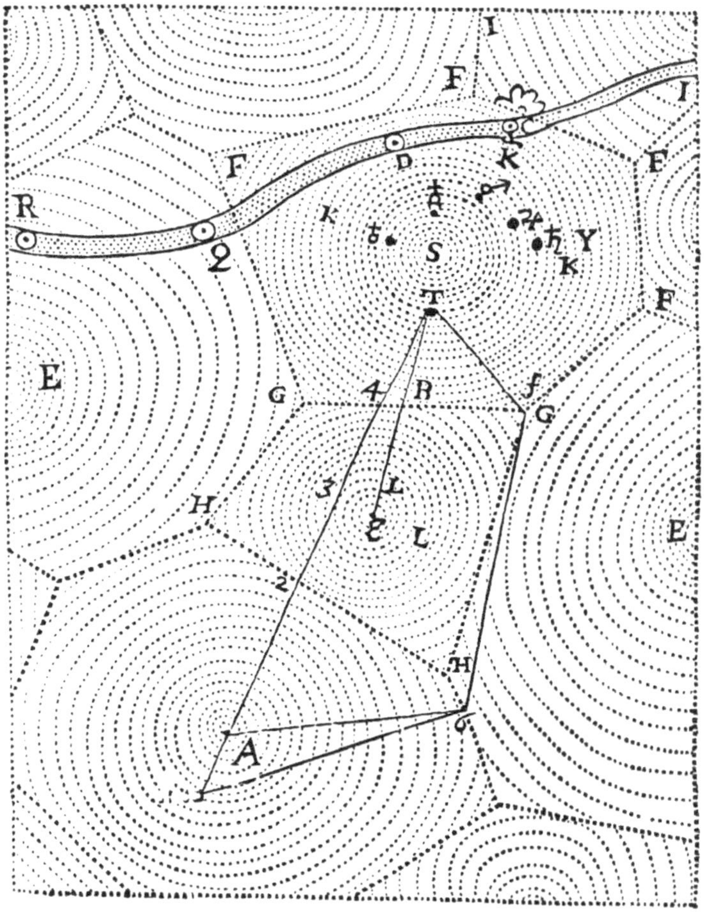

Figure 3.3 Diagram of the Cartesian vortices from Descartes's 1664 *Le Monde*. Image provided by Green Lion Press

Space. . . . From what we know of our own System it may be reasonably concluded that all the rest are with equal wisdom contrived, situated, and provided with accommodations for rational inhabitants. (Ferguson 1794, 4)

With the advent of Newtonian physics, the extended physical universe with numerous planetary systems was firmly ensconced in the scientific worldview.

The common planetary system

By the start of the eighteenth century, what we can call the Copernican principle was heavily in force in speculation about extraterrestrial life, including the question of the existence of planetary systems around other stars. This principle states that the situation of the Earth is typical for the universe as a whole. That statement is, of course, very generalized, and any precise application of the principle – or perhaps we should say "assumption" – was flexible depending on the purposes for which an author used it. In the case of f_p, such an assumption would lead one to assume a value of about 1, with some allowance typically made for the occasional exception to the rule. While this basically describes the situation for the eighteenth and first half of the nineteenth centuries, we shall see that, even in the case of optimistic assessments about the prevalence of planetary systems, the use of the term is anachronistic because planets were not the sole abode of life.

Two especially influential books in the extraterrestrial life debate of the late-seventeenth century were the 1686 *Entretiens sur la pluralité des mondes* (*Conversations on the Plurality of Worlds*) by Bernard le Bovier de Fontenelle (1657–1757) and the posthumously published 1698 *Cosmotheoros*, or *The Celestial Worlds Discover'd,* by Christiaan Huygens (1629–95). Both of these works assumed a universe extended in space, though Fontenelle presented a Cartesian worldview, whereas Huygens advocated for a Newtonian one. Both also spent more time examining the question of what life is like on other planets of the solar system than they did on the nature of the planetary systems of other stars. Nonetheless, both texts speculate on the latter issue, thereby providing us with material relevant to a discussion of f_p.[5]

The first word of Fontenelle's title, "Conversations," refers to the form of the text: a dialogue between a natural philosopher and an intelligent marquise interested in astronomy. The dialogue is broken up into five "evenings," the first four of which are devoted to an explanation of the heliocentric system and the nature of the life that inhabits the various planets of our solar system. It is only on the fifth evening that the philosopher broaches the topic of planetary systems around other stars.

The conversation of that final evening (Fontenelle 1990, 62–73) begins with the philosopher claiming that the stars are immensely far from the Earth and hence must produce their own light. This leads the marquise to the conclusion that he is implying that each star thus represents a sun at the center of a Cartesian vortex, which had been discussed on the fourth evening (1990, 53), and thus each star might itself be the center of a planetary system. After the philosopher points out the similarity these systems have to our own – namely, that the star at the center of the vortex illumines the planets in the system but is only seen from afar by other systems at night – the marquise asks, "may these systems not, despite this similarity, differ in a thousand ways? After all, a basic resemblance doesn't exclude infinite differences" (1990, 65).

The philosopher is then forced to admit ignorance, but at the same time, feels no compunction about speculating:

> One vortex has more planets revolving around its Sun, another has fewer. In one there are subordinate planets which revolve around the larger planets; in another there are none. Here they're gathered around their Sun like a little platoon, beyond which a great void extends, stretching to neighboring vortices; elsewhere they travel around the edges of the vortex and leave the middle empty. I don't doubt that there may be some vortices deserted and without planets, others whose Sun, being off-center, has an orbit itself and carries its planets with it. . . . That's enough for a man who's never left his own vortex. (1990, 65–66)

A little later, the philosopher notes that the Milky Way we see in our sky is "an infinity of small stars . . . so close to one another that they seem to form a continuous whiteness" (1990, 66). The inhabitants of these systems, he tells the marquise, would never have a truly dark night due to the close proximity of so many stars. Moreover, he discusses comets, which he says are planets that pass from one vortex to another; when they enter a new vortex, and regarding that border he notes, "I believe that in passing through there a poor planet [that is, the comet] is pretty violently shaken and its inhabitants don't bear it very well" (1990, 69).

Fontenelle clearly applied the Copernican principle in his speculations about other star systems. Like ours, they constitute a Cartesian vortex, most of which will, again like ours, carry inhabited planets. Though he allows a certain degree of variety among planetary systems, he clearly expects that most stars represent the center of a planetary system, thereby effectively suggesting a value of f_p that must be close to 1.

Huygens, too, in his popular work of the late-seventeenth century, *Cosmotheoros*, spent the majority of his text on the question of extraterrestrial life within our solar system. Liberally employing the Copernican principle, he argues that the other planets of our solar system will be inhabited by creatures much like us, for example, in possessing hands and feet, living in society, and having knowledge of mathematics; in other ways, they might be somewhat different, such as in their "shape."

Huygens comes to the question of other planetary systems only late in the work (like Fontenelle), in book 2, starting with the section entitled "The fix'd Stars so many Suns." After noting that the old theory of immensely large stars has been dismissed due to the realization that "fictitious Rays" in telescopes led to misunderstanding the size of stars (cf. Graney 2012), Huygens states that the stars exist in extended space, "scatter'd and dispersed all over the immense spaces of the Heaven." He asks,

> For then why may not every one of these Stars or Suns have as great a Retinue as our Sun, of Planets, with their Moons, to wait upon them? Nay there's a manifest reason why they should.

He proceeds with a thought experiment: If we imagine ourselves halfway between our Sun and another star, we would find that we could not see planets in either system because the planets are so small. We can see that there would be nothing special about our Sun, which has planets, and the other star. Thus we would be forced to assume that the other star likewise has planets orbiting around it, planets that, like ours, would be full of life. The remainder of the text is then taken up with an analysis of how far away the stars must be, that the universe is infinite in extent, and that the stars' planetary systems are best understood as operating in a Newtonian manner. As it had been for Fontenelle, f_p must for Huygens take a value of about 1: Planetary systems around distant stars are the norm because our system is typical.

These two late-seventeenth-century works thus set the stage for the extensive speculation about extraterrestrial life that endured throughout the eighteenth and much of the nineteenth centuries. Presenting the wide range of authors – scientists, popularizers, and literary and religious figures – would be impossible in this short chapter, and in any event has already been done by Crowe (1986; 2008). It is fair to say, however, that the existence of extraterrestrial life was taken for granted in this period.

In fact, speculation about extraterrestrial life was so prevalent and imaginative that we could, in a sense, attribute a value of greater than 1 to f_p. That is, planets were not the only places where authors of this period expected life to be present. Other locales for life included comets, the Sun, and what we today would call asteroids and dwarf or minor planets.

We have already noted that Fontenelle claimed there would be life on comets and, moreover, that comets could pass from one system to another, thus violating the Drake Equation progression from suitable stars to stars with planetary systems. Fontenelle was not unusual in his claims for life on comets: The index to Crowe (1986, 674–75) lists twenty-one more "advocates of life" on comets. To explain how life could exist on comets, especially once their orbits were understood to pass close to the Sun where their temperatures would be raised immensely, required clever – if not tortured – explanations. For example, Hugh Williamson (1735–1819) argued that heating of an atmosphere was reliant on the "fitness" of the atmosphere to transmit the vibrations caused by the light that hits the particles of air; because comets' atmospheres are made thin by the force of the Sun's light particles driving much of it away from a comet as it nears the Sun, comets' atmospheres are thus less able to transmit heat as they draw closer to the Sun (Crowe 1986, 113–14). Paul Gudin de la Brenellerie (1738–1812) likewise advocates life on comets despite the seemingly inhospitable nature of their changing environments, "using them," as Crowe says, "as an example of the diversity for which nature strives in creating intelligent beings" (Crowe 1986, 180). This reflects the continued use of the Copernican principle that drove speculation on extraterrestrial life: Because we here on Earth are intelligent creatures, and we do not have any special significance in the universe as a whole, intelligent life is expected elsewhere, granting that in different circumstances it will take on different forms suited to its given environment.

Comets indeed were not the only unexpected places where life might develop: the stars themselves could be abodes for extraterrestrial life. Crowe (2011) has presented a survey of various authors who seriously considered the idea of life on the Sun. Generally speaking, authors who claimed life existed on the Sun had to explain how they could exist on a celestial body of such great heat. Two typical answers suggested that the inhabitants of the Sun had a different nature than ours or that the Sun included environments that were not as hot as they might seem. For example, Nicholas of Cusa suggested that the beings who existed on the Sun could be more spiritual in nature, whereas Earthlings are "more gross and material" (Crowe 2011, 169). William Herschel (1738–1822) proposed a model of the Sun in which the source of its heat and light was an outer shell underneath which was a globe not unlike our own, glimpses of which were to be had through sunspots, which were in fact holes in this layer (Crowe 2011, 172). These examples, of which Crowe provides numerous others, do not directly affect an evaluation of the likelihood of planetary systems around other stars – that is, the value we ought to assign to f_p – but they do imply that the term is only inaccurately applied during this period. Planets themselves are not necessary for life to arise – even a star without planets can be the abode of life.

I shall end this section with a brief look at the work of Thomas Dick, specifically his 1837 *Celestial Scenery*. The text presents a broad swath of astronomical knowledge during the first half of the nineteenth century, with chapters on the appearance of the night sky, the arrangement and nature of the celestial bodies within the solar system, and so forth. Dick's ninth chapter contains an extended discussion of the "doctrine of a plurality of worlds," in other words, that life exists in places other than the Earth. But prior to this chapter, Dick had already extensively discussed the population that resides on each body in our solar system. He assumes an average population of 280 persons per square mile, based on the size and current population of England (Dick 1847, 58), for each body. He discusses this within each section in which he discusses a celestial body and then summarizes the information in a table (Dick 1847, 280; see Table 9.2 in this volume). The total population figure of 22 trillion is remarkable enough, but more germane to our consideration here is that he includes not only the planets but also their satellites, the bodies known as Vesta, Juno, Ceres, and Pallas (now defined as asteroids or dwarf planets), and even the rings of Saturn, including the edges of the rings.

Again, we find that our modern conception of the Drake Equation is not congruent with the cosmology of the past. Even a solar system so similar to ours, as envisioned by Dick, contains more possible locations for life than merely the major planets. A far more liberal definition of planet including much smaller bodies than we consider today suggests that, historically speaking, even unusual arrangements of celestial bodies were deemed suitable for life to arise. Like the ancient systems examined above, in which the term f_p was clearly inadequate to account for differences between cosmological worldviews, the astronomy of a century and a half before our own time contained differences that likewise make the term problematic. I suggested earlier in this chapter that a value of greater than 1 for f_p might be most accurate, and while this is clearly problematic in a mathematical sense, it is meant to suggest that the astronomy of the first half of the nineteenth century was extremely optimistic in its assumptions that life exists throughout the universe, even in places that we today would reject as possible locations for life to thrive.

The uncommon planetary system

The second half of the nineteenth century and the beginning of the twentieth century saw increased questioning of the Copernican principle as regards the existence of planetary systems. The work of William Whewell (1794–1866) was a watershed moment for a shift in attitudes about the commonality of extraterrestrial life. In his anonymously published 1853 volume *Of the Plurality of Worlds,* he attacked the notion that extraterrestrial life was common (for various details of Whewell's position, see Crowe 2008, chapter 11). He adduced various arguments to suggest that life on the other planets of the solar system was extremely unlikely, including the lack of water and atmosphere on the Moon; the low density and high gravity of Jupiter; and the relatively narrow region of space that would make up the "Temperate Zone of the Solar System" in which planets could have an environment conducive to life.

Whewell also specifically discussed the likelihood of life around other stars, and herein lie arguments relevant to the question of whether other stars have planetary systems. In chapter 8, "The Fixed Stars," Whewell argued that there is no compelling evidence to think that the stars represent systems like our solar system. At the heart of his argument are two claims: We have no evidence about systems around other stars due to their great distance, and any argument from analogy to our own system is flawed because of the manifold differences between our star and other stars. For example, many stars, he notes, are in fact double star systems, and "a system of planets revolving around or among a pair of stars, which are, at the same time, revolving about one another, is so complex a scheme, so impossible to arrange in a stable manner, that the assumption of the existence of such schemes, without a vestige of evidence, can hardly require confutation" (Whewell 1853, 145). Regarding our lack of observation of planets around other stars, he points out that

> to assume that besides these luminous bodies which we see, that there are dark bodies which we do not see, revolving round the others in permanent orbits, which require special mechanical conditions; and to suppose this, in order that we may build upon this assumption a still larger case, that of living inhabitants of these dark bodies; is a hypothetical procedure, which it seems strange that we should have to combat. (1853, 149)

In other words, Whewell is claiming that, without observational evidence of planets, we cannot assume that they exist; to do so would not be in keeping with scientific method.

Whewell questions the very notion that human natural history – in the long term – forms the basis of a typical natural history for stars throughout the galaxy. Though we see certain similarities in the stars, such as the fact that they are self-luminous and perhaps rotate, we cannot thereby dismiss the obvious differences between our sun and the manifold characteristics of stars, such as the variation in color among the stars, periodic changes in brightness among some stars, or the fact that individual stars have, over the course of recorded human astronomy, changed in appearance. "The assumption that there is anything of the nature of a regular law or order of progress from nebular matter to conscious life,—a law which extends to all the stars, or to many of them,—is in the highest degree precarious and unsupported" (1853, 162–63). That is, Whewell rejects the argument from analogy that our planetary system, and most especially the progress of life to a human-like, intelligent creature, can be taken as typical.

Finally, Whewell concludes that little evidence exists that our planetary system is typical:

> No planet, nor anything which can fairly be regarded as indicating the existence of a planet, revolving about a star, has anywhere been discerned. The discovery of nebulae, or binary systems, or clusters of stars, of periodical stars, of varying and accelerated periods of such stars, all seem to point the other way. (1853, 165)

What value then, would Whewell provide for f_p? It would have to be close to zero, if one demands clear evidence of planetary systems. On the other hand, perhaps Whewell would be forced to admit, by his own logic of not speculating about that for which we lack clear evidence, that we cannot reasonably assign a particular number. In that case, however, he would likewise insist that his contemporaries, with their value close to 1, would have to abandon the certainty with which they propose that planetary systems were the norm.

The first half of the twentieth century also saw the likelihood of planetary systems reduced through an increased questioning of the origins of planetary systems. Early speculations about the origin of planetary systems goes back at least as far as Descartes in the seventeenth century (Williams and Cremin 1968, 40), but two of the most influential theories were presented around 1800 by Pierre Simon Laplace (1749–1827) and William Herschel (1738–1822) (for a full account, see Brush 1996a, 14–36). Laplace put forth what became known as the nebular hypothesis in his *Exposition of the System of the World* in 1796. He suggested that the origin of a stable solar system could come about by a rotating sun cooling and contracting, leaving behind rings of material that coalesced into planets. He claimed Herschel's observations of the nebulae offered evidence of this process in action. Such a theory implied that planetary system formation was common throughout the universe. Though the theory underwent various criticisms and changes (see Brush 1996a, 37–61, 107–117), the general picture it offered was accepted through much of the nineteenth century. Consistent with what we discussed earlier in this chapter regarding the previous century as well as much of the nineteenth century, this theory implied a high value for f_p because it understood solar systems to be commonplace, making implicit use of the Copernican principle by understanding our own system as typical.

Increasing criticism of the nebular hypothesis prompted new theories of planetary formation as the nebular hypothesis was shown to be flawed. Most important in this regard was the work of T. C. Chamberlin (1843–1928) and F. R. Moulton (1872–1952), who proposed shortly after the turn of the century a theory in which planetary systems were formed via close encounters between stars (Brush 1996b, 22–67). Rather than planets accreting from a disk of spinning material, as in the nebular hypothesis, planets formed by material being torn from stars through mutual gravitational attraction as they passed one another. The material then pulled away from the star in a spiral arm, which in turn attracted further material in the region to form planets. Spiral nebulae, which had not yet been determined to be galaxies in their own right, were possible candidates for seeing this theory in action.

James Jeans (1877–1946) would further restrict the expectation of planetary systems when he identified problems with the Chamberlain-Moulton hypothesis and claimed that only a limited range of near collisions and the resulting tidal forces could produce the proper phenomena that would lead to a planetary system like ours (Dick 1998, 72–78). Jeans recognized that his analysis suggested that planetary systems would be quite rare, stating,

> We have absolutely no knowledge as to whether systems similar to our solar system are common in space or not. It is quite possible, for aught we know to the contrary, that our system may have been produced by events of such an exceptional nature that there are only a very few systems similar to ours in existence. It may even be that our system is something quite unique in the whole of space. (quoted in Dick 1998, 74)

At its height of popularity, then, a tidal theory of planetary system formation would imply a value of f_p that was very low, perhaps even approaching zero if our system was indeed unique.

But, as is common in cutting-edge science, the tidal theory of James was itself eventually overturned in the 1940s and 1950s. Not only was the theory itself subject to criticism, a revived nebular hypothesis was developed and observational results claimed to have identified orbital companions to stars (Dick 1998, 78–88). A number of competing theories were brought forth (Williams and Cremin 1968), with the upshot being that planetary systems were now understood to be far more likely than the tidal theory had suggested. By the time of the Green Bank conference at which the Drake Equation was discussed, claims J. P. T. Pearman, f_p probably had an upper limit of 0.5, with a low end of about 0.2, with a reasonable guess based on observational data of about 0.4 (Pearman 1963, 289).

This survey of ideas about the value of a Drake Equation variable has offered an illuminating historical lesson. We have learned that the variable is simply inapplicable in certain cosmologies prevalent in different eras in Western historical thought. In a geocentric universe, popular for close to two millennia, the variable – indeed, the equation itself – does not offer a reliable guide to how philosophers and scientists viewed the question of extraterrestrial life. Even as modern cosmology developed, similar in broad swaths if not in the details, we see that the term doesn't fully encapsulate the thinking about extraterrestrial life. As a heuristic device, an investigation of one term of the Drake Equation has provided an interesting framework to investigate certain issues in the history of the extraterrestrial life debate, a purpose for which the equation was certainly not intended.

References

Aristotle. 1984. *The Complete Works of Aristotle.* 2 vols., ed. Jonathan Barnes. Princeton University Press.

Brush, Stephen. 1996a. *Nebulous Earth: The Origin of the Solar System and the Core of the Earth from Laplace to Jeffreys,* vol. 1 of *A History of Modern Planetary Physics.* Cambridge University Press.

Brush, Stephen. 1996b. *Fruitful Encounters: The Origin of the Solar System and of the Moon from Chamberlin to Apollo,* vol. 3 of *A History of Modern Planetary Physics.* Cambridge University Press.

Crowe, Michael J. 1986. *The Extraterrestrial Life Debate 1750–1900: The Idea of a Plurality of Worlds from Kant to Lowell*. Cambridge University Press. Repr. 1999, as *The Extraterrestrial Life Debate 1750–1900*. Mineola, NY: Dover.

Crowe, Michael J. (ed.) 2008. *The Extraterrestrial Life Debate, Antiquity to 1900: A Source Book*. Notre Dame, IN: University of Notre Dame Press.

Crowe, Michael J. 2011. "The Surprising History of Claims for Life on the Sun." *Journal of Astronomical History and Heritage* 14(3): 169–79.

Dick, Steven J. 1982. *Plurality of Worlds: The Origins of the Extraterrestrial Life Debate from Democritus to Kant*. Cambridge University Press.

Dick, Steven J. 1998. *Life on Other Worlds: The Twentieth-Century Extraterrestrial Life Debate*. Cambridge University Press.

Dick, Thomas. 1847. *Celestial Scenery, or, The Wonders of the Planetary System Displayed: Illustrating the Perfections of Deity and a Plurality of Worlds*. West Brookfield, MA: Merriam and Cooke. Available online via the Hathi Trust Digital Library at http://babel.hathitrust.org/.

Epicurus. 1926. *Epicurus: The Extant Remains*, ed. Cyril Bailey. Oxford: Clarendon Press.

Ferguson, James. 1794. *Astronomy Explained upon Sir Isaac Newton's Principles*. 9th ed. London: T. Longman et al. Available online via the Hathi Trust Digital Library at http://babel.hathitrust.org/.

Fontenelle, Bernard le Bovier de. 1990. *Conversations on the Plurality of Worlds*. Trans. H. A. Hargreaves. Berkeley: University of California Press.

Graney, Christopher M. 2012. "The Work of the Best and Greatest Artist: A Forgotten Story of Religion, Science, and Stars in the Copernican Revolution." *Logos* 15(4): 97–124.

Graney, Christopher M. 2013. "Stars as the Armies of God: Landsbergen's Incorporation of Tycho Brahe's Star-Size Argument into the Copernican Theory." *Journal for the History of Astronomy* 44: 165–72.

Grant, Edward. 1994. *Planets, Stars, and Orbs: The Medieval Cosmos, 1200–1687*. Cambridge University Press.

Huygens, Chistiaan. 1698. *Cosmotheoros, or The Celestial Worlds Discover'd*. Available online via the Institute for History and Foundations of Science at www.staff.science.uu.nl/~gent0113/huygens/huygens_ct_en.htm/.

Kozhamthadam, Job. 1994. *The Discovery of Kepler's Laws: The Interaction of Science, Philosophy, and Religion*. Notre Dame, IN: University of Notre Dame Press.

Macrobius. 1952. *Commentary on the Dream of Scipio*. Trans. William Harris Stahl. New York: Columbia University Press.

Martianus Capella. 1977. *Martianus Capella and the Seven Liberal Arts, vol. 2, The Marriage of Philology and Mercury*. Trans. William Harris Stahl and Richard Johnson with E. L. Burge. New York: Columbia University Press.

Murdoch, John E. 1984. *Album of Science: Antiquity and the Middle Ages*. New York: Charles Scribner's Sons.

Pearman, J. P. T. 1963. "Extraterrestrial Intelligent Life and Interstellar Communication: An Informal Discussion." In *Interstellar Communication: A Collection of Reprints and Original Contributions*, ed. A. G. W. Cameron, 287–93. New York: W. A. Benjamin.

Whewell, William. 1853. *Of the Plurality of Worlds*. Reprinted in a facsimile edition by Chicago: University of Chicago Press, 2001.

Williams, I. P., and A. W. Cremin. 1968. "A Survey of Theories Relating to the Origin of the Solar System." *Quarterly Journal of the Royal Astronomical Society* 9: 40–62.

Notes

1 This chapter is heavily indebted to the work of two noted historians of the extraterrestrial life debate, Michael J. Crowe (1986 and 2008) and Steven J. Dick (1982 and 1998). Those texts provide a thorough understanding of the history of the extraterrestrial life debate.

2 This general picture of the Aristotelian universe can be found primarily in three of Aristotle's works: *Physics*, *On the Heavens*, and *On Generation and Corruption* (Aristotle 1984, 1:315–554). Further natural phenomena that we today consider astronomical but that Aristotle placed beneath the celestial region are considered in his works called *Meteorology* (Aristotle 1984, 1:555–625).

3 This is a bit confusing. The "four elements" apparently refer to the four that surround the earth: water, air, fire, and then the fifth element, or quintessence, which is described in the text that follows.

4 Graney (2012, 105) explains that small telescopes, like those used by early telescopic observers, showed a definite globe for stars. If the Earth were moving but one could not measure parallax among the stars, then the stars would have to be at least a minimum distance away from the Earth. Calculations comparing these distances and the apparent size of the stars led some Copernicans to posit immense sizes for the stars. See also Graney 2013.

5 Note that the question of the habitability of those planets is an issue separate from the existence of life on those planets. See Chapters 5 and 7 for discussions of whether planets around other stars have an appropriate environment for life and indeed have life upon them.

4

Fraction of stars with planetary systems, f_p, 1961 to the present

Chris Impey

Abstract

Of the Drake Equation's seven factors, f_p, the fraction of stars with planets, is the one on which we have made the most progress. Since 1961, we have gone from ignorance to a fairly reliable determination of what this number is. Our progress has occurred in three phases. In the first, which lasted for more than thirty years after 1961, astronomers were limited to speculation and probabilities, based on the Copernican principle and their growing understanding of processes that led to the formation of our solar system. The second phase began in 1995, when the first exoplanet was discovered using the Doppler method, which kicked a new research field into high gear. Excitement mingled with confusion as hot Jupiter-like planets with highly eccentric orbits were found, planets with unprecedented properties. The next decade and a half saw a steady march of the detection limit from Jupiter mass to Neptune mass and more recently to super-Earths, exoplanets with two-to-five times Earth's mass. The census grew from a few to over five hundred. The third phase of exoplanet discovery came with the 2009 launch of the Kepler satellite. Astronomers began to use the transit or eclipse method to detect planets smaller than Earth and, as of 2014, they have found more than 3,000 planetary candidates, about 300 of which are similar to or smaller than Earth. Both the Doppler and the transit methods provide information on potential habitability, and taken together they yield a mean exoplanet density. There is remarkable agreement in the exoplanet count derived from three different detection methods: Doppler, transits, and microlensing. With an uncertainty factor of two, there is an exoplanet for every sequence star, and most of them are small, terrestrial, and orbit red dwarfs. The projected number of terrestrial planets around main sequence stars in the Milky Way is about 200 billion. These recent results put the Drake Equation on

The Drake Equation, ed. Douglas A. Vakoch and Matthew Dowd. Published by Cambridge University Press.

firmer epistemological footing than ever before, a trend that will continue as exoplanet research moves toward the detection of biomarkers and the imprint on exoplanet atmospheres of microbial life.

Introduction

"The Drake Equation is a wonderful way to organize our ignorance." Jill Tarter once used these words in an interview to characterize her longtime colleague's iconic equation (Achenbach 2000). With multiple factors at play in the equation, uncertainty in its product is governed by the most uncertain of them. So what is the state of our ignorance in the early twenty-first century?

The greatest progress in the past half century involves f_p, the fraction of stars with planets. We have understood stellar evolution for nearly a century, so estimating the formation rate of stars that might host planets is a mature field of astrophysics. By contrast, the last three terms in the equation remain stubbornly indeterminate; plausibility arguments can be made for either high or low values. The two factors that precede the final three are also unknown: the number of planets that are habitable, which depends on poorly understood boundary conditions for biology, and the fraction that is actually inhabited, which depends on very challenging measurements of biomarkers that are not yet possible. By contrast, the census of planets around other stars has moved from speculation to a firm statistical basis in just twenty years. Unfortunately, increasingly crisp delineation of f_p does not mitigate uncertainty in the other factors. Drake's metric of cosmic companionship, N, remains as elusive as ever.

Why did it take until 1995 to detect the first exoplanet? Consider a scale model of our solar system. If the Sun is a glowing sphere of plasma ten feet across, Earth is a blue-white marble 400 yards away and Jupiter is a pale yellow sphere the size of a beach ball just over a mile away. On this scale, the solar system is twenty miles across, while the nearest Sun-like star would be another glowing plasma ball a hundred thousand miles away. Looking toward that nearest Sun-like star, a giant planet like Jupiter would reflect a billionth of the star's light and an Earth one-fifth as much, and both would be buried in the glare of the star, since their angular separation is less than the blurring of the star image as seen by a telescope. We can detect planets by the reflex motion they induce on the star they orbit. In the solar system, Jupiter makes the Sun pivot around its edge, a ten-foot wobble that would be imperceptible from thousands of miles away. The periodic Doppler motion induced on the star is also subtle: 11 meters per second for Jupiter and 10 centimeters per second for the Earth; the latter is equivalent to a very slow walking speed. As fractions of the speed of light, these are four parts in a billion for Jupiter and one hundredth that for the Earth. Eclipses or transits offer another detection method, but only when the orbital plane is almost perfectly aligned with the line of sight. Seen from a remote location, Jupiter would dim the Sun's light by 1 percent, but only for five hours every twelve years, and the Earth would dim the Sun by a tiny 0.01 percent. While these techniques for detecting exoplanets have been known for a long time (Struve 1952), all are extremely challenging to carry out in practice (Wright and Gaudi 2012).

The recent "explosion" in the number of known exoplanets may be the most exciting transformation of knowledge in science. Two decades ago, no planets were known beyond the solar system. By early 2014, there are more than a thousand confirmed exoplanets and several thousand additional candidates (Figure 4.1). The detection limit has progressed from Jovian gas giants to Earth-like terrestrial planets. Multi-planet systems are common. The pace of progress and discovery has been dizzying even for experts in the field.

Speculation

For centuries, scientists and philosophers speculated about the existence of planets around other stars. The Copernican revolution transformed Earth from the center of the universe to a rocky body orbiting a typical star, and a "principle of mediocrity" suggested other solar systems should exist. By extension, this heuristic suggests the existence of planets similar to ours, and fuels expectations of life beyond Earth and with it the whole subject of astrobiology.

In the period from 1961 to 1995, astronomers had no good evidence for exoplanets, but there was some confidence that they existed. Solar nebula theory had been developed during the eighteenth century in separate works by Swedenborg, Kant, and then LaPlace. The process leading to the formation of a disk that condenses into a star and planets in circular orbits was assumed to be universal, so nebular theory predicted planets around most or all single stars. However, nebular theory was unable to explain why the planets in the solar system carry 99 percent of the system's angular momentum, so by the 1960s, it had been largely abandoned (Williams and Cremin 1968). Victor Safronov (1972) achieved a breakthrough when he single-handedly formulated and solved many questions about planet formation. He described a process whereby dust grains aggregated into rocky material, after which gravity took over and grew planetary embryos. By the early 1990s, theorists thought that they could explain most of the attributes of our solar system and, by extension, remote solar systems as well.

Meanwhile, the observational search for exoplanets was causing frustration and false hopes. Historically, the best hope for exoplanet detection was astrometry of nearby stars to look for anomalies or "wobbles" caused by an orbiting planet.

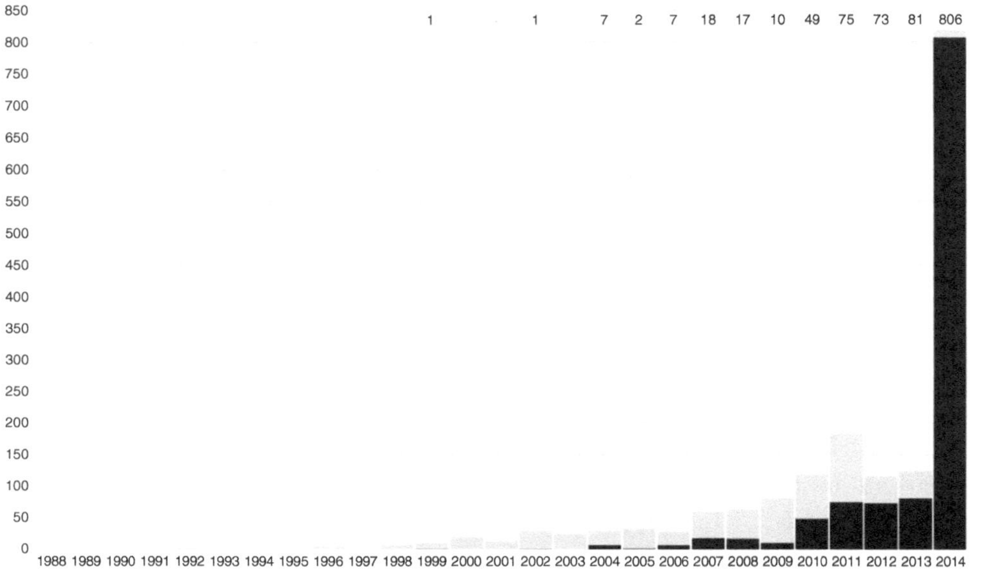

Figure 4.1 The explosive growth in the number of confirmed exoplanets in a histogram of number versus discovery date. The dark shading shows exoplanets discovered by the transit method, almost entirely from NASA's Kepler mission; the pale shading shows exoplanets discovered by all other methods, mostly using radial-velocity data. Kepler stopped taking high-quality data in 2013, so the pace of discovery will slow, but several thousand more candidate exoplanets will eventually be confirmed.
Credit: The Open Exoplanet Catalog and Aldaron/Wikimedia Commons

Claims of an exoplanet associated with the binary star 70 Ophiuchus date back to the mid-nineteenth century and resurfaced in the mid-twentieth, but Heintz (1988) later discredited this evidence. Peter van de Kamp (1969) used photographic plates to claim to detect one, and then two, Jupiter-mass planets around the nearby red dwarf Barnard's Star, as part of an astrometric campaign to find "dark" companions to nearby stars (van de Kamp 1986). Recent astrometric and radial-velocity observations have finally ruled out his claim (Choi et al. 2013).

Exoplanets induce a reflex motion in the parent star that manifests as a periodic Doppler shift. Advances in spectrograph design and the replacement of photographic plates by couple-charged devices in the 1970s finally gave observers the tools they needed to bring exoplanets within reach. The first claim of a radial-velocity detection of an exoplanet was a planetary companion to Gamma Cephei of about two times Jupiter's mass (Campbell, Walker, and Yang 1988). But the interpretation of the evidence was called into question, and it was fifteen years before an exoplanet was confirmed in this system (Hatzes et al. 2003). Soon after, an object eleven times Jupiter's mass was found in orbit around HD 114762

(Latham et al. 1989), a mass implying a probable brown dwarf. This detection was also subsequently confirmed (Butler et al. 2006). A third detection of a three-Jupiter-mass object around Beta Gemini needed twenty-five years of data to be confirmed (Hatzes and Cochran 1993; Hatzes et al. 2006).

The first exoplanet discovery came from an unexpected direction. Pulsars are collapsed, rapidly spinning remnants of massive stars. Their rotation is so irregular that anomalies can be measured to a precision of one part in a trillion, allowing detection of orbiting planets as slight as a tenth of the Earth's mass. There were two pulsar planet false alarms, in 1979 and 1991, with the latter claim receiving a lot of publicity. The following year, Wolszczan and Frail (1992) used very accurate timing measurements to detect two Earth-mass planets around the millisecond pulsar PSR B1257+12. That claim has stood the test of time, including the subsequent discovery of a third Moon-mass body. Pulsar planets have proven to be rare (Lorimer 2005), but it must be vexing for radio astronomers to find the first planet-mass objects beyond the solar system yet have the discovery relegated to the status of an exotic anomaly.

Success

Exoplanet discovery was formally ushered in on October 6, 1995, when Michel Mayor and Didier Queloz of the Geneva Observatory announced the discovery of an exoplanet half Jupiter's mass orbiting the G star 51 Pegasi (Mayor and Queloz 1995). While the Jupiter-mass exoplanet was unseen and detected only by the Doppler method, the star at a distance of fifty light years was bright enough to be visible to the naked eye. Mayor and Queloz researched fast binary stars, which have orbital periods of hours or days, so they observed their candidates frequently. The competing Californian group of Geoff Marcy and Paul Butler ran their experiment for eight years and had removed 51 Pegasi from the sample due to an error in the star catalog. Since the Doppler method could barely detect a Jupiter-mass planet and since Jupiter orbits the Sun in twelve years, they anticipated having to collect data for a long time before being able to detect any exoplanets. Although they missed out on the initial discovery, Marcy and Butler's team found three exoplanets by the end of 1995 and were part of the exciting early wave of this new field (Marcy and Butler 1998). These two groups have led the field in terms of Doppler detection ever since, using innovations in spectrograph design and instrumental technique to drive the detection limit from Jupiter mass toward Earth mass (Figure 4.2).

Success was accompanied by confusion. The first exoplanet 51 Peg b was a Jupiter-mass planet much closer to its star than Mercury is to the Sun, whipping around a complete orbit in just over four days, at a scorching temperature of

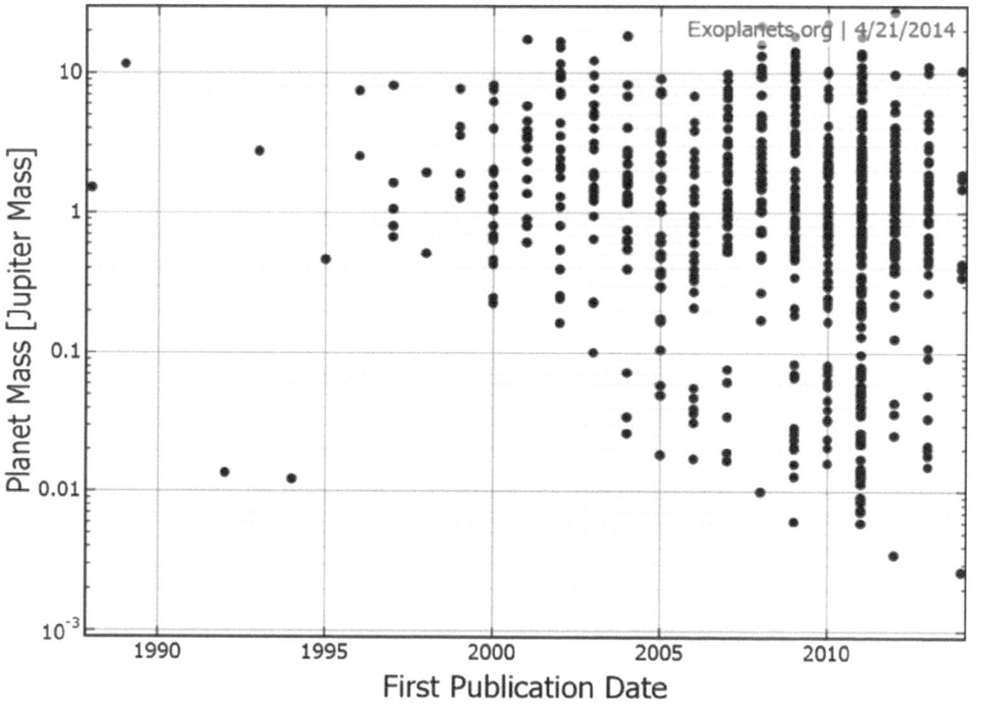

Figure 4.2 Impressive progress in the detection limit on exoplanet mass by Doppler detection can be seen in this graph of mass versus year of publication. The limit is now at an Earth mass, which is 300 times less than Jupiter mass, but most newly discovered exoplanets continue to be gas giants. There is an upper bound to planetary mass around ten times Jupiter mass. Note that there were five "precursor" discoveries before 51 Pegasi in 1995, but only two pulsar planets were confirmed before 1995.
Credit: Data from the Exoplanet Orbit Database and the Exoplanet Data Explorer at Exoplanets.org

1,000 degrees Celsius (1,800 °F). Discoveries were announced at a rate of about one a month for the first few years, accelerating to one a week by 2005. Properties of exoplanets are governed by an observational filter: you can detect only planets accessible to your detection technique. Statistical properties of the Doppler method sample are always skewed toward high mass and short period, the objects that require less data of lower quality to detect. But the first few dozen exoplanets were surprising because they were so massive and so close to their parent stars, and most of them had orbital eccentricities larger than any of the planets in our solar system. "Hot Jupiters" were completely unexpected.

Giant planets cannot form so close to their stars; there isn't enough material at those distances in the proto-planetary disk and the temperature is too high (Lin, Bodenheimer, and Richardson 1996). Instead, they migrate inward due to interactions with each other and with material in the disk. Inward motion must happen quickly since it takes only about a million years to grow by accretion from dust grain into planet embryos or planetesimals. These embryos are Moon- or Mars-sized in the inner regions and several times Earth's size beyond the snow line. One migration mode involves subtle resonance interactions between the embryo and gas in the disk, and another occurs after an embryo has grown to near-Jupiter mass and clears a gap in the disk, after which both the planet and the gap migrate to smaller distances. But that's not the whole story, as recent observations show that many hot Jupiters have highly inclined orbits, and some even go around their stars in the opposite direction to the star's rotation (Albrecht et al. 2012). The details are complex, and planets interact violently with each other and can migrate inward or outward depending on circumstances. Theory is not yet mature enough to predict exoplanet properties.

After the initial surprise of the hot Jupiters, planet hunters patiently extended their time baseline and lowered their detection thresholds. After nearly twenty years, it's still too early to measure the abundance of normal gas giants on orbits like those in the solar system, although some proxies have finally been detected (Howard et al. 2010). Observational biases still favor detection of the most massive and rapidly orbiting exoplanets, but we're gradually getting a sense of the underlying population (Figure 4.3). The Doppler method has detected several dozen super-Earths, rocky planets with three to ten times Earth's mass, but a true Earth clone is just beyond reach, although the recently discovered Kepler 186f comes close (Mayor et al. 2009; 2013; Quintana et al. 2014). Physical properties beyond mass and size can sometimes be measured (Charbonneau et al. 2002; 2005; Baraffe, Chabrier, and Barman 2010).

A sampling of the exoplanet "zoo" will clarify how diverse these other worlds are in their physical properties. The Methuselah planet, or PSR B1620–26b, is 12,400 light years away and 12.7 billion years old. It orbits a pulsar and a white dwarf, and most likely formed around a Sun-like star, but when they entered the dense environment of the M4 globular cluster, the planet was captured by a neutron star and its companion while the planet's original host was ejected from the system (Sigurdsson et al. 2003). HD 80606b has an eccentricity of 0.94, which is almost as high as Halley's comet. It travels from an Earth-like distance from its star to closer than Mercury, getting blasted as if by a blowtorch every four months (Laughlin et al. 2009). Other exoplanets are scorched all the time. Corot-7b is five times Earth's mass and has a tight twenty-hour orbit with its star-facing side at 2,330 degrees Celsius (4,220 °F) and its outward-facing side at −220 degrees Celsius (−370 °F). With an atmosphere of sodium and oxygen, the hot side probably has magma raining down from the sky (Léger et al. 2011). SWEEPS-10 is even closer to its parent star, a red dwarf. The Jupiter-mass planet

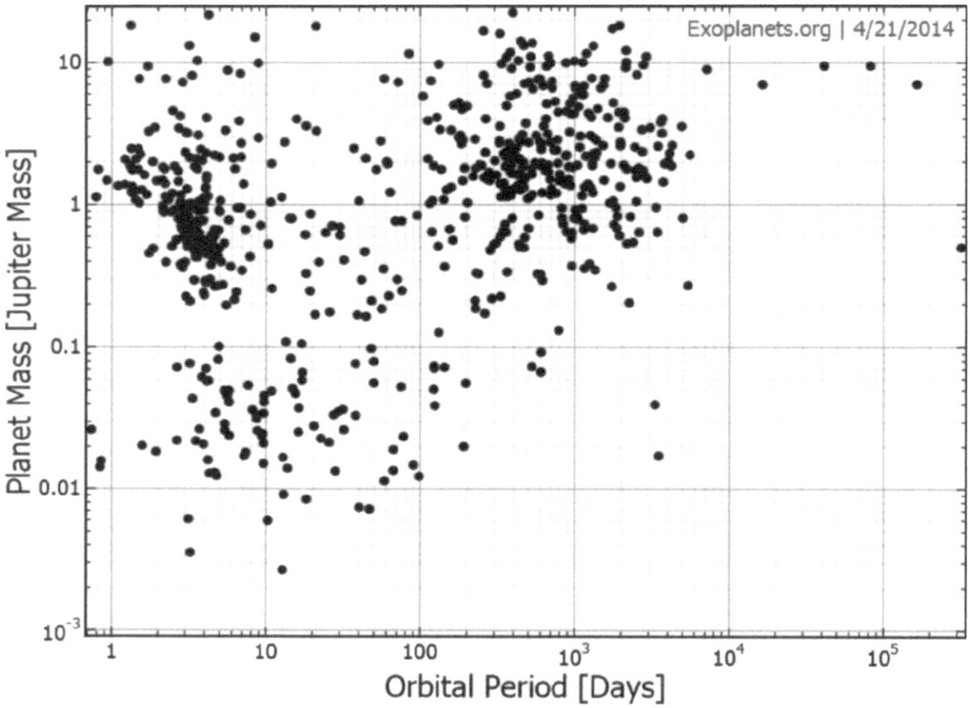

Figure 4.3 Orbital period versus mass for exoplanets discovered by the Doppler method. Mass sensitivity is limited by experimental technique and signal-to-noise, so the lack of low-mass planets does not reflect the underlying population. Similarly, the lack of gas giants with periods more than 10^4 days reflects the fact that data has been collected only since 1995. Conversely, the orbital distribution of giant planets with periods up to a few years is complete, so the clump of "hot Jupiters" at periods of a few days to a week is significant. The paucity of Uranus- and Neptune-mass planets is also probably real.
Credit: Data from the Exoplanet Orbit Database and the Exoplanet Data Explorer at Exoplanets.org

whips around in ten hours, 200 times faster than Mercury (Sahu et al. 2006). Kepler 16b orbits twin red dwarfs and is near the edge of the habitable zone. Double sunsets would be visible, as from the fictional planet Tatooine in *Star Wars* (Doyle et al. 2011). In addition to these extremes, there are dozens of hot and icy giants, water worlds, rocky super-Earths, and even free-floating (or rogue) planets (Debes and Sigurdsson 2013).

Formally, Drake Equation factor f_p is the fraction of stars with planetary *systems,* so we must measure the multiplicity of each system. In 1992, the pulsar PSR B1257+12 was found to have two Earth-mass planets, and multiple objects were soon detected in the growing haul of exoplanets. In Fourier analysis of Doppler spectroscopy of a parent star, each planet contributes a separate sinusoid, so the data can be simultaneously fit for multiple planets, in a modern version of Kepler's "harmony of the spheres." Teasing out multiple planets from noisy data is tricky and care has to be taken that the orbits are dynamically reasonable (Horner et al. forthcoming). As of early 2014, there were about 460 multi-planet systems known, including about 100 systems with three planets, 49 with four, and 18 with five or more (Schneider 2014). The Sun-like star HD 10180 has at least seven, and possibly as many as nine planets, thereby rivaling the solar system in richness (Tuomi 2012). The architectures of many of these solar systems are quite unlike our own. For example, the exoplanet currently closest to Earth's size orbits the habitable zone of a dwarf star outside four gas giants on tight, rapid orbits (Quintana et al. 2014).

Another aspect of life in the universe not explicitly captured in the Drake Equation is the fraction of planets with habitable moons. Moons are expected to be common around gas giant planets, and since the solar system contains at least three with the potential for life – Europa, Titan, and Enceladus – searches for exomoons are crucial (Canup and Ward 2006). Exomoons are at the limit of detection with current techniques, but one good candidate has been announced (Bennett et al. 2014) and several systematic searches are underway (Kipping et al. 2012; 2014). To see the dramatic progress that took place in a decade of exoplanet discovery, compare the *Annual Reviews of Astronomy and Astrophysics* articles of Marcy and Butler (1998) and Udry and Santos (2007), and see Perryman (2011). Over this time, the field moved from a handful of exoplanets with no sense of the underlying population to large statistical samples with hundreds of times more sensitive detection limits in mass. For interviews with leading exoplanet hunters and a nontechnical overview of the growth of astrobiology, see Impey (2010; 2011).

Census

The most recent phase of exoplanet discovery began on March 7, 2009, when NASA launched its Kepler spacecraft on a Delta II rocket from Cape Canaveral, Florida. Although the Doppler method still yields many confirmed exoplanets, it was "eclipsed" by Kepler's transit method, which has generated a deluge of exoplanet candidates in the past few years. Very few exoplanet systems are aligned suitably for a transit, so the strategy is to "stare" at a patch of sky containing a large number of stars. Kepler's modest one-meter mirror measured the brightness of 170,000 stars in the direction of the Cygnus constellation every seven minutes for four-and-a-half years. Three transits have to be observed to confirm an exoplanet. If the size has been measured with a transit, a subsequent Doppler detection can be used to measure mass and thereby characterize the exoplanet's mean density.

Kepler has blown the lid off the search for low-mass planets. The team announced over 1,200 candidates in early 2011, over fifty of which were in their habitable zones, and five of which are probably less than twice the Earth's size (Borucki et al. 2011). By early 2012, the number of candidates had grown to over 2,300, nearly 250 of which were less than 1.3 times the Earth's size (Batalha et al. 2012). In early 2014, the Kepler team announced that the number of confirmed exoplanets exceeded 1,000, with over 2,900 additional candidates. It's just a matter of time before Earth-like planets are found in Earth-like orbits. Mission leader Bill Borucki and his team pitched the project to NASA Headquarters in 1992, but it was rejected as being technically too difficult. In 1994 they tried again, but this time it was rejected as too expensive. In 1996 and then again in 1998, the proposal was rejected on technical grounds, even though lab work had proved the concept and exoplanets had recently been discovered. By the time the project was given the go ahead as a NASA Discovery class mission, the first transit had been detected from the ground (Charbonneau et al. 2000).

There's a bittersweet coda to the Kepler success story. NASA extended the three-and-a-half year mission in 2012 because of greater-than-anticipated noise in the stars and the detectors. But failure of two of the four reaction wheels used for pointing meant that NASA had to abandon the primary mission in 2013. Analysis of previously collected data should lead to a confirmed exoplanet yield of about 3,200. NASA has solicited proposals for scientific use of the compromised spacecraft, and as of mid-2014 is considering proposals to continue studying exoplanets with lower photometric precision but over a much larger area of sky.

The census of exoplanets is subject to selection effects, regardless of the method used. This field is not yet mature enough to fully understand and model those selection effects and produce a global value for f_p. But general trends are well understood. For Doppler detection, the radial-velocity variation is inversely proportional to the square root of the orbital distance and proportional to the planet mass times the sine of the inclination angle of the orbit. Because of the uncertainty in inclination, a minimum mass is measured, and for any sample of planet systems with random orientations the masses will, on average, be underestimated by a factor of two. In transit detection, the signal scales with the square of the ratio of planet radius to star radius. For normal Jupiter–Sun systems with random orientations, the transit depth is 1 percent, but the odds of a transit alignment are only one in a thousand. This rises to one in ten for hot Jupiters. Analogs of Earth orbiting Sun-like stars have a transit depth of 0.01 percent and a one-in-200 probability of suitable alignment (Wright and Gaudi 2012). Transit geometries are more favorable for planets around small red dwarfs than around solar-type stars.

Direct imaging complements Doppler and transit detection because it is most sensitive to massive planets with large semi-major axes around the nearest stars.

An image with an exoplanet separated from its star provides compelling and direct evidence of an orbiting body, and the orbit can be traced by multiple observations. High-contrast imaging is required because the reflected light from a giant planet is swamped by hundreds-of-millions-of-times-brighter starlight. For a giant planet orbiting 5 AU from a star at a distance of 50 parsecs, the maximum angular separation is 0.1 arcseconds. The planet's visibility is ten times better in the near-infrared than in the optical regime, though the actual contrast depends on the composition and scattering properties of the planetary atmosphere (Seager 2010). Exoplanets were first imaged a decade after they were discovered (Chauvin et al. 2004), and the number successfully imaged is still less than fifty. Advances in achieving better contrast through adaptive optics led to the first image of a multiple-planet system just a few years later (Marois et al. 2008).

A fourth method for detecting exoplanets employs microlensing. When any star passes directly in front of another star, general relativity predicts a brightening of the background star by about 30 percent as the intervening star magnifies its light. The light curve is achromatic and time-symmetric, so microlensing is easily distinguishable from other types of stellar variability. No image splitting is seen because the gravity deflection angle is very small and unresolved. If the foreground star has an orbiting exoplanet, it can cause a secondary brightening; the duration is inversely proportional to the planet's mass. Microlensing succeeded around the same time as direct imaging (Bond et al. 2004) and it has the sensitivity to detect Earth-like planets (Gaudi et al. 2008). Unfortunately, the incidence rate of microlensing events is only one in a million and the events are not repeatable, limiting what we can learn. As of early 2014, about two dozen exoplanets have been detected using microlensing.

In the past five years, the first reliable estimates of f_p have emerged. Doppler surveys in the pre-Kepler era showed that about 10 percent of Sun-like stars have planets, with indications that the true fraction might be much higher and that rocky terrestrial planets might outnumber gas giants (Marcy et al. 2005). Recent research has moved this estimate upwards, as both Doppler and transit methods reach sensitivities where Earth proxies are detectable.

Our expectations are increasingly guided by computer simulations. The astrophysics of planet formation is complex, but recent simulations are able to model core accretion within gravitating disks and estimate the internal structure of planets that form, as well as the ensemble properties of the population (Mordasini, Alibert, and Benz 2009; Alibert et al. 2013). An alternative approach is to sacrifice the modeling of internal planet structure but follow a large number of planetary embryos (Thommes, Matsumura, and Rasio 2008). All simulations naturally form rich planetary systems, often with high multiplicity. Synthetic

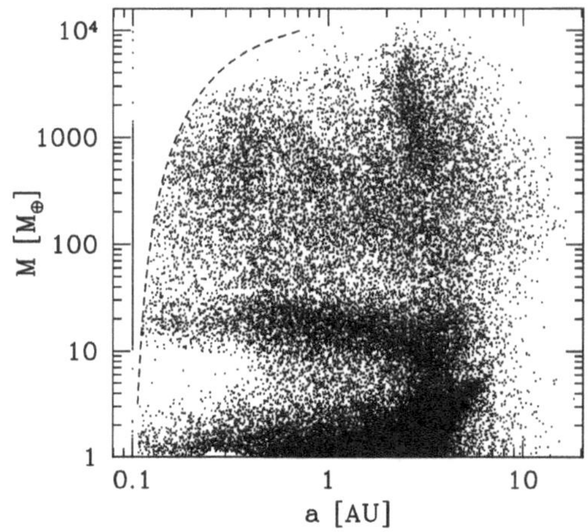

Figure 4.4 The results of a theoretical core-accretion model applied to synthesize the exoplanet population. This is the distribution of orbital period versus mass for over 50,000 synthesized exoplanets. The dashed line represents a "feeding limit," where if planets migrate to smaller distances than the line they are no longer followed within the simulation. The vast majority of the Earth-like planets are near the bottom of the diagram, and many of the small gas giants 10 to 30 times Earth's mass are censored by the limits of observational data so are seriously underrepresented in exoplanet catalogs.
Credit: Christoph Mordasini and EDP Sciences

exoplanet studies suggest a vast population of Earth and super-Earth planets at distances of 0.3 to 3 AU, which is still strongly censored by observational limits (Figure 4.4).

The Drake Equation projects a census of communicable civilizations in the Milky Way, and while early exoplanet surveys concentrated on stars similar to the Sun in spectral type and mass, the biggest leverage on N comes from red dwarfs. Main sequence stars under 0.5 solar masses outnumber stars more massive than the Sun by a factor of a hundred. Tuomi et al. (forthcoming) carried out a Bayesian analysis of high-quality Doppler spectroscopy of nearby M dwarfs and deduced an incidence rate of ~1.4 planets per star, of which ~1.1 are between three and ten Earth masses, and ~0.2 are in their stars' habitable zones (Figure 4.5). Comparison with Kepler results can be difficult because Kepler observed more massive dwarfs farther from the galactic plane; transits measure size, not mass, and the mass-radius relationship for super-Earths is poorly understood. However, transit surveys give consistent results, with ~0.9 planets per star between one half and four times the Earth's radius, of which ~0.2 are in their dwarf star's habitable zones (Dressing and Charbonneau 2013; Morton and Swift 2014). The incidence rate is slightly lower for Sun-like stars (Figure 4.6), with ~0.5 planets between one-half and four-times Earth's radius (Marcy et al. 2014). Both techniques reveal a rapid

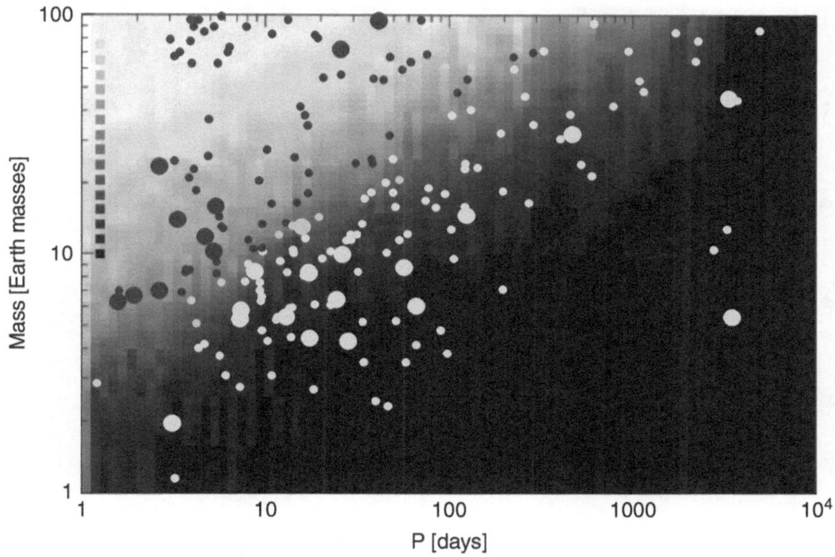

Figure 4.5 Planet detection probability shown as shading from black (0 percent) to pale grey (80 percent) for Doppler data to look for exoplanets around nearby M dwarfs. The Bayesian analysis accounts for all noise sources and for correlations in the data. Large dots are planets and candidate planets around M dwarfs, and smaller dots are planets around all other types of stars. The analysis indicates more than one Earth mass or larger exoplanet per red dwarf star, about one in five of which are in the habitable zone of the star.
Credit: Mikko Tuomi and Oxford Journals

rise in the incidence rate of terrestrial planets toward the low end of the mass or size function (Howard 2013).

A crucial third census comes from microlensing surveys. Although microlensing has discovered the fewest exoplanets, it is the only technique that is not strongly skewed toward massive planets close to their stars. Microlensing can deliver planets all the way down to Mars mass in the separation range 0.5 to 10 AU. Statistical analysis of six years of data gives perhaps the most complete census, an average of ~1.6 planets per star (Cassan et al. 2012). Moreover, microlensing is able to detect free-floating planets or nomads. Chaotic dynamics in solar systems can lead to the rearrangement of planetary orbits or even their ejection from the system entirely. Observations toward the dense star fields of the galactic bulge suggest ~1.8 Jupiter-mass planets per star (Sumi et al. 2011), and there may be as many as 10^5 unbound objects more massive than Pluto per main sequence star (Strigari et al. 2012). Microlensing surveys are still subject to poor statistics and modeling uncertainties, but these results suggest that the golden age of exoplanets is just beginning.

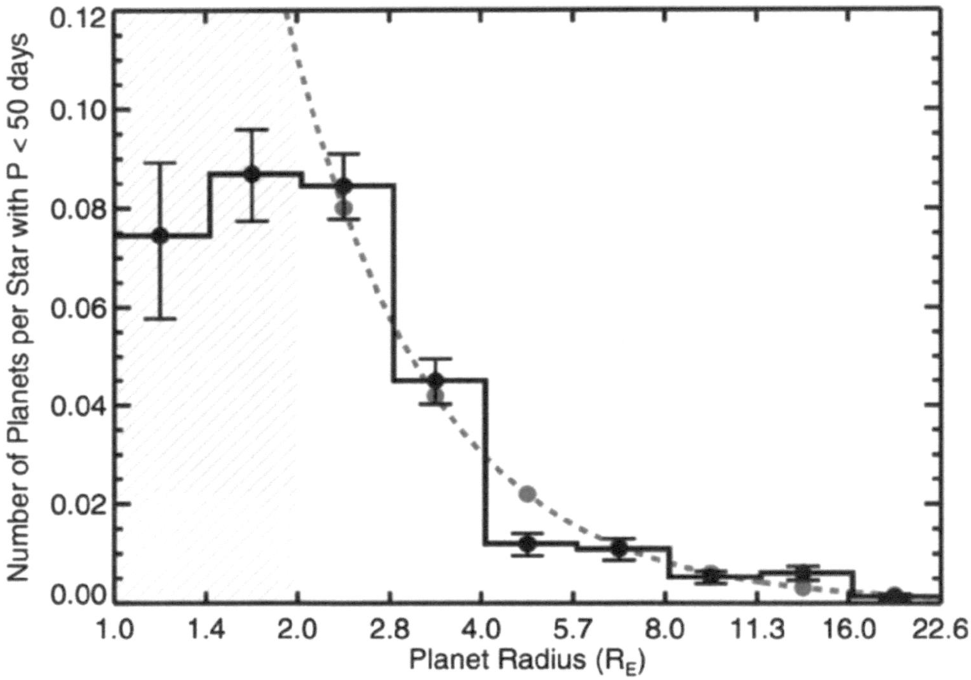

Figure 4.6 The occurrence rate of exoplanets from Kepler transit data for periods of less than fifty days. The error bars show statistical errors and the cross-hatched region marks where the detection statistics are incomplete. The dashed line indicates a power law fit to the data. Recent analysis of Kepler data indicates that the plateau or flattening of the incidence rate for small planets is real. Eventually, Kepler data will be used to extend the occurrence rate to orbital periods of a year or more.
Credit: Andrew Howard and IOP Science

Conclusion

Among the billions of planets in the Milky Way, attention naturally focuses on the most Earth-like. Kepler has already identified several hundred within a factor of two of the Earth in mass, about a dozen of which are located in the habitable zones of their parent stars. In early 2014, Kepler 186f was shown to be just 10 percent bigger than the Earth, and in the habitable zone of a red dwarf half the Sun's mass at a distance of 500 light years (Quintana et al. 2014). While the search for an Earth "twin" understandably captures headlines, the more prosaic work of conducting surveys and understanding their selection biases continues. In the past five years, three different methods have converged on the conclusion that $f_p \sim 1$.

Going the next step and converting the exoplanet incidence rate into a total for the Milky Way galaxy is not a trivial calculation. Volume-limited surveys of stars down the main sequence fusion limit are complete only to 100 parsecs, so it is a large extrapolation to project the mass function across our entire stellar system (Lépine and Gaidos 2013). The mass of the galaxy is better determined than its total star count, so deriving the main sequence census means accounting for dark matter and abundant stellar remnants like white dwarfs. A rough estimate is 200 billion stars, about 75 percent of which are red dwarfs. A substantial fraction of them are likely to be habitable, and red dwarfs will sustain life for hundreds of billions of years. The number of exoplanets in the universe is a factor of 10^{11} larger.

In terms of both time and space, the real estate available for astrobiology is staggering. All of the raw materials likely to be required for life in the universe – carbon, water, a rocky planet, and a local energy source – are abundant. In a seminal science fiction novel from almost a century ago, Olaf Stapledon (1937) wrote about alien civilizations in the Milky Way, including one that inhabited a tidally locked planet in a red dwarf system. In the year that the Kepler mission was launched, the film director James Cameron released *Avatar*. For just over $400 million, it's possible to make a fantasy movie about life on a moon of a giant exoplanet, or it's possible to actually do the search. Reality is catching up with imagination.

References

Achenbach, J. 2000. "Life Beyond Earth." *National Geographic* 197 (Jan): 24.

Albrecht, Simon, Joshua N. Winn, John A. Johnson, Andrew W. Howard, Geoffrey W. Marcy, R. Paul Butler, Pamela Arriagada, Jeffrey D. Crane, Stephen A. Shectman, Ian B. Thompson, Teruyuki Hirano, Gaspar Bakos, and Joel D. Hartman. 2012. "Obliquities of Hot Jupiter Host Stars: Evidence for Tidal Interactions and Primordial Misalignments." *The Astrophysical Journal* 757: 18–43.

Alibert, Yann, F. Carron, A. Fortier, S. Pfyffer, W. Benz, C. Mordasini, and D. Swoboda. 2013. "Theoretical Models of Planetary System Formation: Mass vs. Semi-Major Axis." *Astronomy and Astrophysics* 558: 109–22.

Baraffe, Isabelle, Gilles Chabrier, and Travis Barman. 2010. "The Physical Properties of Exoplanets." *Reports on Progress in Physics* 73: 16901–40.

Batalha, Natalie M., Jason F. Rowe, Stephen T. Bryson, Thomas Barclay, Christopher J. Burke, et al. 2012. "Planetary Candidates Observed by Kepler. III. Analysis of the First 16 Months of Data." *The Astrophysical Journal Supplements* 204: 24–45.

Bennett, David P., V. Batista, I. A. Bond, C. S. Bennett, D. Suzuki, et al. 2014. "MOA-2011-BLG-262Lb: A Sub-Earth-Mass Moon Orbiting a Gas Giant Primary or a High Velocity Planetary System in the Galactic Bulge." *The Astrophysical Journal* 785: 155–68.

Bond, Ian, A. Udalski, M. Jaroszynski, N. J. Rattenbury, B. Paczynski, et al. 2004. "OGLE 2003-BLG-235/MOA 2003-BLG-53: A Planetary Microlensing Event." *The Astrophysical Journal Letters* 606: 155–58.

Borucki, William J., David G. Koch, Gibor Basri, Natalie Batalha, Timothy M. Brown, et al. 2011. "Characteristics of Planetary Candidates Observed by Kepler. II. Analysis of the First Four Months of Data." *The Astrophysical Journal* 736: 19–40.

Butler, R. Paul, J. T. Wright, G. W. Marcy, D. A. Fischer, S. S. Vogt, C. G. Tinney, H. R. A. Jones, B. D. Carter, J. A. Johnson, C. McCarthy, and A. J. Penny. 2006. "A Catalog of Nearby Exoplanets." *The Astrophysical Journal* 646: 505–22.

Campbell, Bruce, Gordon Walker, and Stevenson Yang. 1988. "A Search for Sub-stellar Companions to Solar-type Stars." *Astronomical Journal* 331: 902–21.

Canup, Robin, and William Ward. 2006. "A Common Mass Scaling Relation for Satellite Systems of Gaseous Planets." *Nature* 441: 834–39.

Cassan, Arnaud, D. Kubas, J.-P. Beaulieu, M. Dominik, K. Horne, et al. 2012. "One or More Bound Planets per Milky Way Star from Microlensing Observations." *Nature* 481: 167–69.

Charbonneau, David, Timothy M. Brown, David W. Latham, and Michel Mayor. 2000. "Detection of Planetary Transits across a Sun-like Star." *The Astrophysical Journal Letters* 529: 45–48.

Charbonneau, David, Timothy M. Brown, David W. Latham, and Michel Mayor. 2002. "Detection of an Extrasolar Planet Atmosphere." *The Astrophysical Journal* 568: 377–84.

Charbonneau, David, Timothy M. Brown, David W. Latham, and Michel Mayor. 2005. "Detection of Thermal Emission from an Extrasolar Planet." *The Astrophysical Journal* 626: 523–29.

Chauvin, Gael, A.-M. Lagrange, C. Dumas, B. Zuckerman, D. Mouillet, I. Song, J.-L. Beuzit, and P. Lowrance. 2004. "A Giant Candidate Planet around a Young Brown Dwarf: Direct VLT/NACO Observations using IR Wave-front Sensing." *Astronomy and Astrophysics* 425: L29–32.

Choi, Jieun, Chris McCarthy, Geoffrey W. Marcy, Andrew W. Howard, Debra A. Fischer, John A. Johnson, Howard Isaacson, and Jason T. Wright. 2013. "Precise Doppler Monitoring of Barnard's Star." *The Astrophysical Journal* 764: 131–43

Debes, John H., and Steinn Sigurdsson. 2013. "The Survival Rate of Ejected Terrestrial Planets with Moons." *The Astrophysical Journal* 668: L167–70.

Doyle, Lawrence R., Joshua A. Carter, Daniel C. Fabrycky, Robert W. Slawson, Steve B. Howell, et al. 2011. "Kepler-16: A Transiting Circumbinary Planet." *Science* 333: 1602–6.

Dressing, Courtney D., and David Charbonneau. 2013. "The Occurrence Rate of Small Planets around Small Stars." *The Astrophysical Journal* 767: 95–116.

Gaudi, B. Scott, Joseph Patterson, David S. Spiegel, Thomas Krajci, R. Koff, et al. 2008. "Discovery of a Very Bright, Nearby Gravitational Microlensing Event." *The Astrophysical Journal* 677: 1268–77.

Hatzes, Artie P., and William D. Cochran. 1993. "Long Period Radial Velocity Variations in Three K Giants." *The Astrophysical Journal* 413: 339–48.

Hatzes, Artie P., William D. Cochran, Michael Endl, Barbara McArthur, Diane B. Paulson, Gordon A. H. Walker, Bruce Campbell, and Stephenson Yang. 2003. "A Planetary Companion to γ Cephei A." *The Astrophysical Journal* 599: 1383–94.

Hatzes, Artie P., William D. Cochran, Michael Endl, Barbara McArthur, Diane B. Paulson, Gordon A. H. Walker, Bruce Campbell, and Stephenson Yang. 2006. "Confirmation of the Planet Hypothesis for the Long Period Radial Velocity Variations of Beta Gemini." *Astronomy and Astrophysics* 457: 335–41.

Heintz, Wulff D. 1988. "The Binary Star 70 Ophiuchi Revisited." *Journal of the Royal Astronomical Society of Canada* 82: 140–45.

Horner, Jonathan A., Robert A. Wittenmyer, Chris G. Tinney, Paul Robertson, Tobias C. Hinse, and Jonathan P. Marshall. Forthcoming. "Dynamical Constraints on

Multi-Planet Exoplanetary Systems." In the 12th Annual Australian Space Science Conference, Melbourne.

Howard, Andrew W. 2013. "Observed Properties of Extrasolar Planets." *Science* 340: 572–76.

Howard, Andrew W., John Asher Johnson, Geoffrey W. Marcy, Debra A. Fischer, Jason T. Wright, David Bernat, Gregory W. Henry, Kathryn M. G. Peek, Howard Isaacson, Kevin Apps, Michael Endl, William D. Cochran, Jeff A. Valenti, Jay Anderson, and Nikolai E. Piskunov. 2010. "The California Planet Survey. I. Four New Giant Exoplanets." *The Astrophysical Journal* 271: 1467–81.

Impey, Chris D. 2011. *The Living Cosmos: Our Search for Life in the Universe*. Cambridge University Press.

Impey, Chris D., ed. 2010. *Talking About Life: Conversations on Astrobiology*. Cambridge University Press.

Kipping, David M., Gáspár Á. Bakos, Lars A. Buchhave, David Nesvorny, and Allan Schmitt. 2012. "The Hunt for Exomoons with Kepler. I. Description of a New Observational Project." *The Astrophysical Journal* 750: 115–34.

Kipping, David M., Gáspár Á. Bakos, Lars A. Buchhave, David Nesvorny, Allan Schmitt. 2014. "The Hunt for Exomoons with Kepler. IV. A Search for Moons Around Eight M Dwarfs." *The Astrophysical Journal* 784: 28–42.

Latham, David W., Robert P. Stefanik, Tsevi Mazeh, Michel Mayor, and Gilbert Burki. 1989. "The Unseen Companion of HD 114762: A Probable Brown Dwarf." *Nature* 339: 38–40.

Laughlin, Gregory, Drake Deming, Jonathan Langton, Daniel Kasen, Steve Vogt, Paul Butler, Eugenio Rivera, and Stefano Meschiari. 2009. "Rapid Heating of the Atmosphere of an Extrasolar Planet." *Nature* 457: 562–64.

Léger, Alain, O. Grasset, B. Fegley, F. Codron, A. F. Albarede, P. Barge, et al. 2011. "The Extreme Physical Properties of the CoRoT-7b Super-Earth." *Icarus* 213: 1–11.

Lépine, Sébastien, and Eric Gaidos. 2013. "The Northern Census of M Dwarfs within 100 pc and Its Potential for Exoplanet Surveys." *Astronomische Nachrichten* 334: 176–79.

Lin, Douglas N. C., Peter Bodenheimer, and Derek C. Richardson. 1996. "Orbital Migration of the Planetary Companion of 51 Pegasi to its Present Location." *Nature* 380: 606–7.

Lorimer, Duncan R. 2005. "Binary and Millisecond Pulsars." *Living Reviews in Relativity* 8: 7–18.

Marcy, Geoffrey W., and R. Paul Butler. 1998. "Detection of Extrasolar Giant Planets." *Annual Reviews of Astronomy and Astrophysics* 36: 57–97.

Marcy, Geoffrey W., R. Paul Butler, Debra Fischer, Steven Vogt, Jason T. Wright, Chris G. Tinney, and Hugh R. A. Jones. 2005. "Observed Properties of Exoplanets: Masses, Orbits, and Metallicities." *Progress of Theoretical Physics Supplement* 158: 24–42.

Marcy, Geoffrey W., R. Paul Butler, Debra Fischer, Steven Vogt, Jason T. Wright, Chris G. Tinney, and Hugh R. A. Jones. 2014. "Occurrence and Core-Envelope Structure of 1–4x Earth-sized Planets around Sun-like Stars." *Proceedings of the National Academy of Science* 111(35): 12655–60.

Marois, Christian, Bruce Macintosh, Travis Barman, B. Zuckerman, Inseok Song, Jennifer Patience, David Lafrenière, Rene Doyon. 2008. "Direct Imaging of Multiple Planets Orbiting the Star HR 8799." *Science* 322: 1348–52.

Mayor, Michel, and Didier Queloz. 1995. "A Jupiter-Mass Companion to a Solar-Type Star." *Nature* 378: 355–59.

Mayor, Michel, X. Bonfils, T. Forveille, X. Delfosse, S. Udry, J.-L. Bertaux, H. Beust, F. Bouchy, C. Lovis, F. Pepe, C. Perrier, D. Queloz, and N. C. Santos. 2009. "The HARPS Search for Southern Extrasolar Planets. XVIII. An Earth-Mass Planet in the GJ 581 Planetary System." *Astronomy and Astrophysics* 507:487–94.

Mayor, Michel, X. Bonfils, T. Forveille, X. Delfosse, S. Udry, J.-L. Bertaux, H. Beust, F. Bouchy, C. Lovis, F. Pepe, C. Perrier, D. Queloz, and N. C. Santos. 2013. "The HARPS Search for Southern Extrasolar Planets. XXXIV. Occurrence, Mass Distribution and Orbital Properties of Super-Earths and Neptune-Mass Planets." *Astronomy and Astrophysics* 555: 410–23.

Mordasini, Christoph, Yann Alibert, and Willy Benz. 2009. "Extrasolar Planet Population Synthesis. I. Method, Formation Tracks, and Mass-Distance Distribution." *Astronomy and Astrophysics* 501: 1139–60.

Morton, Timothy D., and Jon Swift. 2014. "The Radius Distribution of Small Planets around Cool Stars." *The Astrophysical Journal* 791: 10–22.

Perryman, Michael A. C. 2011. *The Exoplanet Handbook.* Cambridge University Press.

Quintana, Elisa V., Thomas Barclay, Sean N. Raymond, Jason F. Rowe, Emeline Bolmont, et al. 2014. "An Earth-Size Planet in the Habitable Zone of a Cool Star." *Science* 344: 277–80.

Safronov, Victor S. 1972. *Evolution of the Protoplanetary Cloud and Formation of the Earth and the Planets.* Jerusalem: Israel Program for Scientific Translations.

Sahu, Kailash C., Stefano Casertano, Howard E. Bond, Jeff Valenti, T. Ed Smith, et al. 2006. "Transiting Extrasolar Planetary Candidates in the Galactic Bulge." *Nature* 443: 534–40.

Seager, Sara. 2010. *Exoplanet Atmospheres: Physical Processes.* Princeton University Press.

Schneider, Jean. 2014. "The Extrasolar Planets Encyclopedia." An interactive catalog maintained online at http://exoplanet.eu/catalog/.

Sigurdsson, Steinn, Harvey B. Richer, Brad M. Hansen, Ingrid H. Stairs, and Stephen E. Thorsett. 2003. "A Young White Dwarf Companion to Pulsar B1620–26: Evidence for Early Planet Formation." *Science* 301: 193–96.

Stapledon, Olaf. 1937. *Starmaker.* London: Methuen. Repr. 2008, Mineola, NY: Dover.

Strigari, Louis E.,Matteo Barnabe, Philip J. Marshall, and Roger D. Blandford. 2012. "Nomads of the Galaxy." *Monthly Notices of the Royal Astronomical Society* 423: 1856–65.

Struve, Otto. 1952. "Proposal for a Project of High-Precision Radial Velocity Work." *The Observatory* 72: 199–200.

Sumi, Takahiro, K. Kamiya, A. Udalski, D. P. Bennett, I. A. Bond, et al. 2011. "Unbound or Distant Planetary Mass Population Detected by Gravitational Microlensing." *Nature* 473: 349–52.

Thommes, Edward W., Soko Matsumura, and Frederic A. Rasio. 2008. "Gas Disks to Gas Giants: Simulating the Birth of Planetary Systems." *Science* 321: 814–17.

Tuomi, Mikko. 2012. "Evidence for 9 Planets in the HD 10180 System." *Astronomy and Astrophysics* 543: 52–64.

Tuomi, Mikko, Hugh R. A. Jones, John R. Barnes, Guillem Anglada-Escudé, and James S. Jenkins. Forthcoming. "Bayesian Search for Low-mass Planets Around Nearby M Dwarfs: Estimates for Occurrence Rate Based on Global Detectability Statistics." *Monthly Notices of the Royal Astronomical Society.*

Udry, Stephane, and Nuno C. Santos. 2007. "Statistical Properties of Exoplanets." *Annual Reviews of Astronomy and Astrophysics* 45: 397–39.

van de Kamp, Peter. 1969. "Alternate Dynamical Analysis of Barnard's Stars." *Astronomical Journal* 74: 757–59.

van de Kamp, Peter. 1986. "Dark Companions of Stars: Astrometric Commentary on the Lower End of the Main Sequence." *Space Science Reviews* 43: 211–327.

Williams, Iwan P., and A. W. Cremin. 1968. "A Survey of Theories Relating to the Origin of the Solar System." *Quarterly Journal of the Royal Astronomical Society* 9: 40–62.

Wolszczan, Alexander, and Dail A. Frail. 1992. "A Planetary System around the Millisecond Pulsar PSR1257+12." *Nature* 355: 145–47.

Wright, Jason T., and B. Scott Gaudi. 2012. "Exoplanet Detection Methods." In *Planets, Stars, and Stellar Systems*, ed. Terry D. Oswalt, Linda M. French, and Paul Kalas, 489–540 Dordrecht: Springer.

5

Number of planets, per solar system, with an environment suitable for life, n_e, pre-1961

Florence Raulin Cerceau

Abstract

The Drake Equation factor n_e, the number of planets per planetary system with an environment suitable for life, has recently received increased attention as astronomers discover many planetary systems within the Milky Way. This factor is closely associated with the notion of planetary habitability, which is not new. During the second part of the nineteenth century, astronomers such as Richard Proctor (1837–1888) and Camille Flammarion (1842–1925) studied planetary habitability in our solar system a century before this concept was updated by the systematic exploration of the solar system and the detection of exoplanets. The question was tackled in scientific terms while studies of the Martian surface were intensifying. The term "habitability" was used mainly for Mars, which was regarded as a sister planet of Earth and looked like a possible abode for life. More generally, studies during this period attempted to compare Earth with other planets in the solar system, in spite of the lack of accurate data on their environments. Analogy soon became the basic way in which to calculate the level of habitability of each planet. In the meantime, spectroscopy was developing quickly and began to provide more data about planetary atmospheres.

At the dawn of the space age, Hubertus Strughold (1898–1987) suggested that we examine the planets of the solar system using the concept of "planetary ecology." This notion was very similar to the concept of habitability employed earlier by nineteenth-century pioneers. Strughold also coined the term "ecosphere" to describe the region around a star that has conditions suitable for life-bearing planets, a concept equivalent to the habitable zone. A decade later, Stephen Dole redefined the concept of habitability as the planetary conditions suitable for human life. He used probability methods to estimate the number of planets in the galaxy

The Drake Equation, ed. Douglas A. Vakoch and Matthew Dowd. Published by Cambridge University Press.

where human life might be possible. This chapter reviews the main historical contributions up to the 1960s regarding the concept of planetary habitability and its different meanings over time.

Introduction

The continuous discovery of exoplanets since 1995 has led us to reconsider the question of planetary habitability. Astronomers usually adhere to a conventional and conservative definition of habitability: the zone around a star within which water can exist in stable liquid form on the surface of a rocky planet (Impey 2013). Newly discovered characteristics of habitable zones in our galaxy have given this term new meanings and nuances, yet we have to be careful when defining the habitable zone in our solar system or in others (Marcy 2014). Despite recent discoveries, it remains unclear whether our solar system is "typical" (Impey 2013; Chapter 4). The discovery of more and more planetary systems and the presumption that every star in our galaxy could have at least one planet have enlarged our understanding of planetary diversity. Nowadays, we have to contemplate studying an "exoplanet zoo" (Impey 2013) where planets are distributed everywhere in each system. In Chapter 4, Impey gives a sampling of this exoplanet zoo to clarify how diverse these other worlds are in their physical properties. From an exobiological viewpoint, the most interesting part of these discoveries concerns the possibility of exo-life and Earth-like planets. If we are surrounded by millions of Earth-like planets, it becomes realistic to study the alien worlds that populate the galaxy and their habitability.

However, habitability is not a new concept. This notion was tackled during the seventeenth century (e.g., Fontenelle 1686; Huygens 1698), when the plurality of worlds was first openly debated. The Copernican theory (as presented by Nicholas Copernicus in *De revolutionibus orbium coelestium*, 1543) dethroned the Earth from the central place in the universe, in spite of the many difficulties created by this new paradigm. In this model of the universe, every planet in the solar system rotated around the Sun. Hence, the Earth became no more than one planet among many and was no longer the center of the universe. It was, in fact, a planet like any other in the solar system. This made it conceivable that other planets could be inhabited as well. During the seventeenth century, however, very few people supported the idea that worlds similar to ours could exist. For instance, Kepler, in his *Somnium* (*The Dream*), talked about the Moon's habitability but believed that major difficulties remained in asserting that our planet was like any other in the solar system.

Planetary habitability began to be defined in scientific terms and widely discussed in the astronomical community during the second part of the nineteenth century, which was a period of active Martian observations. In addition, biological evolution and spectroscopy represented contemporaneous breakthroughs in, respectively, theory and technique (Dick 1996). Spectroscopy confirmed the unity of nature by observational methods, which led to the detection of similar molecules in other planetary or stellar environments. This new scientific technique strengthened the idea that the materials used by living beings were common in the universe. Darwin's theory of evolution (Darwin 1859) provided a scientific context in which the physical evolution of the universe became conceivable, along with mechanisms for evolution from inorganic matter to life (Dick 1996).

On the question of planetary habitability, this chapter will highlight first the pioneering late-nineteenth-century views of the English astronomer Richard A. Proctor (1837–1888), and the Frenchmen Camille Flammarion (1842–1925) and Jules Janssen (1824–1907), who were well known personalities active on this topic. The period was very prolific in studies about habitability, and other scientists – such as Howe and McFarland – also gave their opinion about that, as discussed by Briot and Schneider in Chapter 6.

Nearly a century later at the dawn of the space age, the question of habitability reappeared in a completely different context. While the first programs for launching artificial satellites were revolutionary, the question of life (or more specifically of human life) in space and of life elsewhere began to be explored in concrete terms. We also present in this chapter space-medicine pioneer Hubertus Strughold's views about planetary ecology, a concept similar to planetary habitability. Finally, we detail Stephen H. Dole's ideas about planet habitability, which applied to humanity and was quantified with the help of many factors in a similar way as the Drake Equation.

Fictional and poetical views about habitability during the seventeenth century

During the first part of the seventeenth century, the question of habitability was mainly focused on the possibility of life on the Moon. In November 1609, Galileo first pointed his new refracting telescope (with 20x magnification) at our satellite, with the goal of making a detailed study of its surface. From November 30 until December 18, he examined and drew the Moon and described his discoveries in his book *Sidereus Nuncius*, published in 1610. Galileo concluded from his observations that the Moon's surface consisted of valleys, plains, and mountains and looked very much like the surface of the Earth (Galilei 1989, 48–49). The dark spots were shadows cast by these mountains and valleys as the Sun's light fell on them. These observations raised the question of the Moon's habitability, also strengthened by the Copernican theory.

The German astronomer Johannes Kepler (1571–1630) is often considered a precursor of science fiction with his book *The Dream*, or *Somnium, sive opus posthumum de astronomia lunaris* (Kepler 2003), posthumously published by his son in 1634.[1] In this story, Kepler's main aim was very serious: giving arguments to support Copernican theory. The story takes place on the Moon, especially chosen to demonstrate physical laws seen from our satellite instead of our planet. By analogy, Kepler changes the astronomical reference system and explains what an observer would be able to see from the Moon.

Kepler also describes the Moon as inhabited. Plants grow very quickly, and animals resemble running or flying reptiles. But beings do not live long on our satellite, since they are gigantic. Lunarians are human-like and live hidden in caves on the far side of the Moon, where they are sheltered from the harsh cold of the night and the heat of the day (each of which lasts for two weeks). Kepler's Lunarian world is fiction, but he provides many explanations of an astronomical nature in numerous footnotes, which are very often longer than the text itself.

Fifty years later, Bernard le Bovier de Fontenelle (1657–1757), a French philosopher and writer, published *Entretiens sur la Pluralité des Mondes*, or *Conversations on the Plurality of Worlds* (Fontenelle 1686). This time, the question of habitability focused on the whole solar system. The book was a runaway success, going through hundreds of editions in many different languages (Knevitt 2011).

This influential piece of science popularization comes in the form of a pleasant, elegant dialogue between a philosopher and a marchioness (*la Marquise de la Mésengère*) who is curious about the natural world. Through this dialogue, Fontenelle expounds on the Copernican world system (see Chapter 3) and speculates about the inhabitants of other planets in the solar system. Every evening for five nights – six in Fontenelle's second edition of 1687 – the two interlocutors are drawn into a lively conversation about the Moon and planets. The philosopher explains that the Earth is one of many planets orbiting the Sun, and the Sun is one of many stars in the universe, all of them subject to the same forces that govern the whole universe, as explained in "The Fifth Evening's Conversation. That the fixed Stars are so many Suns, every one of which gives Light to a World" (Fontenelle 1715, 133).

A vivid discussion follows about whether there could be life on the Moon. They discuss each planet, but curiously Mars has no special interest for the philosopher. Why? Incredibly, because Mars is too much like our planet: "Mars has nothing curious that I know of" (Fontenelle 1715, 111). In Fontenelle's description, even the four moons of Jupiter – discovered by Galileo in 1609 – are inhabited. Finally, as they continue to enjoy themselves, the marchioness and the philosopher speculate on what the inhabitants of the planets would be like (Knevitt 2011).

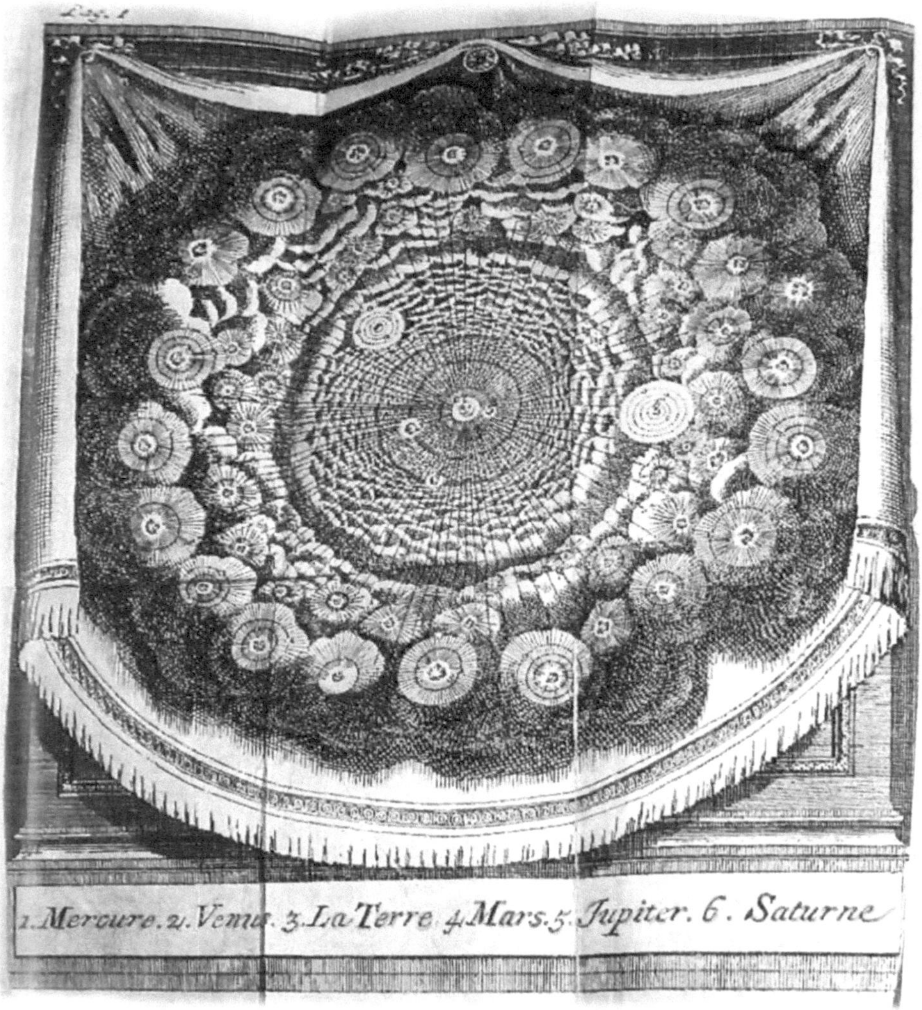

Figure 5.1 The solar system and the stars, from *Entretiens sur la pluralité des mondes* (*Conversations on the Plurality of Worlds*) (Fontenelle 1686).
Credit: Gallica, Bibliothèque Nationale de France

In this popular-science book, one of the first of its kind, Fontenelle supported the Copernican system and cheerfully conversed on the way planets could be inhabited. This book was intended above all to divert the reader. Fontenelle's background was philosophy and poetry, and of course, scientific arguments were not his key concern. However, he said in his preface:

I have fancied nothing concerning the inhabitants of the many Worlds, which must have been wholly Fabulous and Chimerical; I have said all that can be reasonably thought of them, and the Visions which I have added, have some real Foundation; what is true, and what is false are mingled together, but so as to be easily distinguished. (Fontenelle 1715)

As implied here, the question of habitability is tackled only in a poetic way and cannot be associated with scientific methodology or data.

In these same years, the Dutch astronomer and mathematician Christiaan Huygens (1629–1695) wrote a treatise entitled *Cosmotheoros: or, Conjectures Concerning The Planetary Worlds, their Inhabitants and Productions*, posthumously published by his brother Constantine (Huygens 1698). This book presents Huygens's speculations on other planetary systems and how these planets could be inhabited. His viewpoint is based on Copernican theory but also on a strong belief in the power of God. The supposed "innumerable multitude of stars" is the greatest part of God's creation (Huygens [1762] 2010, 10), including the planets and their inhabitants.

In this book, Huygens points out that water played an important role in growing and nourishing life, and also helped spark its origin. He wonders if water is present on other planets: "Since it is certain that the earth and Jupiter have their water and clouds, there is no reason why the other planets should be without them" (Huygens 1698, 24). He thought that the distance to the Sun was a decisive factor for such a possibility: "Every planet therefore must have its water of such a temper,[2] as to be proportioned to its heat" (Huygens 1698, 24).

While Huygens asserts that a great variety of different creatures could exist on other planets, he remains rather anthropocentric, proposing that "rational creatures" on other planets would have the same mind, body, and senses as people on the Earth. Likewise, plants and animals living on other planets rotating other stars would be very similar to terrestrial ones:

What a wonderful and amazing scheme have we here of the magnificent vastness of the universe! So many suns, so many earths, and every one of them stocked with so many herbs, trees, and animals, and adorned with so many seas and mountains! (Huygens 1698, 115)

It should be noted that Huygens and Fontenelle, in their respective writings, have mainly thought about the way planets could be inhabited. It is not exactly talking about "habitability" because in that case the discussion focused on the different creatures that populate the planets. As we will see in the following section, Flammarion distinguished "habitability" from "habitation."[3]

In Chapter 3, Dowd shows that Fontenelle, as well as Huygens, proceeded through an analysis such that the factor f_p could have taken a value of about 1: According to these authors, planetary systems around distant stars are the norm because our system is typical. It is more difficult to propose a value for n_e.

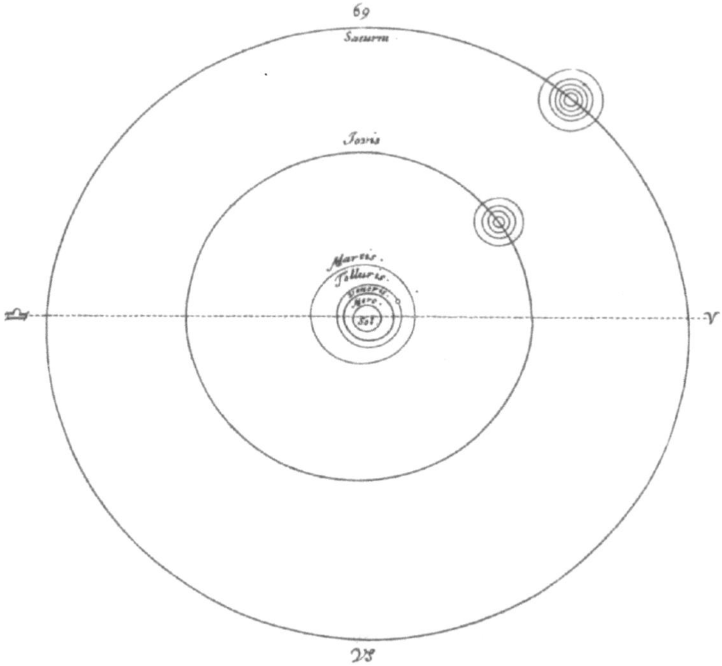

Figure 5.2 *Diagram of the solar system, from Cosmotheoros.*
Credit: Huygens (1698, 12)

If for these two authors, as noted by Dowd, planets are like ours, then they are full of life. If we follow this line of reasoning, the product of the factors f_l, the fraction of suitable planets on which life actually appears, and n_e, the number of planets per solar system with an environment suitable for life, could reach a value of 5, 6, or more, depending on the number of planets around each star. Of course, these (anachronistic) conclusions are very subjective and based only on speculations themselves extrapolated from what could be imagined about the planets of our solar system and beyond, which remained very unclear at that time.

Populating other planets was mainly a way to support Copernican theory, as Fontenelle actively demonstrated. One thing is certain: Too many uncertainties do not allow an accurate evaluation of n_e, if we set ourselves back to Fontenelle's period. As stated by the marquise in the *Conversations on the Plurality of Worlds* when talking about the Moon and its inhabitants with a touch of humor, it would be "a great deal of Ignorance upon a very little Knowledge" (Fontenelle 1715, 77).

The scientific foundations of planetary habitability during the second part of the nineteenth century

The Martian period

Life on the Moon was still accepted as possible during the first part of the nineteenth century. The German astronomer Franz von Paula Gruithuisen (1774–1852) published many papers about lunar life, one of the most important of which is "Discovery of Many Distinct Traces of Lunar Inhabitants, Especially of One of Their Colossal Buildings" (1824, paper in German). His observations of the lunar surface led him to conclude that climates and corresponding forms of vegetation were present on our satellite (Crowe 1986, 203). He also extrapolated from his observations – though he was aware that they had reached the limits of his telescope's possibilities – that the surface of the Moon was covered by cities, fortifications, and even temples, which implied religious activity. Gruithuisen's pluralistic ideas were later extended to other planets. In the 1830s, he published a series of papers arguing for life on Mercury, Venus, and even comets (Crowe 1986, 204). However, his hasty conclusions were severely criticized by his contemporaries, among them Carl F. Gauss and Wilhelm Olbers, who attributed his deductions to the power of his imagination, even as they supported the idea of lunar life.

In the 1860s, the Moon began to be seen as a "dead" world without any atmosphere, and was no longer the center of interest for speculations about life elsewhere. Improvements in telescopes led to increased observations of Mars, which seemed to be very similar to Earth. Distinguished astronomers attempted to penetrate the secrets of its surface. Beer and Mädler produced the first maps of Mars in the 1830s, and an intense period of mapping of the planet's surface began in the second part of the nineteenth century. The canals controversy, launched in 1877 by Giovanni Schiaparelli (1835–1910) and considerably developed by Percival Lowell (1855–1916) at the turn of the twentieth century (see Lowell 1909), intensified the importance attached to the study of the surface of the red planet. Mars, which was widely supposed to be older than Earth, represented a stirring world for speculations about life elsewhere. In part, this reflected an understanding that Mars, like Earth, had seasons, seas, continents, vegetation, and perhaps an advanced civilization.

Could Mars be a home for life? The title of Lowell's book *Mars as the Abode of Life* (Lowell 1909) reflects his belief in Martian life – and he took an extreme stand on that question during his years at the Flagstaff Observatory. He was convinced of the presence of complex life on the red planet and examined what he called "Martian ecology." Because he imagined a kind of model of "sustainable development" taking place on the red planet, he can rightfully be regarded as the first "exoecologist," as asserted by Markley (Markley 2005).

The possibility of another form of intelligent life in our solar system was a great subject of discussion in the astronomical community during the canals controversy, even as Lowell's interpretation of canals was widely disputed. In contrast, the existence of Martian vegetation was largely accepted. This was the case because variations in ground cover on Mars were obvious from the middle of the nineteenth century and were interpreted as evidence of seasons correlated with some kind of vegetation. The challenge was then to determine the *minimum* parameters required to allow the existence on Mars of some simple forms of life. The problem of Martian vegetation remained unsolved until the age of space exploration.

Planetary habitability has been a scientific field worth studying in detail since the second part of the nineteenth century. This chapter presents the selected viewpoints of three astronomers: Richard Anthony Proctor, Jules Janssen, and Camille Flammarion, who were pioneers in this area.

Richard Anthony Proctor

British astronomer Richard Anthony Proctor – famous for his first detailed map of the planet Mars (1867) and his talent in popularizing astronomy – published *Other Worlds Than Ours* (1870), in which he systematically examined the planets of our solar system and stated that habitability was a determining factor in answering the question of the existence of other life forms in the universe. More specifically, his study focused on criteria for planetary habitability and their physical and environmental parameters, such as climate, seasons, atmosphere, geology, and gravity. According to Proctor, defining planetary habitability was a very difficult task but could be overcome by considering possible analogies with our planet. Proctor based his methodology on analogy by comparing terrestrial environmental parameters with those of each planet. He believed that the existence and diversity of life forms depended on conditions specific to the surface of each planet.

One important point in Proctor's study of habitability was that he integrated Darwin's theory of evolution (Darwin 1859). He examined the question of adaptation throughout the book, especially in the chapter entitled "What Our Earth Teaches Us." According to Proctor, Darwin's works demonstrated a correlation between environmental changes (and their rhythm and intensity) in a specific habitat with the survival (or demise) of species in this habitat (Proctor 1870, 24). The conclusion of this observation was that *specific* conditions in the environment could be appropriate only to *specific* species. If many other worlds exist, Proctor thought they would be very different from ours (as the title of his book makes explicit). Creatures on their surfaces could be very unusual and, for instance, delight in environments inhospitable to terrestrial life. According to Proctor, these other worlds shelter life in other ways.

Proctor examined the bodies in our solar system: Mercury, Venus, Mars, Jupiter, Saturn, Uranus, and Neptune, as well as the Moon and other satellites, meteors, and comets. He gave special attention to Mars – "the miniature of our Earth," in Proctor's words (Proctor 1870, 90). Many physical analogies with our planet could be found on Mars: continents, seas, straits, even water on the surface. The atmosphere contained water vapor in a cycle equivalent to the terrestrial one. The Martian world described by Proctor would allow any sort of living being, from the simplest forms of vegetation to much more complex life forms.

At the time of the publication of *Other Worlds Than Ours*, Proctor thought that even Mercury and Venus, which offer physical conditions vastly different from our planet, could have life on their surfaces. However, since the environmental parameters are so dissimilar from Earth, these planets could shelter unfamiliar forms of life – including *microscopic creatures* on Mercury, for instance. The same argument was applied to the giant outer planets, especially Jupiter. Proctor assumed that Jupiter was not yet a suitable abode for living creatures, but suggested that one day could be a living world very different from those we know. Living creatures on Jupiter would be much smaller than inhabitants of Earth, and might possibly be "the most favored races existing throughout the whole range of the solar system" (Proctor 1870, 115).

However, a few years later Proctor expressed some doubts about intelligent life in the solar system (see Chapter 9). His book, *Our Place among Infinities*, published in 1875, contained a chapter titled "A New Theory of Life in Other Worlds" that rejected the existence of intelligent extraterrestrials not only on most planets in our solar system but also on other stellar systems (Crowe and Dowd 2013).

Camille Flammarion

In France, the famous astronomer Camille Flammarion published his first book in 1862, *La pluralité des mondes habités: Étude où l'on expose les conditions d'habitabilité des terres célestes discutées au point de vue de l'astronomie, de la physiologie et de la philosophie naturelle* (*The Plurality of Inhabited Worlds*), which quickly became famous for its support for the plurality of worlds. It was so successful that many editions of this book were published. In this chapter, we have used most often the 1868 publication that has been significantly augmented compared to the first one. In that book, Flammarion described some facts related to habitability (Flammarion 1868, 255):

- Earth, as a planet, is unremarkable.
- The other planets of the solar system are likely to have different conditions for habitability leading to life forms very different from terrestrial ones.
- Living beings on each world match the "physiological" state of the planet.
- The degree of habitability can be defined by analogies and understanding the differences between each world.
- The conditions of habitability on Mars and on Earth could be very similar. Climatic environment, physical features, and atmospheric conditions would be close enough to establish a parallel between the two planets. In line with this assumption, he believed that the inhabitants of Mars could be similar in many ways to those of Earth.

Flammarion's early ideas about planetary habitability can be found in most of his subsequent writings. In *Les Terres du Ciel* (*The Lands of the Sky*) published in 1884, Flammarion advocated the diversity of life and held that various adaptations on Earth were directly connected to changing environments (as proposed by Proctor, following Darwin's theory). On Earth, different habitats also led to different forms of life. This observation can be extended to other planets in our solar system and even to (presumed) numerous inhabited worlds across the galaxy. In *Les Terres du Ciel*, Flammarion highlighted the various conditions in which life could exist in each world and the large diversity likely to exist in the universe. This is the key point raised by Flammarion.

Flammarion fervently developed the topic of habitability in a two-volume work devoted to Mars, entitled *La planète Mars et ses conditions d'habitabilité – Synthèse générale de toutes les observations* (*The Planet Mars and Its Conditions of Habitability – General Synthesis of the Whole Observations about Mars*) (Flammarion 1892–1909). Between Flammarion's first work (1862) and this one (1892–1909), the canals controversy had strengthened interest in Martian habitability, even though Flammarion considered the canals to be natural. In these two volumes, he offered a synthesis of Martian observations carried out until that date, mainly concerning surface structures, atmosphere, and climate.

The methodological approach used to study habitability is clearly described in Flammarion's books, particularly *The Plurality of Inhabited Worlds*. Like Proctor, he assumed that reasoning by analogy was necessary to carry studies about habitability through to a successful conclusion. He believed that analogy was able to proceed from the "known" to the "unknown" and systematically considered the planets of the solar system to examine their similarities and differences. However, if he used the principle of analogy to study habitability on other planets, he did not support the principle of similarity, which is quite different. He believed that it was a big mistake to take our world as the unconditional model for the universe. We can't determine the biological organization of other living beings in the universe by relying solely on similarities with our planet (Flammarion 1862, 18).

Figure 5.3 Portrait of Flammarion by Dagron (1884), "Has more travelled between the stars than on our planet" ("A plus voyagé dans les étoiles que sur notre planète").
Credit: Gallica, Bibliothèque Nationale de France

He admitted that the question of habitability therefore remained very uncertain given humanity's state of knowledge at the end of the nineteenth century. It mainly consisted of formulating plausible conjectures, and even this remained a challenge. Eventually, he came to the conclusion that analogy, even if sure and fruitful, had

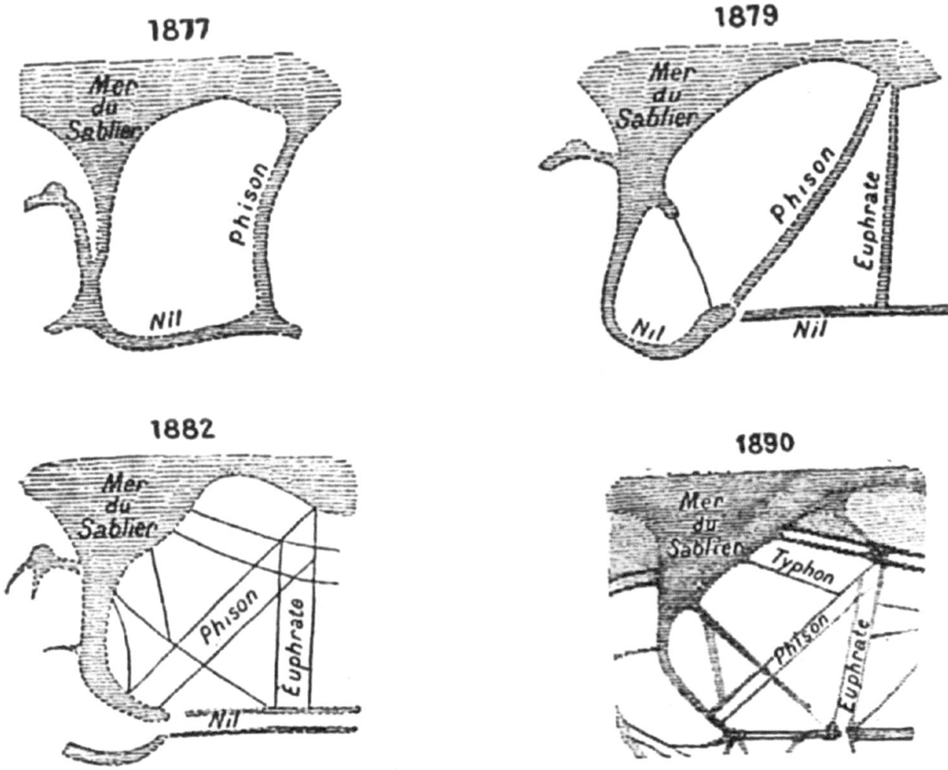

Figure 5.4 The supposed canals on Mars seemed to be subject to many changes, following successive observations made at the end of the nineteenth century. Flammarion tried to find an explanation attesting their natural origin.
Credit: La planète Mars et ses conditions d'habitabilité, Flammarion (1892, 1:571)

limits and could not be applied to the search for the specific characteristics inherent to each world (Flammarion 1868, 251). One remarkable point is that Flammarion did not agree with *anthropomorphism.* He noted that most authors attempted to define the nature of the inhabitants of other worlds by comparing them to humans. According to him, this approach was far too widely accepted even in a scientific context.

In addition, he differentiated "habitability" from "habitation" (Flammarion 1862)[4]:

- *Habitability* concerns correlations between the presumed physical and environmental conditions of the planets and their physiological conditions (compatible with existence of living forms).
- *Habitation* concerns the mental and physical state of each "mankind" supposed to live on other planets.

The universe could be then filled with various "mankinds," each living in harmony with the characteristics of its planet. This viewpoint drastically contrasts with that formulated by Huygens two centuries earlier (Huygens 1698). Through his numerous writings, Flammarion exerted considerable influence over the debate on habitability and the plurality of worlds, especially in France.

Jules Janssen

After the end of the canal controversy in 1909 (see Antoniadi 1930), the question of habitability reemerged with the birth of spectral analysis of planets. Jules Janssen (1824–1907), who founded the Meudon Observatory-France, contributed to the development of this new scientific field and strongly supported its use to detect atmospheric compounds in planetary atmospheres. In his view, it allowed us to search for water vapor, one of the main conditions required for terrestrial life (Janssen in Dehérain, 1929–30). In 1867, he stated that he had discovered by spectroscopy the presence of water vapor in the atmospheres of Mars and Saturn (Janssen 1867),[5] but he was in fact looking at terrestrial signatures.[6] Despite this error, he made effective use of spectroscopy to study the possibility of life on other planets. When combined with the new methods of physical astronomy that came with the birth of astrophysics, he seemed to have found in spectroscopy a possible key to answering the question of extraterrestrial life.

Janssen believed that the question of habitability was one of the most interesting ones facing humans (Janssen, in Dehérain 1929–30, 2:116). He understood that a strong link existed between the environmental conditions on a planetary surface – especially the presence of water vapor in the atmosphere and liquid water on the surface – and the possibilities for life to appear and exist. Spectroscopy could help determine the parameters defining the plausible conditions for life, once we knew the chemical constitution of planetary atmospheres.

At the same time, spectroscopy also limited the possibility for life on other planets. In identifying the presumed components of planetary atmospheres, the technique eliminated planets whose atmospheres did not contain water vapor, despite doubts about its detection. Conclusions about habitability were very hard to draw. But planetary spectroscopy confirmed, as noted by Janssen, the material unity of the universe, since molecules analogous to the terrestrial ones were detected elsewhere in the universe. According to Janssen, nature is uniform in its laws (Janssen, in Dehérain 1929–30, 1:380).

The problem of qualitatively and quantitatively determining the chemical composition of the Martian atmosphere remained partly unsolved until the 1940s, even as it appeared to be key to determining planetary habitability. And, even if no advanced civilization lived on Mars, many doubts still remained about the existence of vegetation. This uncertainty persisted until the beginning of space exploration in the 1960s. Spectroscopy could help, remotely, to answer this question through the detection of planetary atmospheric components. Early experiments in this field were carried out from the 1940s to 1960s.

Undoubtedly, the second part of the nineteenth century marked the start of a methodological approach to tackle the question of habitability on other planets. Through the studies of (what was known about) the characteristics and parameters of the planets in the solar system, speculations about plausible environments for life could be extrapolated from analogies and differences between our planet and other ones. In addition, spectroscopy became a wonderful tool to explore their atmospheres and then to select (and at the same time to limit) environments suitable for life. While the methodological approach became more and more scientific, however, habitability studies were restrictive to the solar system (especially focused on Martian environment).

Before the age of space exploration

Belarusian astronomer Gavriil Adrianovich Tikhov (1875–1960) studied the optical properties of terrestrial plants in extreme environments and aimed to detect by spectroscopy primitive Martian vegetation (see Briot 2013). Observations and field works led him to combine astronomical spectroscopy with botany. He coined the term "astrobotany" and became the founder and director of the section of the same name in the Kazakhstan Science Academy. The scientific activity of the Astrobotany Section came to an end in 1964, four years after his death, and his work remained poorly recognized.

In the 1940s, the Netherlands-born American astronomer Gerard P. Kuiper (1905–73) and later the French-born American astronomer Gérard de Vaucouleurs (1918–95) carried out spectrographic observations of Mars. They questioned whether there was vegetation on its surface, perhaps in the form of lichen or some other "primitive" species. However, they both excluded the presence on Mars of green plants with chlorophyll (Raulin Cerceau and Bilodeau 2011).

Less than a decade before the first probes orbited Mars, a period of excitement reemerged after a series of detections. William M. Sinton (1925–2004) used infrared spectroscopy and found absorption bands that seemed to match typical bands of terrestrial plants (Sinton 1957). However, these results, which depended on the interpretation of spectrograms (Dick 2013), have never been confirmed. Controversy about Martian vegetation lasted more than a century and faded in 1964 when the space probe Mariner 4 sent the first images of a barren and arid surface.

In the USSR, Russian rocket specialist Konstantin E. Tsiolkovsky (1857–1935) made significant contributions to our understanding of the origin and evolution of intelligent life in the universe. Tsiolkovsky has long been widely known for significant contributions to spaceflight theory, but his philosophical writings are less recognized. He believed that extraterrestrial intelligence was common in the universe (Lytkin, Finney, and Alepko 1995). His popular essay, "The Planets are Occupied by Living Beings" (1933, in Russian), outlines how other suns could have planets capable of sustaining life, how life could arise on some of these and evolve into higher forms, and so on, using a logic similar to that of the Drake Equation (Finney, Finney, and Lytkin 2000). Tsiolkovsky also attempted to quantify the number of habitable planets:

> Every sun has about ten big planets and thousands of small ones. At least one of them is similar to Earth: in temperature, size, weight, water, air and so on. So how can one deny organic life on them? (Tsiolkovsky 1933, qtd. in Finney, Finney, and Lytkin 2000, 748)

After a series of astronomical considerations, he concluded: "Every sun has not one but probably several populated planets" (Tsiolkovsky 1933, qtd. in Finney et al. 2000, 748).

It should be underlined that until the 1930s, many Western astronomers were still heavily influenced by James Jeans's hypothesis, which held that formation of planetary systems was relatively rare (Finney, Finney, and Lytkin 2000). The inconsistency of Jean's hypothesis gradually became evident in the 1930s (Shklovskiy 1963). After the collapse of this idea, pioneers developed new viewpoints about exoplanets.

The mid-twentieth century: Planetary ecology and habitability for humans

At the dawn of the space age, German physiologist Hubertus Strughold (1898–1987) proposed an idea very close to the concept of habitability previously defined by Proctor, Janssen, and Flammarion. Strughold was one of the pioneers of space medicine.[7] He coined the term *planetary ecology* to describe "the study of the planetary conditions necessary for life" and developed his theory in *The Green and Red Planet: A Physiological Study of the Possibility of Life on Mars* (1954). In this book, he provided fresh insights into the topic of habitability with new ideas: combining physical planetary data with physiological ones, in the light of what was known at the time about terrestrial life. His viewpoint was inspired by Percival Lowell's book, *Mars as the Abode of Life*, which he described as "the most impressive, most original book" about the possibility of life on other planets (Strughold 1954, 6).

As a physiologist, he defined methods for a "biological" study of planets (Strughold 1954). He said it was necessary to raise the question of life on other planets at the biological level. We must note that, until then, only astronomers had considered the concept of habitability.

Strughold starts his book with a survey of the physiological foundations of "life as we know it" on Earth. Then he justifies his assumptions about planetary ecology with some well-established principles of ecology and physiology known on Earth. Strughold's definition of planetary ecology is quite similar to that of planetary habitability: "the science which studies all the planets, including the earth, with regard to their comparative fitness as a biological environment" (Strughold 1954, 2). He was the first to combine physical and environmental parameters with biological ones. As with his predecessors, Strughold proceeded by analogy and started his study with comparisons involving biological and planetary parameters. Of course, many advances had been obtained in planetology and biology since the first concepts formulated by pioneers at the end of the nineteenth century.

In terms of planetary habitability or ecology, Strughold based his arguments on two definitions (Strughold 1954, 2):

- The definition of *physical ecology*: "Ecology is that science which treats of the physical environment of a place or region, with regard to its fitness as a site for the existence and development of living things."
- The definition of *physiological ecology*: "Ecology deals also with the adaptive reactions or responses of living things to their environment, in order to make their existence easier wherever they might be."

According to Strughold, the astronomical discoveries made during the first part of the twentieth century in planetary atmospheres provided a lot of data that could be used by biologists. In this way, he sought to remove barriers between astronomy and biology, allowing biologists to enter the discussion about life elsewhere in the universe. One of his arguments was to define the limits imposed on living organisms. On Earth, he stated, only specific organisms could survive in extreme environments, characterized in particular by extreme temperatures. This conclusion could be extrapolated to other planets and lead to parameters that matched planetary ecology. Applied to the celestial bodies of our solar system, he came to the conclusion – based especially on temperature – that nearly all planets should be excluded, except Mars and Venus:

> From the standpoint of temperature alone Mars and perhaps Venus are the only planets, aside from the Earth, which at present possess the prerequisites for living matter as we know it. All the other planets are excluded, for their temperatures lie far outside the range of active life. (Strughold 1954, 31)

As was the case with many of his contemporaries and predecessors, Strughold conferred special attention on Mars as a potential *biological environment*. He said that the presence of molecular oxygen was a crucial parameter for the existence of living organisms. However, in his opinion, since molecular oxygen (O_2) had still not been detected in the Martian atmosphere, the habitability of this planet might be very limited. In spite of this, he believed that the absence of molecular oxygen did not preclude some possible forms of primitive Martian life, like lichens or bacteria.

It must be pointed out that Strughold proposed distance to the Sun as the decisive determinant of the possibilities of life on each planet in the solar system. He described the pioneering concept of "thermal ecosphere of the sun," including planets capable of supporting life similar to ours (Strughold 1954, 36). This early definition has been studied by Dole (Dole 1964) and is comparable to the *habitable zone*[8] defined twenty-five years later by Hart (Hart 1979).

In 1959, in a paper entitled "The Problem of Life in the Universe and the Mode of Star Formation," Su-Shu Huang also tackled the question of the habitable zone (Huang 1959), considering the spectral types of the stars and their evolution on the

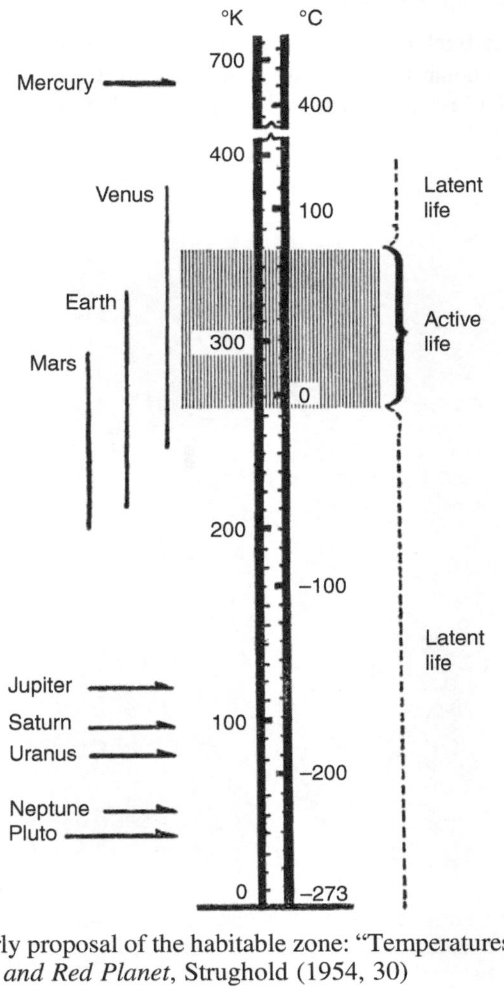

Figure 5.5 An early proposal of the habitable zone: "Temperatures of the Planets." From *The Green and Red Planet*, Strughold (1954, 30)

main sequence. According to Huang, "the best candidates are the hypothetical planets of main sequence F (preferably late F), G, and K (preferably early K) stars." He concluded that, within five parsecs from Earth, "only three stars (the sun, ε Eridani, and τ Ceti) have good chance for supporting advanced living beings on their planets, if such planets exist" (Huang 1959, 422). Frank Drake targeted these stars (ε Eridani and τ Ceti) in the OZMA project, which are the closest stars (about 11 light-years) from the Sun after the triple-star system Alpha Centauri.

A few years later, Stephen H. Dole (1916–2000) also examined the concept of planetary habitability. Dole's formulation of planetary habitability followed the Green Bank Conference (1961). However, his ideas are relevant in this chapter because he tackles the problem in a different way than Drake's conceptualization, but with parameters comparable to Drake's. Dole's book *Habitable Planets for Men* (1964) proposes the following definition:

> The use of the term "habitable planet" is meant to imply a planet with surface conditions naturally suitable for human beings, that is, one that does not require extensive feats of engineering to remodel its atmosphere or its surface so that people in large numbers can live there. (Dole 1964, 4)

Figure 5.6 Chinese-American astrophysicist Su-Shu Huang (1915–77), pioneer in the concept of habitable zone.
Courtesy: Northwestern University Archives

In this case, habitability chiefly concerned the planetary conditions suitable *for human life*, even if these conditions were also suitable to other forms of (terrestrial) life. Dole attempted to delineate the astronomical circumstances (mass of the planet, period of rotation, age, axial inclination, level of illumination, orbital eccentricity, mass of the star) that produce these conditions. Then he evaluated the probabilities of finding these conditions elsewhere in the galaxy. From that, he deduced the number of habitable planets for humans in the galaxy following an equation somewhat similar to the Drake Equation of 1961, except for the important fact that the Drake equation deals with the number of communicating civilizations. Dole's equation can be expressed as the following product of factors (Dole 1964, 82):

$$N_{HP} = N_s P_p P_i P_D P_M P_e P_B P_R P_A P_L$$

where:

- N_s, prevalence of stars in the suitable mass range, 0.35 to 1.43 solar masses;
- P_p, probability that a given star has planets in orbit;
- P_i, probability that the inclination of the planet's equator is correct for its orbital distance;
- P_D, probability that at least one planet orbits within an ecosphere;
- P_M, probability that the planet has a suitable mass, 0.4 to 2.35 Earth masses;
- P_e, probability that the planet's orbital eccentricity is sufficiently low;
- P_B, probability that the presence of a second star has not rendered the planet uninhabitable;
- P_R, probability that the planet's rate of rotation is neither too fast nor too slow;
- P_A, probability that the planet is of the proper age;
- P_L, probability that, all astronomical conditions being suitable, life has developed on the planet.

Dole was aware that any parameter in this equation was likely to be imprecise, since not all factors were known accurately, or at all. Nevertheless, the result obtained by Dole was, roughly, 600 million habitable planets in our galaxy (Dole and Asimov 1964, 171). Of the planets in the solar system, only Earth and Venus were considered to be in the range of habitability, as Mars fell well below the assessed minimum (Dole and Asimov 1964, 94).

Dole redefined the term "ecosphere," previously proposed by Strughold:

> "Ecosphere" will be used to mean a region in space, in the vicinity of a star, in which suitable planets can have surface conditions compatible with the origin, evolution to complex forms, and continuous existence of land life and surface conditions suitable for human beings, along with the ecological complex on which they depend. The ecosphere lies between two spherical shells centered on the star. Inside the inner shell, illuminance levels are too high; outside the outer shell, they are too low. (Dole 1964, 64)

If we think in terms of concepts, the Drake equation factor n_e seems to be more equivalent to the pioneering views on habitability formulated by Proctor, Flammarion, and Janssen, and also to Strughold's viewpoint, than to Dole's. One major difference between the concept of habitability used in the Drake equation and in Dole's concept of habitability is the additional parameter dealing with human life: "Free oxygen in a planetary atmosphere is essential for a planet to be considered habitable" (Dole and Asimov, 1964, 162–63).

Dole developed his concept of habitability, as applied to humans, while carrying out studies on the physical and physiological requirements of human beings in spacecraft. Human destiny was in that case a key preoccupation:

> If new planets offer new kinds of environments for man, man may well respond by developing new varieties of himself. (Dole and Asimov 1964, 209)

However, more general conclusions could be drawn from this study. The analysis of the ten factors proposed by Dole and Asimov (1964, 141–73) to describe their equation provides a lot of information, which can be more generally applied for the search for (terrestrial-like) life in a planetary environment.

Conclusion

Today, the number n_e of planets suitable for life per planetary system can be estimated scientifically (see Chapter 6) because we know more about the parameters of Earth-like exoplanets and their potential habitability. Furthermore, space exploration has recently shown that the large satellites of giant planets could also be relevant targets for the search for life elsewhere. This point was highlighted by Dole and Asimov as early as 1964 (Dole and Asimov 1964, 130–33). The definition of the habitable zone could then be extended to large natural satellites of exoplanets that are not necessarily in the "standard" habitable zone. Studying habitability requires various astronomical, geophysical, and geochemical parameters that are connected with the nature of the planetary surface, sub-surface, and atmosphere. As earlier proposed by Strughold, today's criteria for habitability also integrate biological parameters, which are supposed to identify limits in which (terrestrial) living forms could survive. Extreme environments on Earth are in that case very significant, since they could provide data that could be extrapolated to other celestial bodies.

During the pioneering period of the study of habitability (Proctor, Janssen, Flammarion), the question of habitability was mainly related to the solar system, taking into account the physical and environmental conditions necessary for life. This period makes for the first time a clear distinction between the question of "habitability" (the possibility to be inhabited) and that of "habitation" (being inhabited). When nineteenth-century pioneers began to take a deep interest in the notion of habitability, studies of the Martian surface were intensifying. However, the habitability of other planets in the solar system, such as Venus, was not excluded. Nevertheless, with a lack of data on these environmental parameters, it was difficult to draw solid conclusions about the possibility for other planets to be inhabited. Nothing was known with certainty about the environment of our neighboring planets and beyond. However, planetary habitability was, as we have noted, correlated with permanent liquid water on surfaces, a view very close to our current one.

The first scientific approaches to study planetary habitability have a strong common point: the methodological choice of *analogy*. Proctor, Flammarion, and Janssen used it to discuss habitability. As highlighted by Flammarion, analogy has limits, even if it remains (up to now) a unique and tangible way to estimate the possibilities of life on other planets. For instance, analogy is still used today when looking for (past) life on Mars. It is clear that habitability depends on many complex criteria, including those required *at a minimum* for the presence of life as it was suggested by the pioneers, and on the definition of life itself – a point that was not evident before the birth of molecular biology.

On the one hand, modern views of habitability have significantly restricted the possibilities of life in the solar system compared to the possibilities considered at the end of the nineteenth century. In that respect, it could be assumed that the value of the factor n_e has decreased since the pioneering views on habitability. On the other hand, modern views have clarified the range of possibilities for other environments suitable for life and have revealed new habitable environments such as the subsurface oceans of satellites of the giant outer planets. Of course, the spectacular increase in discoveries of exoplanets offers many other possibilities that could bring to reality the dreams of our pioneers. Beyond the solar system, new prospects are now in view with the coming detection of biosignatures on other planetary systems.

References

Antoniadi, Eugène. 1930. *The Planet Mars*. Translated from the French by Patrick Moore. Chatham: W & J Mackay, 1975.

Briot, Danielle. 2013. "The Creator of Astrobotany, Gavriil Adrianovich Tikhov." In *Astrobiology, History and Society: Life Beyond Earth and the Impact of Discovery*, ed. Douglas A. Vakoch, 175–85. Heidelberg: Springer.

Crowe, Michael J. 1986. *The Extraterrestrial Life Debate 1750–1900: The Idea of a Plurality of Worlds from Kant to Lowell*. Cambridge University Press.

Crowe, Michael J., and Matthew F. Dowd. 2013. "The Extraterrestrial Life Debate from Antiquity to 1900." In *Astrobiology, History, and Society: Life Beyond Earth and the Impact of Discovery*, ed. Douglas A. Vakoch, 3–56. Heidelberg: Springer.

Darwin, Charles M. A. 1859. *On The Origin of Species by Means of Natural Selection.* London: John Murray.
Dehérain, Henri. 1929–30. *Jules Janssen, Œuvres Scientifiques. 2 vols.* Paris: Editions de la Société d'éditions géographiques, maritimes et coloniales.
Dick, Steven J. 1996. *The Biological Universe: The Twentieth-Century Extraterrestrial Life Debate and the Limits of Science.* Cambridge University Press.
Dick, Steven J. 2013. "The Twentieth-Century History of the Extraterrestrial Life Debate: Major Themes and Lessons Learned." In *Astrobiology, History, and Society: Life Beyond Earth and the Impact of Discovery*, ed. Douglas A. Vakoch, 133–73. Heidelberg: Springer.
Dole, Stephen H. 1964. *Habitable Planets for Men.* New York: Blaisdell.
Dole, Stephen H., and Isaac Asimov. 1964. *Planets for Man.* New York: Random House.
Finney, B., L. Finney, and V. Lytkin. 2000. "Tsiolkovsky and Extraterrestrial Intelligence." *Acta Astronautica* 46(10–12): 745–49.
Flammarion, Camille. 1862. *La pluralité des mondes habités: Étude où l'on expose les conditions d'habitabilité des terres célestes, discutées au point de vue de l'astronomie, de la physiologie et de la philosophie naturelle.* Paris: Mallet-Bachelier.
Flammarion, Camille. 1868. *La pluralité des mondes habités: Étude où l'on expose les conditions d'habitabilité des terres célestes, discutées au point de vue de l'astronomie, de la physiologie et de la philosophie naturelle.* Paris: Gauthier-Villars.
Flammarion, Camille. 1884. *Les Terres du ciel: Voyage astronomique sur les autres mondes et description des conditions actuelles de la vie sur les diverses planètes du système solaire.* Paris: Marpon & Flammarion.
Flammarion, Camille. 1892–1909. *La planète Mars et ses conditions d'habitabilité: Synthèse générale de toutes les observations. 2 vols.* Paris: Gauthiers-Villars.
Fontenelle, Bernard (le Bovier de). 1686. *Entretiens sur la pluralité des mondes.* Available online at http://abu.cnam.fr/BIB/auteurs/fontenelleb.html.
Fontenelle, Bernard (le Bovier de). 1715. *Conversations on the Plurality of Worlds.* Translated by William Gardiner from the Last Paris Edition. London: A. Bettesworth.
Galilei, Galileo. 1989. *Sidereus Nuncius or the Sidereal Messenger.* Translated with by Albert Van Helden. University of Chicago Press. Original publication 1610.
Hart, Michael H. 1979. "Habitable Zones About Main Sequence Stars." *Icarus* 37(1): 351–57.
Huang, Su-Shu. 1959. "The Problem of Life in the Universe and the Mode of Star Formation." *Publications of the Astronomical Society of the Pacific* 71(422): 421–24.
Huygens, Christiaan. 1698. *Cosmotheoros (The Celestial Worlds Discovered: or, Conjectures Concerning the Inhabitants, Plants and Productions of the Worlds in the Planets).* London: Timothy Childe.
Huygens, Christiaan. [1762] 2010. *Cosmotheoros: or, conjectures concerning the inhabitants of the planets.* Translated from the Latin of Christian Huygens. A new edition, corrected. Glasgow: Gale ECCO Prints Editions.
Impey, Chris. 2013. "The First Thousand Exoplanets: Twenty Years of Excitement and Discovery." In *Astrobiology, History, and Society: Life Beyond Earth and the Impact of Discovery*, ed. Douglas A. Vakoch, 201–12. Heidelberg: Springer.
Janssen, Jules. 1867. *Annuaire de la Société météorologique de France, Bulletin des Sciences, 11 juin 1867*, p. 218.
Kepler, Johannes. 2003. *Kepler's Somnium, The Dream or Posthumous Work on Lunar Astronomy.* Translated by Edward Rosen. New York: Dover Publications. Original publication 1634.

Knevitt, Oliver. 2011. "Fontenelle's *Conversations* (1686): Popular Science Writing at Its Very Best." Available online at www.science20.com/between_death_and_data/fontenelle's_conversations_1686_popular_science_writing_its_very_best-76400/.

Lowell, Percival. 1909. *Mars as the Abode of Life*. New York: MacMillan.

Lytkin, Vladimir, Ben Finney, and Liudmila Alepko. 1995. "Tsiolkovsky, Russian Cosmism and Extraterrestrial Intelligence." *The Quarterly Journal of the Royal Astronomical Society* 36: 369–76.

Marcy, Geoffrey. 2014. *Paper presented at the 5th Conference "Searching for Life Signatures."* UNESCO, Paris, March 20–21.

Markley, Robert. 2005. *Dying Planet: Mars in Science and the Imagination*. Durham: Duke University Press.

Proctor, Richard A. 1870. *Other Worlds Than Ours: The Plurality of Worlds Studied under the Light of Recent Scientific Researches*. New York: A. L. Burt.

Raulin Cerceau, Florence, and Bénédicte Bilodeau. 2011. *D'autres planètes habitées dans l'univers?* Paris: Ellipses.

Sheehan, William. 1996. *The Planet Mars: A History of Observation and Discovery*. Tuscon: University of Arizona Press.

Shklovskiy, Iosif S. 1963. "Is Communication Possible with Intelligent Beings on Other Planets?" In *Interstellar Communication, A Collection of Reprints and Original Contributions*, ed. A. G. W. Cameron, 5–16. New York: W. A. Benjamin.

Sinton, William M. 1957. "Spectroscopic Evidence for Vegetation on Mars." *The Astrophysical Journal* 126, no. 2: 231–39.

Slipher, Vesto M. 1908. "The Spectrum of Mars." *Astrophysics Journal* 28: 397–404, and plate 36.

Strughold, Hubertus. 1954. *The Green and Red Planet: A Physiological Study of the Possibility of Life on Mars*. London: Sidgwick and Jackson.

Notes

I would like to thank Danielle Briot for the in-depth discussions we had that were very helpful to improve the manuscript.

1 Published in 1634, but work in progress from 1609 to 1630.

2 Frozen or evaporated.

3 Concerning the Drake Equation, the way planets could be inhabited would correspond with the factor f_l.

4 For our purpose in examining the Drake Equation, this differentiation is similar to the distinction between the factor n_e and the factor f_l.

5 The spectroscopic detection of water vapor on Mars was also announced by William Huggins in England and Hermann Vogel in Germany (Sheehan 1996).

6 William Wallace Campbell denied the Martian result in 1894 after his spectroscopic observations at the Lick Observatory (Sheehan 1996). However, the controversy still continued with Slipher, who claimed in 1908 the spectrographic detection of Martian water vapor and oxygen (Slipher 1908).

7 Strughold was considered the "Father of Space Medicine," but he also was in charge of medical research for the Luftwaffe during the Second World War. He emigrated to the United States after the Second World War.

8 In Chapter 9, Crowe notes an even earlier view of an equivalent notion to the habitable zone (the "temperate zone") coming from William Whewell in 1853.

6

Number of planets, per solar system, with an environment suitable for life, n_e, 1961 to the present

Danielle Briot and Jean Schneider

Abstract

One of the Drake Equation factors that has changed the most since 1961 is n_e, the average number of planets per star that can potentially support life. This factor is still evolving.

The definition of conditions for developing life is related to the definition of a circumstellar habitable zone. It is generally defined as the zone in which physical conditions make presence of liquid water possible. As a first approximation, it implies a temperature between 0 and 100 degrees Celsius, the so-called Goldilocks condition. This preliminary estimate is based on the temperature of the star: in other words, its spectral type and the distance between the star and its planet. Changes in the n_e term are principally due to unexpected characteristics on one hand of many exoplanets and on the other hand of some objects in the solar system.

We first discuss the basic condition for life: liquid water. Stellar spectral types where life is most likely to emerge and exist are reviewed, and less favorable conditions of hot and cold stars are discussed. We also look at planets orbiting one member of a binary star, or around both stars ("circumbinary planets"). We point out the physical conditions necessary for planets to shelter life, including mass and other physical parameters. Exomoons are interesting objects and are discussed as well. Relations between the star and its planetary system are reviewed. No longer is distance the only parameter to be considered. In the case of terrestrial exoplanets with large eccentricity that cross the habitable zone, life with some phases of hibernation may be possible. Some terrestrial exoplanets orbit so close to their star that they are co-rotating, keeping one face to the star at all times, which implies that a temperate annular zone may exist between the very

The Drake Equation, ed. Douglas A. Vakoch and Matthew Dowd. Published by Cambridge University Press.

hot face in front of the star and the very cold face on the opposite side. Characteristics of some satellites in the solar system discovered since the space age show that some tidal effects are liable to extend the habitable zone, as can be seen by the detection of oceans flowing below the icy surface of Europa or internal water springing from the geysers of Enceladus. It would be interesting to search for and study as a source of possible life moons of giant exoplanets located in the habitable zone of their star. As a conclusion, the continuously increasing number of small rocky planets provides great encouragement to search for extraterrestrial life. Indeed, they show a high rate per star and satisfy the conditions necessary for producing life.

Introduction

"It is difficult to treat in a small space of a subject on which so much has been written, unless, indeed, one is confined to a mere statement of known facts." This is the first sentence of a paper by Herbert A. Howe, titled "The Habitability of Other Worlds" and written as early as 1885.

The term n_e is the third one in the Drake Equation, which is also known as the Green Bank Equation. The definition of the term n_e is not exactly the same in different sources. For each star defined by the second term in the Drake Equation, f_p, or the fraction of stars that form planets, the term n_e is sometimes defined as the number of planets hospitable to life, or the number of planets similar to Earth, or the average number of planets in the ecosphere of the star. When the Drake Equation was written, all of these definitions were equivalent. But today, as we shall see, it is possible that planets with properties very different from Earth's are good candidates for sheltering life. In this chapter, we will use the definition of the average number of planets that can potentially support life per star that has planets.

The term n_e, together with the term f_p, is probably the one about which our knowledge has most rapidly increased since 1961. It is closely linked to what is now called the habitable zone. The word habitability was initially used about planets in the solar system. Su-Shu Huang probably used the word habitable for the first time in relation to extrasolar planets in 1959, before the creation of the Drake Equation (Huang 1959). Huang, whose studies are surprisingly ahead of their time, was a member of the group for which the Drake Equation was developed. The habitable zone has also been called the ecosphere. (e.g., Dole 1964; Shklovski and Sagan 1966). The inclusion of this term in the Drake Equation triggered studies about this subject, especially in the 1970s (e.g., Hohlfeld and Terzian 1977).

Studies on the habitable zone of extrasolar planets were very rare when the Drake Equation was formulated. The number of publications on this topic is continuously increasing: whereas only very few papers appeared in the 1970s, starting from the 2000s, several tens were published each year and are still increasing. Currently, one scientific paper about this subject appears about every two days, and often more than that. A good source for papers on this subject is the Extrasolar Planets Encyclopedia at Exoplanet.eu (Schneider et al. 2011). This is now a deeply studied subject, and knowledge about it is changing fast. Some studies examine the subject globally, whereas others are devoted to a specific planet, object, or process. Two main reasons justify scientific interest and studies in this subject: recent discoveries of the properties of some moons in the solar systems outside of the habitable zone, and the discovery of many exoplanets in the habitable zone that have unexpected properties. The problem of the habitable zone has to be seen from a broader view. The classical definition of habitability is based on the assumption about the possibility of liquid water on the planet. This definition is probably due to Michael Hart (1979), and this hypothesis is discussed below. At first sight, this hypothesis implies at least that the temperature of the planet ranges between 0 and 100 degrees Celsius – around 273 to 373 Kelvin – for an Earth-mass planet. For super Earths – that is to say, rocky exoplanets of two to approximatively ten or twenty times Earth mass – the ground atmospheric pressure may be much higher, leading to a higher boiling temperature. Underwater vents in terrestrial oceans have shown that life may exist up to 200 degrees Celsius. The habitable zone is therefore defined by a temperature range in which the planet has to lie. This condition is sometimes named the Goldilocks condition, or we speak of Goldilocks planets. This expression is derived from a fairy tale in which Goldilocks makes some tests in the Three Bears house and continually selects the mean condition; in one particular case, the breakfast has to be neither too hot nor too cold. As a first approximation, the habitable zone for planets orbiting various spectral type stars is defined by taking into account the star's temperature and the decrease of temperature due to the distance between the star and the planet. Usual figures show a habitable zone whose distance from the star and broadness are defined as a function of the star's temperature. Figure 6.1 shows an example of this habitable zone corresponding to one Earth-mass planet as a function of the stellar mass. In this figure, we can see also that the habitable zone of the Sun includes the planets Earth and Mars, with the latter very near the zone's outer edge.

Such a definition of the habitable zone involves the liquid-water hypothesis, properties of the star, properties of the planets, and the distance and relationships between star and planet. We shall successively study the various elements that come into play in this definition. However, as Robert McFarland wrote, "The error consists in virtually assuming that the climate of a place depends solely on its

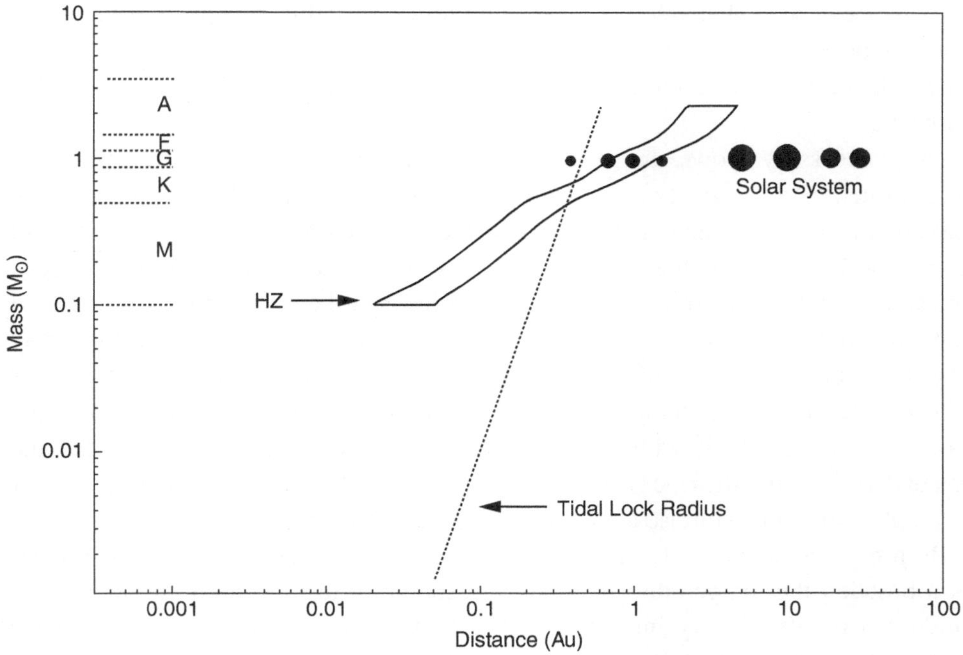

Figure 6.1 Diagram showing the habitable zone (solid curves) as a function of the stellar mass, from Kasting, Whitmire, and Reynolds (1993, 124). The dotted line represents the distance at which an Earth-like planet in a circular orbit would be locked into synchronous rotation within 4.5 billion years as a result of tidal damping.

distance from the *Sun* – whereas this is only one of a hundred causes" (McFarland 1883, 217). Though written in 1883 and dealing only with the planets of the solar system, this sentence could also be applied to the exoplanets. Indeed, studies of some objects in the solar system show that other physical processes are relevant, and so the definition of the habitable zone has to be broadened to take into account these factors.

Definition of the habitable zone and hypotheses for other forms of life

As noted above, the fundamental conditions generally adopted for habitability is the presence of liquid water. Obviously, this condition is based on the only form of life that we know and the conditions necessary for the emergence of this form of life on the planet Earth. In fact, we don't really know how to define what is life, and whether a totally different form of life would be possible on another planet located far from the solar system. Life on Earth is based on water and carbon chemistry. Liquid water is often considered indispensable to life or at least to the emergence of life. There are many different reasons for this: mainly, that biochemical reactions require a fluid, and water remains liquid in a very large temperature interval, is a very good solvent, and is one of the most abundant molecules in the universe.

It is necessary, however, to investigate whether other solvents are possible. Titan, satellite of the giant planet Saturn, is the only satellite in the solar system that has an atmosphere, which makes it a very interesting case study regarding different forms of hypothetical life. Thus Titan was observed by the Cassini-Huygens mission, which discovered a very surprising result: it has standing lakes of liquid methane and ethane, as seen in Figure 6.2. It has been suggested that they may shelter a special form of life for which methane would play the role that water plays on Earth. Living beings would inhale dihydrogen, H_2, instead of dioxygen, O_2, then metabolize it with acetylene instead of glucose, and exhale methane instead of carbon dioxide. Other molecules can also be good solvents – for example, ammonia, sulfuric acid, and formamide. Ammonia seems to be a particularly promising solvent. It can be used as a cooler in water-ammonia composites and broaden the temperature interval corresponding to the liquid phase. Formamide can be also a very interesting solvent. Its liquid phase has an even larger temperature interval with limits of 0 and 222 degrees Celsius (Leitner et al. 2008).

It is also possible to imagine some completely different forms of life that would be nonchemical. For instance, Jean Schneider proposed a crystalline physiology

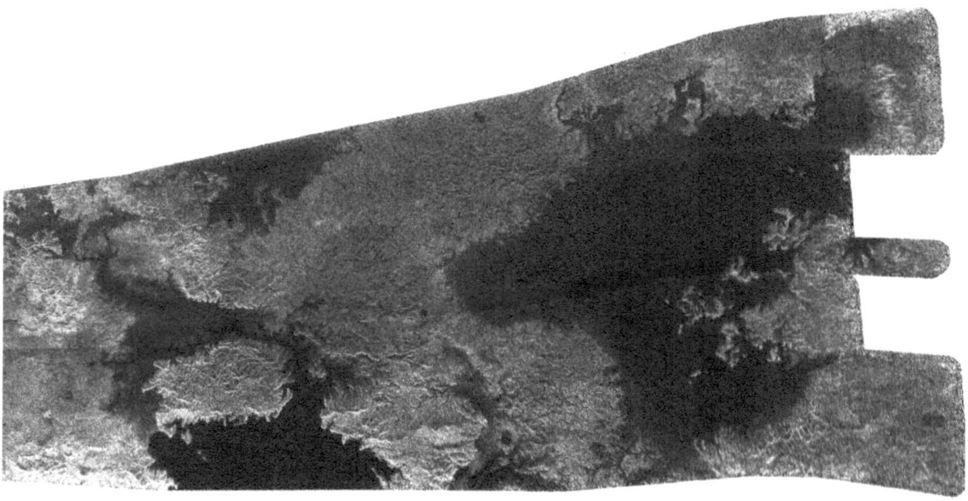

Figure 6.2 Lakes on Titan. This image of Saturn's larger moon, Titan, obtained by the radar instrument on the spacecraft Cassini features dunes and lakes that are thought to consist of liquid methane and ethane.
Credit: NASA/JPL

for a hypothetical lifeform (Schneider 1977). This form of life would satisfy many conditions: stability, rich information content, and diffusion into the surrounding environment.

Generally speaking, life on extrasolar planets is conceived as a copy or at most an extrapolation of the only form of life that we know, which is the terrestrial life. The question is whether nature enforces these limits or is much more imaginative and elastic.

Physical properties of stars

Some words about physical properties of stars

Stars considered "normal" are classified as a function of their temperature, determined by examining their spectra. This classification is indicated by a capital letter. From the hottest to the coldest stars, the list of spectral types is O, B, A, F, G, K, and M. All astronomers know the mnemonic sentence for remembering this list: Oh, Be A Fine Girl, Kiss Me. Each spectral type is divided in ten sub-spectral types indicated by an Arabic numeral. A Roman numeral from I to V is added to indicate the luminosity class of the star. Our Sun is a G2 V star. The surface temperature of the hottest stars, O stars, is about 35,000 Kelvin, and of the coldest stars, M stars, is lower than 3,000 Kelvin. So the habitable zone for cold stars is nearer the star. The classical habitable zone of the Sun encompasses the orbits of Earth and Mars. Hot stars are blue, massive, and evolve very quickly. Cold stars are red, less massive, and evolve much more slowly. Our Sun is a yellow star of intermediate mass.

Some necessary physical conditions of stars for the emergence of life

A certain time interval is obviously necessary for life to emerge. The continuously habitable zone (CHZ) is defined as the region within which a planet can keep a moderate surface temperature during a time long enough for life to appear and develop (Hart 1979). The time needed for life to emerge is difficult to determine. The time span of the CHZ is generally considered to be between 1 and 5 Gyr (Gyr = gigayear = one billion [10^9] years). So only stars in a stable stage of their life have to be considered. The longest stable phase occurs when the star is of luminosity class V. Let us consider the Hertzsprung–Russell diagram (H-R diagram), which is the fundamental diagram of stellar astrophysics. In this diagram, the luminosities of stars are plotted versus their temperature. The luminosity class V stars are on the main sequence, and this phase corresponds to 90 percent of the life of the star. The other phases of a star's life are much shorter. During these phases, properties of stars change drastically and too rapidly at the astronomical time scale needed for life to develop to a high degree of complexity. Furthermore, in some cases, stars can undergo very rapid (hours to months or years) variations in size and temperature, which implies that their ability to sustain life becomes very problematic.

Our Sun, which is an intermediate mass star, has been staying on the main sequence for about 5 Gyr. Such a large time interval allowed life to emerge and develop to an extreme level of complexity on the planet Earth. The Sun is expected to remain on the main sequence for about another 5 Gyr. So a solar-type star, a type known as able to shelter complex life, will spend 10 Gyr on the main sequence of the H-R diagram. However, even during the long and most stable phase of its life, a star undergoes some luminosity changes. Since the Sun first arrived on the main sequence, its luminosity has increased, thereby moving the habitable zone outward. It is possible that Venus was located in the habitable zone in the past. After its 10 Gyr long stage on the main sequence, the Sun will become a red giant: red because its temperature decreases and giant because its radius increases tremendously. Its radius will increase tenfold, and its surface will reach the Earth's orbit. Even if life on Earth continues up to this moment, it will then disappear.

Let us now see what happens for stars hotter and cooler than the Sun.

Life conditions for a planet orbiting a hot star

The hotter and more massive stars are, the faster they evolve. Hot stars stay on the main sequence only for some millions of years before exploding as a supernova. This time interval is shorter than the one needed for life to evolve in the CHZ. For life to develop and become as complex as life on Earth, hot stars are not good candidates.

Furthermore, massive, blue, hot stars radiate principally in ultraviolet wavelengths. So if life developed on a planet orbiting a hot star, it would need an atmosphere opaque to ultraviolet radiation. Finally, let us note that hot stars are very rare in the Sun's neighborhood. Because they are very bright, they can be seen from a greater distance. However, in this case, the detection of life on these planets would be especially difficult. Thus, very hot stars do not seem very good candidates for the emergence of life or for their observation from Earth.

Life conditions for a planet orbiting a cold star

The M stars, which are the coldest stars on the main sequence, are the most numerous in our galaxy, the Milky Way. They represent three-quarters of all main sequence stars. They can be considered very good candidates to look for planets in the habitable zone (see, for example, Quintana et al. 2014).

This is due to several factors. A very successful method for detecting extrasolar planets is the detection of transits – that is to say, observing the regular periodic decrease of a star's luminosity when a planet passes between the star and the observer, therefore occulting a part of the luminous surface of the star. In the case of a cold and small star, the habitable zone is close to the star, and the ratio of planet–star surfaces is larger. This increases the probability to observe transits. In addition, the evolution of cold stars is very slow. Physical conditions that are stable for a very long time would allow more time for life to develop to a complex level.

If the habitable zone is too close to the star, however, the planet moves in co-rotation; in other words, the rotational period of the planet is the same as its orbital period. A good example of such a motion is that of the Moon in relation to the Earth. In such a case, the planet always faces the same side toward its star. This side is permanently illuminated by the star, whereas the other side is always in darkness. It is difficult to determine the temperature of each side of the planet. In the absence of any atmosphere, obviously an inappropriate situation for life, one side of the planet will be very hot and the other side very cold. In the case of an atmosphere with turbulent movements, heat transfer can occur from one face to the other. The annular part of the planet, the terminator, which is located at the edge between the brightened and the dark sides, can be also considered. The temperature of this intermediate area might be suitable for life, and perhaps some forms of life can develop here. Furthermore, if the planet has a libration movement, as does the Moon, the "temperate" area is enlarged. Some studies of extremophiles on Earth demonstrate that life can exist in conditions harder than we may expect in these areas.

Double and multiple stars

In our galaxy, large rates of stars are actually binary or multiple stars. By some estimates, 70 percent of stars in the vicinity of the Sun fall into this class. If some planets exist in the habitable zones of binary or multiple stars, the possibility of discovering habitable planets in our galaxy is much greater. As early as 1960, Huang (1960) studied the habitable zones of binary systems, reviewing the several cases of binary stars, star separation, and elliptical orbits. However, he explicitly regretted not being able to determine if the formation process of binary stars also allows for planetary formation. As an example, he considers the triple system α Centauri and discusses the possibility that a planet exists in its habitable zone. Let us recall that the three stars forming the α Centauri are the three stars nearest to the Sun. It is thus the stellar system for which the distance is the most favorable to observe and study its planets by direct imaging. This makes it very important to determine its habitable zone, investigate whether some planets exist in this system, and learn whether those planets are located in the habitable zone. In 2012, a planet was discovered orbiting α Centauri B; unfortunately, it is too close to the star to be in the habitable zone, since its period is four terrestrial days (Dumusque et al. 2012).

Two main types of orbits can be distinguished for planets in binary stars systems: S-type, in which the planet orbits only one of the two stars, and P-type, in which the planet orbits both stars. This second type is also known as circumbinary movement. There is also another type called T-type, in which a planet may orbit close to one of the Lagrangian equilibrium points L4 and L5 of the binary system.

The first planet in a double star system, 16 Cyg B b, was detected in 1996, which was only one year after the discovery of the first exoplanet. It is an S-type planet orbiting around the star 16 Cyg B, which is quite similar to the Sun. Currently, more planets with S-type movement than planets with a P-type movement are known. It is difficult to give a precise number for planets in each type because new ones are discovered constantly.

The problem of habitable planets orbiting multiple stars was studied in 1977 by Hohlfeld and Terzian (1977). This study was driven by the Green Bank Equation, another name for the Drake Equation as mentioned earlier in this chapter. Even before the discovery of the first extrasolar planet, some studies focused on the stability of orbits of exoplanets in binary systems, because a stable orbit is fundamental to the emergence of life and to the determination of habitable zones in the cases of double or multiple stars (see, e.g., Benest 1991). As in other fields connected to the habitability of planets, the number of such studies has increased drastically since then.

Habitable planet properties

Planet mass: Terrestrial or gaseous planet?

The solar system has eight planets. Starting outward from the Sun, there are first four terrestrial, or rocky, planets: Mercury, Venus, Earth, and Mars. Next, there are four giant and gaseous planets: Jupiter, Saturn, Uranus, and Neptune. Rocky planets are less massive and gaseous planets more massive. Gaseous planets have an iron core surrounded by a very thick and opaque gaseous layer. Life on a Jupiter-like planet is very difficult to imagine. Up to now, only rocky planets were thought to shelter life. Because those planets are less massive, however, they are the most difficult to detect, so the first planets detected were giant planets. In our solar system, the mass of the most massive terrestrial planet, Earth, is 5.974×10^{27} grams, and the mass of the least massive gaseous planet,

Uranus, is 86.8×10^{27} grams. The most massive planet, Jupiter, is 1898.7×10^{27} grams, and the star Sun is 1.989×10^{33} grams. To know if life can develop on an exoplanet, it is fundamental to determine whether this planet is rocky or gaseous. This determination is complicated by the fact that we do not know exactly what the limit mass between terrestrial and gaseous planets is, or even if such a limit mass exists. At this moment, it would appear that some exoplanets with masses intermediate between the rocky and gaseous planets in our solar system could be gaseous or rocky exoplanets. So even if we know the mass of an exoplanet, it is not always possible to determine whether or not the planet is liable to shelter life.

However, the present situation can only be improved. Results from NASA's Kepler space telescope show that the mean mass of newly discovered planets decreases with time as more and more small mass planets are discovered. It is likely that more small mass planets exist than large mass ones. Future instruments will greatly increase the number of discovered small mass planets.

Other physical planet conditions

A planet must have an atmosphere to shelter life, and this atmosphere must be neither too dense nor too thin. Remember that the limits of the habitable zone, where water is liquid – above the freezing point and below the boiling point – depend on the atmospheric pressure. It can be very different with various planets. In particular, for ten to twenty Earth-mass super Earths, the boiling temperature may be well beyond 100 degrees Celsius. The planet's mass plays an important role in the presence of atmosphere and its physical characteristics. For example, the mass of the planet Mercury is too small to have and keep an atmosphere. But the planet's mass is not the only parameter that affects the presence of an atmosphere compatible with life. The mass of Venus, 4.869×10^{27} grams, is smaller than the Earth's mass, 5.974×10^{27} grams, and nevertheless the atmosphere of Venus is very opaque and dense. The atmospheric pressure on the surface of Venus is nearly 100 times that on Earth. The presence of this atmosphere causes a very strong greenhouse effect, so that the temperature at the surface of Venus is around 465 degrees Celsius or 738 Kelvin. Life on Venus is really difficult to imagine, except perhaps something very unlike what we know on Earth that could exist in the upper part of the atmosphere. The extremely different conditions in the atmospheres of Venus and Earth are probably due to the location of Venus, which is closer to the Sun than Earth, even very near the inner edge of the habitable zone. Furthermore, it is possible that for life to exist, tectonic activity is needed to reprocess carbonate rocks end return carbon dioxide in the atmosphere (Kasting, Whitmire, and Reynolds 1993).

Exomoons

In our solar system, the terrestrial planets are closer to the Sun and the giant planets are farther away. According to models of planet formation from a disc surrounding the central star, giant planets can form only beyond the ice line, and so beyond the habitable zone. This is observed in the solar system. However, since the discovery of the first extrasolar planet, 51 Pegasi b, by Mayor and Queloz in 1995 (Mayor and Queloz 1995), we know that giant extrasolar planets can exist quite close to their star. One of the greatest surprises of this discovery was this planet's period, which is 4.23077 days. This result implies that the planet 51 Peg b is very close to its star, even though it is a giant planet because its mass is 0.468 ($\pm$ 0.007) times the mass of Jupiter. This observational result is explained by migration processes: giant planets form beyond the ice line, then move by some migration processes in the protoplanetary disc, which leads the planets very close to the star. These planets are called hot Jupiters. The migration processes have been deeply studied since the discovery of 51 Peg b.

Other giant planets can be located in the habitable zone. They are not habitable because they are gaseous, but they can have moons, called exomoons, which are rocky objects and where life is possible (Williams, Kasting, and Wade 1997). Habitable zones of exomoons are considered to be the same ones as habitable zones of their planet. So life may exist on moons of a giant planet located in the habitable zone. Moreover, from properties of some moons in the solar system, we shall see that habitable zones corresponding to exomoons can extend habitable zones for planets. It is important to detect such objects because they can play a very interesting role in the search for life. Exomoons have become a next step in exoplanetology (see Schneider, Lainey, and Cabrera 2015).

Connections between the planet and its star

The classical definition of habitable zone is based on the temperature of the star and the distance between the star and the planet. For the emergence of life, a planet must also have certain properties. It is necessary also to look at the relationships between star and planet.

Ellipticity of the planet trajectory

As we saw earlier in this chapter, one surprising result of the discovery of the first extrasolar planet was that a giant planet was so close to its star. As more and more planets are discovered, they present other unexpected physical properties. Whereas the eight planets of the solar system have orbits that are nearly circular, many extrasolar planets have highly elliptical orbits. The probability of a highly elliptical orbit increases with the distance between the star and the planet. How can we define a habitable zone for a planet whose distance from its star varies by a large amount? We can imagine a life with hibernation times connected to some phases of the planet orbit, as in many terrestrial animals. We do not really know if a constant temperature corresponding to a circular orbit during some amount of time is necessary for the emergence of life.

Variations of obliquity of planets

The only planet that we know shelters life, Earth, has an essentially stable obliquity, exhibiting only small variations around the mean value 23°26'14.89". However, Laskar, Joutel, and Robutel (1993) have shown that in the absence of the Moon, the obliquity would be chaotic and would vary from nearly 0 degrees up to about 85 degrees. These extended variations would imply dramatic changes in climate unfavorable for the development of life. The most widely accepted model for the formation of the Earth and the Moon implies an impact with some body during the formation of the solar planetary system. Such an event does not happen often, so the probability that an exoplanet has a moon with the same properties as ours is low. This point argues against the presence of complex forms of life being typical on exoplanets.

Properties of co-rotating planets

As noted earlier, many planets, some terrestrial, have been discovered very close to their star. In these cases, they are co-rotating. If the star is hot, a planet close to its star is not in the habitable zone, but a part of the planet could be at a temperature suitable for life. The part of the planet facing the star is superheated but the part opposite is hardly reached by the star's heat. Obviously, the diffusion of heat from the very hot zone facing the star to the opposite zone depends on the properties of the planet's atmosphere and on the heat transfer induced by its rotating and turbulent motions. Between hot and cold planet's faces, a temperate zone can exist and may be habitable. The planet CoRoT 7 b is a good example of such a planet, and many studies have been performed about conditions of a possible habitability on this object.

Tidal effects producing liquid water

There is an area where knowledge has been greatly expanded since the formulation of the Drake equation. On-site observations by spacecrafts of some objects in the solar system have demonstrated that liquid water can be found on objects whose temperatures are determined by factors other than their distance to the Sun. Other physical processes can also be efficient at providing heat. Some tidal influences caused by nearby objects with a large mass can increase the temperature of the object concerned. In the solar system, the giant planets are outside the habitable zone. However, observations of some of their satellites obviously show heating, and even in several cases, liquid water is detected.

Tidal effects for Jupiter satellites

Tidal interaction exercises great influence on the Galilean satellites of Jupiter: Io, Europa, Ganymede, and Callisto. These objects were first observed by Galileo Galilei in 1610 (Galilei 1610). They played a very important role in the history of human knowledge. Whereas the Earth was then thought of the center of the universe, and every celestial body was supposed to orbit the Earth, they are the first objects that were observed orbiting Jupiter – that is, an object other than Earth. The finite speed of light was discovered from observations of these objects (Roemer 1676). When the Voyager spacecrafts made their first close-up images of these satellites, observers could see that the four satellites are very different from each other. Io, the closest to Jupiter, displays a very active volcanism. This activity had been predicted by Peale, Cassen, and Reynolds (1979). Obviously a planet or a moon similar to Io would not be habitable, but this object provides the strongest proof we have of huge internal heating due to the tidal forces caused by proximity to Jupiter. The possibility of an exoplanet or an exomoon similar to Io was first studied in 2008 by Briot, Lellouch, and Schneider, who coined the words Super-Io and Hyper-Io (Briot, Lellouch, and Schneider 2011; Briot and Schneider 2010).

While the investigation for an object similar to Io would be very interesting in extrasolar systems, another Galilean satellite is more directly connected to research for extraterrestrial life in our solar system: the second Galilean satellite, Europa. It has a frozen surface, which was first observed and studied by spacecrafts Voyager 1, Voyager 2, and Galileo. Some examples of features observed on the surface of Europa are displayed in Figure 6.3. The question of liquid water on Europa was first proposed in 1979 (Cassen and Reynolds 1979), and thanks to the presence of an ocean of liquid water under the ice layer, the habitability of Europa has been studied since 1983 (Reynolds et al. 1983). Very recently, water vapor plumes have been detected. Nowadays, Europa is one of the most studied and favorite candidates for research about life elsewhere in the solar system. Several space missions are planned to study its icy crust and ocean.

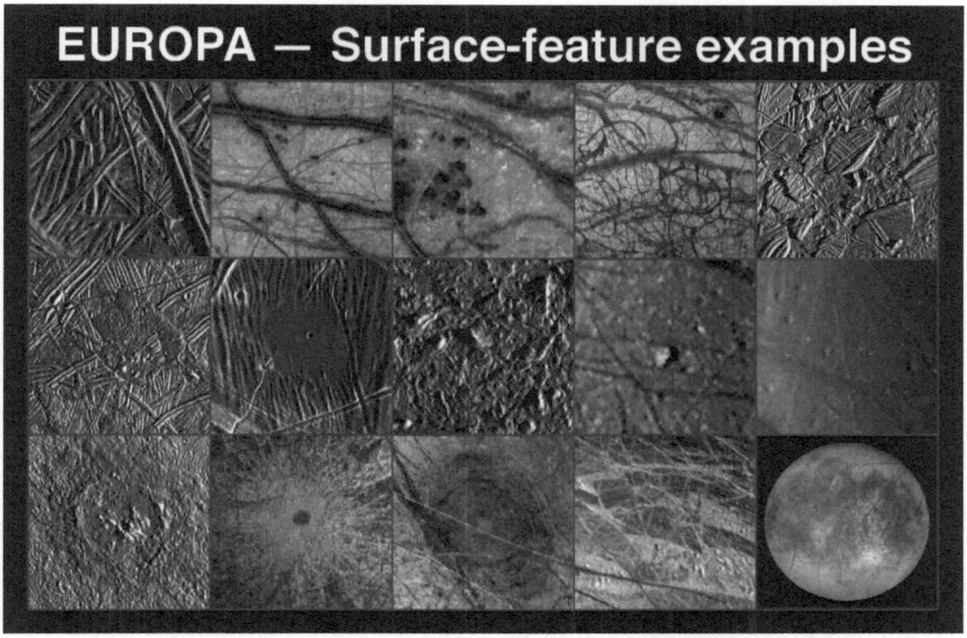

Figure 6.3 Jupiter's icy moon Europa is shown at the bottom right of the figure. The other fifteen frames show the great variety of its surface features, indicating the possible presence of liquid water beneath the ice.
Credit: NASA/JPL/DLR

Tidal effects for Saturn satellites

Liquid water was also discovered in an object located even farther outside the habitable zone defined by the distance of the Sun. Enceladus, a Saturn satellite, was observed by the orbiter Cassini. This object is covered with ice. Some areas are cratered, whereas other zones are without craters due to impacts with small bodies. It can be deduced that the zones without craters are younger than the zones with craters, so the evolution of Enceladus is still active. In 2005, ice and water vapor plumes were observed emerging from long fractures on this moon's south pole, as can be seen in Figure 6.4. The last observations show the presence of liquid water, maybe an ocean, under the thick ice layer. Fractures at the south pole are heated from Enceladus' interior. The heating source could be due to tidal forces or radioactivity. Some chemical compounds in internal water, such as salt, can depress its freezing point. Enceladus is another promising body in the solar system to search for extraterrestrial life.

The detailed knowledge we have acquired of the only planetary system that we can study in detail shows that the habitable zone can be extended much more outside of the classical definition. Moons of giant planets can have liquid water far

Figure 6.4 Enceladus shows off its beautiful plume to the Cassini spacecraft's cameras.
Courtesy: NASA/JPL-Caltech

outside the classical habitable zone. These results are obviously encouraging for research on extraterrestrial life in other planetary systems, especially researching and studying exomoons.

Conclusion

We have demonstrated how the concept of the possibility of the development of life on an extrasolar planet has greatly evolved from 1961, on the one hand because of the properties of thousands of extrasolar planets detected since 1995, and on the other hand because of the properties of objects in the solar system viewed from and studied by spacecrafts. All that can be said today on this subject might be made obsolete or invalid almost instantly. The number of extrasolar planets in our galaxy is sometimes estimated to exceed the number of stars – that is, 200 or 300 billion stars. We already know several thousand planets and many more are expected to be discovered soon. Among those planets, a very interesting rate is supposed to satisfy the conditions necessary for producing life. Knowing giant planets are the easiest to discover, we can expect that more and more small planets will be very likely discovered in the near future. We demonstrated how the conditions fixed initially for the habitability zone can be extended. If we keep in mind the ratio observed in the solar system, where two of eight planets are in the habitable zone, we can tentatively presume that our galaxy may contain several billions of planets liable to shelter life.

References

Benest, Daniel. 1991. "Habitable Planetary Orbits around α Centauri and Other Binaries." In *Bioastronomy: The Search for Extraterrestrial Life: The Exploration Broadens*, ed. Jean Heidmann and Michael J. Klein, 44–47. Proceedings of the Third International Symposium on Bioastronomy 1990, Lecture Notes in Physics 390.

Briot, Danielle, and Jean Schneider. 2010. "Occurrence, Physical Conditions, and Observations of Super-Ios and Hyper-Ios." In *Pathways Towards Habitable Planets*, ed. Vincent Coudé du Foresto, Dawn M. Gelino, and Ignasi Ribas, 409–10. Astronomical Society of the Pacific Conference Series 430.

Briot, Danielle, Emmanuel Lellouch, and Jean Schneider. 2011. "What Could be Observed in the Case of Super-Ios and Hyper-Ios ?" In *Molecules in the Atmospheres of Extrasolar Planets*, ed. J. P. Beaulieu, S. Dieteres, and G. Tinetti, 155–58. Astronomical Society of the Pacific Conference Series 450.

Cassen, Patrick M., and Ray T. Reynolds. 1979. "Is There Liquid Water on Europa?" *Geophysical Research Letters* 6: 731–34.

Dole, Stephen H. 1964. *Habitable Planets for Man*. New York: Blaisdell.

Dumusque, Xavier, Francesco Pepe, Christophe Lovis, Damien Ségransan, Johannes Sahlmann, Willy Benz, François Bouchy, Michel Mayor, Didier Queloz, Nuno Santos, and Stéphane Udry. 2012. "An Earth-Mass Planet Orbiting α Centauri B." *Nature* 491: 207–11.

Galilei, Galileo. 1610. *Sidereus Nuuncius*. Venice: Thomas Baglioni.

Grasset, Olivier, M. K. Dougherty, A. Coustenis, E. J. Bunce, C. Erd, D. Titov, M. Blanc, A. Coates, P. Drossart, L. N. Fletcher, H. Hussmann, R. Jaumann, N. Krupp, J.-P. Lebreton, O. Prieto-Ballesteros, P. Tortora, F. Tosi, and T. Van Hoolst. 2013. "Jupiter Icy Moons Explorer (JUICE): An ESA Mission to Orbit Ganymede and Characterise the Jupiter System." *Planetary and Space Science* 78: 1–21.

Hart, Michael H. 1979. "Habitable Zones around Main Sequence Stars." *Icarus* 37: 351–57.

Hohlfeld, Robert Grant, and Yervant Terzian. 1977. "Multiple Stars and the Number of Habitable Planets in the Galaxy." *Icarus* 30: 598–600.

Howe, Herbert Alonzo. 1885. "The Habitability of Other Worlds." *Sidereal Messenger* 4: 294–98.

Huang, Su-Shu. 1959. "The Problem of Life in the Universe and the Mode of Star Formation." *Publications of the Astronomical Society of the Pacific* 71: 421–24.

Huang, Su-Shu. 1960. "Life Supporting Regions in the Vicinity of Binary Systems." *Publications of the Astronomical Society of the Pacific* 72: 106–14.

Kasting, James F., Daniel P. Whitmire, and Ray T. Reynolds. 1993. "Habitable Zones around Main Sequence Stars." *Icarus* 101: 108–28.

Laskar, Jacques, Frédéric Joutel, and Philippe Robutel. 1993. "Stabilization of the Earth's Obliquity by the Moon." *Nature* 361: 615–17.

Leitner, Johannes J., R. Schwarz, B. Funk, E. Pilat-Lohinger, M. G. Firneis, R. Dvorak, V. Eybl, S. Eggl, and M. Aittola. 2008. "Alternative Solvents as a Basis for Life Supporting Zones in Planetary Systems." European Planetary Science Congress. Proceedings of the Conference Held 21–25 September 2008 in Münster, Germany.

Mayor, Michel, and Didier Queloz. 1995. "A Jupiter-Mass Companion to a Solar-type Star." *Nature* 378: 355–59.

McFarland, Robert White. 1883. "Habitability of the Planets." *Sidereal Messenger* 2: 217–18.

Peale, Stanton J., Patrick M. Cassen, and Ray T. Reynolds. 1979. "Melting of Io by Tidal Dissipation." *Science* 203: 892–94.

Quintana, Elisa V., Thomas Barclay, Sean N. Raymond, Jason F. Rowe, and Emeline Bolmont. 2014. "An Earth-Sized Planet in the Habitable Zone of a Cool Star." *Science* 344: 277–80.

Reynolds, Ray T., Steven W. Squyres, David S. Colburn, and Christopher P. McKay. 1983. "On the Habitability of Europa." *Icarus* 56: 246–54.

Roemer, Ole. 1676. "Démonstration touchant le mouvement de la lumière trouvé par M. Romer de l'Academie Royale des Sciences." Journal des Sçavans, 233–36. Académie royale des sciences, 16 décembre 1676.

Schneider, Jean. 1977. "A Model for a Non-Chemical Form of Life: Crystalline Physiology." *Origins of Life* 8: 33–38.

Schneider, Jean, Cyrill Dedieu, Pierre Le Sidaner, Renaud Savalle, and Ivan Zolotukhin. 2011. "Defining and Cataloging Exoplanets: The exoplanet.eu Database" *Astronomy & Astrophysics* 532: A79–A89.

Schneider, Jean, Valéry Lainey, and Juan Cabrera. 2015. "A Next Step in Planetology: Exomoons." *International Journal of Astrobiology* 14: 191.

Shklovskii, Iosif Samuilovic, and Carl Sagan. 1966. *Intelligent Life in the Universe*. San Francisco: Holden-Day.

Williams, Darren M., James F. Kasting, and Richard A. Wade. 1997. "Habitable Moons around Extrasolar Giant Planets." *Nature* 385: 234–36.

Note

It is a pleasure to thank Florence Raulin-Cerceau, Philippe Robutel, and Siegfried Eggl for their very useful and interesting discussions.

7

Fraction of suitable planets on which life actually appears, f_l, pre-1961

Stéphane Tirard

Abstract

An understanding of the f_l factor of the Drake Equation rests on the notion of a planet suitable for the development of life. The goal of this chapter is to present the transformation of this notion as it has evolved over the last two centuries. Three periods can be distinguished. The first period, which begins toward the end of the nineteenth century, is characterized by the conviction that life is present throughout the entire universe. Some scientists imagine other worlds, often comparable to our earth. Others develop the theory of panspermia and the dispersion of germs of life to all of the new planets. The second period, from the 1920s to the 1950s, is marked by the formulation of complete interdisciplinary hypotheses about the origins of life on earth. These hypotheses were notably focused on understanding the conditions in which life appeared on earth, the only known example of the development of life. The third period, from the 1950s to the present, has studied the possible probiotic chemical pathways that may have pertained on earth. There has been broad scientific acceptance of the idea that life may exist elsewhere in the universe, though the idea has been contested. However, such opposition has been based more on philosophical conception than on scientific argument.

Introduction

Our ideas about suitable planets on which life could appear have evolved greatly over the two last centuries. This chapter attempts to provide an overview of scientific works that explored, inspired, or shaped our views. It argues that before the formulation of the Drake Equation, our conception of suitable planets on which life could appear depended on what we knew about the origin of life on Earth.

The Drake Equation, ed. Douglas A. Vakoch and Matthew Dowd. Published by Cambridge University Press.

First, this chapter will emphasize that belief in the widespread presence of life in the universe was common during the second part of the nineteenth century, and will do this using two examples. The first is the view held by astronomers such as Camille Flammarion, who popularized astronomy and described life on different planets of the solar system. The second is the case of the panspermists, from William Thomson (1872) to Arrhenius at the beginning of the twentieth century, who believed that the germs of life are present everywhere in the universe and could develop on each planet.

Next, this chapter examines how thoughts during the first part of the twentieth century on the origin of life on the Earth (Bernal 1951; Haldane 1929; Oparin 1924; 1938) indirectly constituted a crucial step for beliefs about the possibility of life on other suitable planets.

In the third and last part, this chapter focuses on the emergence of prebiotic chemistry, which led during the 1950s to theories of how life emerged on the Earth, which could apply to other planets. The significance of considerations from prebiotic chemistry carries past 1961, when the Drake Equation was proposed, and so these sections will examine how the main ideas of the 1950s and 1960s on the origin of life were discussed by the French biologist Jacques Monod in his book, *Le hasard et la nécessité* (1970).

Life on Earth and life on other planets: Reasoning by analogy by the scientists at the end of the nineteenth century

During the second part of the nineteenth century, the possibility of life on other planets was linked to two contrasting theories. On one hand, the origin of life was explained by the evolution of inert matter to living matter, a process by which life evolved independently on each planet. On the other hand, the origin of life on planets was explained by panspermia and its arrival in the form of germs from space.

Evolution and analogy

Astronomers at the end of the nineteenth century described planets of the solar system in detail. Some scientists, astronomers, and biologists claimed that life could appear on all of the telluric planets of the solar system, a development they saw as the logical implication of the evolutionary theories of Darwinism and Lamarckism.

The first example is the French astronomer Jules Janssen (1824–1907), who founded the physical astronomy observatory at Meudon, near Paris, and was very interested in the possibility of life on other planets. According to him, astronomers have to seek evidence for plants on other planets, and physical astronomy could help. Direct observations were not possible, but if conditions were analogous to Earth's (especially the presence of water vapor) the presence of life became plausible (Raulin-Cerceau 2006, 16–17).

Camille Flammarion (1842–1925) was a nonacademic French astronomer who had his own observatory and was an important author of popular science texts. He published numerous books that were successful in France (e.g., Flammarion

...Peut-être ne serions-nous pas très dépayses en arrivant devant un paysage de Vénus.

Figure 7.1 "Perhaps, we would not be so disoriented in front of a Venus landscape." View of Venus from Flammarion (1884).
Image courtesy Stéphane Tirard

1877; 1884; 1892) and claimed that life exists everywhere in the universe, including on the closest planets, Mercury, Venus, and Mars. According to him, all of these telluric planets can be compared to the Earth and had histories conditioned by their distance from the Sun. As the title of his book *Les Terres du ciel* (1884) expresses very well, Flammarion's descriptions were founded on the hypothesis that some conditions analogous to Earth exist on the other planets. Therefore, life follows the pattern found on the Earth, and is more advanced on Mars and less on Venus.

His case shows the importance of concepts of biological evolution. Indeed, he believed that the evolution of life on the other planets is linked to the Lamarckian conception of evolution, which was dominant in France at this time (Loison 2011). Lamarckism, and later neo-Lamarckism, were founded on concept of plasticity and maintained that the environment explains the transformation in species. Therefore, life necessarily appears on all planets with more or less the same conditions as the Earth.

The last lines of the book reveal his philosophical conception of life: "Life grows endlessly in space and time; it is universal and eternal; it fills the infinite with its harmony, and reigns over the centuries of centuries, during the endless eternity."[1] In this book, Flammarion exceeded the limits of astronomy. According him, there would be a sort of continuous process from the formation of planets to the evolution of human beings. Applying these principles, he described the evolution of "human" societies on these planets. In doing so, it is clear that Flammarion's intention was more philosophical than rigorously scientific.

From these examples, we can see that belief in the presence of life in the universe can occur in several ways. Janssen represents the pure scientist and suggests one method for detecting life on other planets. Flammarion, a popular scientist, uses speculation to offer pleasant dreams to his readers.

Regarding the notion of suitable planets, it is interesting to note that in these hypotheses Earth was the central reference. The concept of suitable planets seems to depend on understanding the origin and evolution of life here. In these examples, we observe how life on the Earth is obviously the best model, because it is the only one. The important consequence is that conditions on other planets are supposed to be equivalent to the Earth's. Moreover, the Lamarckian conception of evolution explained how analogous conditions could lead to the same results, that is to say, to analogous living beings.

Panspermia: Life everywhere

Panspermia theories share a belief that life is present across the universe and that each new planet is seeded by germs from space. Panspermia is also linked to evolution, but for different reasons according to different authors.

Figure 7.2 "Here, comes down the sky another light, here, bloom plants that are not plants." View of Mars from Flammarion (1884).
Image courtesy Stéphane Tirard

William Thomson (Lord Kelvin, 1824–1907), a leading panspermia theorist during the second part of the nineteenth century, endorsed it in opposition to Darwinism. He used it to explain how life began on Earth without evolution and proposed a very simple model in which different germs (seeds or moss, for example) arrived here on meteorites. He claimed: "The hypothesis that life originated on this Earth through moss-grown fragments from the ruins of another world may seem wild and visionary; all I maintain is that it is not unscientific"

(Thomson 1872, cv). He concluded his piece by directly attacking Darwinism and endorsing the role of the Creator.

However, panspermia did not remain in simple opposition to evolution. Philippe van Tieghem (1839–1914), a leader of French botany at this time and a convinced panspermist, illustrates a Darwinist opinion. In 1891, he wrote:

The vegetation of the Earth is only a very small part of the vegetation of the Universe. Once able to accept vegetable life, the Earth is populated by plants, as today an island or a rock are populated by seeds coming from their environment. The only objection, which can be argued, is the factual isolation of the Earth in space. But not everybody accepts this isolation. The fall of meteorites denies it. It would have sufficed that, once or a few times, some seeds enclosed in a meteorite, or by any other means, arrived on the terrestrial globe after its cooling. On the Earth seeded, everything would be themselves developed from the positive germs. The vegetation of the Earth had a beginning and will have an end; but the vegetation of the Universe is eternal as the universe itself.[2]

Therefore, in van Tieghem's proposal as in Thomson's, one central idea is the eternity and universality of life. They have the same explanation for the arrival of life on Earth, even though they disagreed about its destiny, with van Tieghem as evolutionist and Thomson as what we can all a fixist.

The most developed theory of panspermia was proposed by Svante Arrhenius (1859–1927), a Swedish chemist who explained the conditions of the circulation in space of microscopic germs and their distribution on planets:

Finally, we perceive that, according to this version of the theory of panspermia, all organic beings in the whole universe should be related to one another, and should consist of cells which are built up of carbon, hydrogen, oxygen, and nitrogen. The imagined existence of living beings in other worlds in whose constitution carbon is supposed to be replaced by silicon or titanium must be relegated to the realm of improbability. Life on other inhabited planets has probably developed along lines which are closely related to those of our earth, and this implies the conclusion that life must always recommence from its very lowest type, just as every individual, however highly developed it may be, has by itself passed through all the stages of evolution from the single cell upward. (Arrhenius 1908, 229–30)

Arrhenius admitted that on every planet where life arrived, it would evolve according to local circumstances.

Two important points must be underlined in these different conceptions of panspermia. First, the panspermian theory mobilized two main concepts: eternity and infinity. Indeed, according to the panspermists, life has existed since eternity and is found everywhere in the infinity of the universe. Secondly, germs of life must be present in space. Therefore, not only must planets be suitable for life, but also space itself.

In 1910, the Frenchman Paul Becquerel (1879–1955) challenged panspermia, using results from experiments on the resistance of germs in hypothetical space conditions. He claimed that life could not survive UV rays (Becquerel 1910, 87–88), and his conclusion helped bring about the decline of panspermian theory. At this moment, the more realistic scenario for the origin of life seemed to be evolution, and this meant that life in the universe was dispersed in multiple locations.

The notion of suitable planets requires an understanding of the conditions under which life appeared on Earth

During the first part of the twentieth century, several important hypotheses emerged about the origin of life on Earth, all of which had in common research on conditions on the primitive Earth at the time of life's origin. Understanding these conditions was tightly linked to comparative planetological data. This period tied questions of the origin of life to scientific theories about the formation and transformation of planets.

Primitive conditions and the evolution of the planets

The concept of evolution from inert to living matter was progressively developed by scientists during the late nineteenth and early twentieth centuries. But assumptions about conditions on primitive Earth were not very accurate. Scientists were more interested in biological processes than in an accurate portrayal of the primitive Earth. Complete models of the origin of life in relation to its environment came only in the 1920s, and they were an important step in the emergence of the concept of suitable planets.

It is very well known that during the 1920s, the two main theories on the origins of life were those of Alexandre I. Oparin (1894–1980) and John B. S. Haldane (1892–1964), developed independently in the USSR in 1924 and Great Britain in 1929. These two assumptions focused on how life emerged on Earth, and also on the context in which this happened – that is to say, conditions on the primitive planet at that time. The two authors needed data not only from chemistry and biology, but also from geology and planetology, and successfully put the origin of life into a broader geological and planetological context.

In 1936, Oparin, who became a specialist on life's origin, published in Russian a book later titled in English *The Origin of Life*, which described the evolution of Earth and the evolution of conditions on the planet (Oparin 1938). His description, as with his text of 1924, was broadly interdisciplinary.

The first two chapters concern "Theories of spontaneous generation of life" and "The theories of the continuity of life," which respectively eliminated the possibilities for spontaneous generation and panspermia. In his third chapter,

Oparin presented the "Theories of the origin of life at some distant period of the earth's existence." In this chapter, he discussed Pfluger's, Leduc's, and Haeckel's theories. In the fourth chapter, Oparin described his own theories and his plan for the rest of the book, in which he develops his ideas about the process of life's origin:

Chapter 4: "Primary forms of carbon and nitrogen compounds"
Chapter 5: "Origin of organic substances, primary proteins"
Chapter 6: "The origin of primary colloidal systems"
Chapter 7: "Origin of primary organisms"
Chapter 8: "Evolution of primary organisms"

It is interesting to note that chapter four is essentially about planetology, or the effect of stars on the formation of planets. He demonstrated the particular position of the Earth in the solar system and the emergence of carbon and nitrogen compounds on the young planet. Indeed, Oparin's goal was to explain the appearance of these fundamental chemical compounds, a process that he believed leads to the appearance of carbon and nitrogen on Earth:

> Carbon made its first appearance on the Earth's surface not in the oxidized form of carbon dioxide but, on the contrary, in the reduced state, in the form of hydrocarbon. . . . Thus, it can be assumed with a high degree of probability that nitrogen, like carbon, first appeared on the Earth's surface in its reduced state, in the form of ammonia. (1938 102, 104)

In this chapter, Oparin presented his ideas about the reduced state of the primitive atmosphere, and focused on the common history of the Sun and Earth. His topic is not the problem of life in the universe, but he accurately described the conditions that a planet has to present for the appearance of life. His emphasis was on Earth's suitability.

Carbon dioxide or not?

It is possible to study how discussions about the origin of life on Earth focused on some very particular primitive conditions or very specific points of the suitability. The case of the presence or absence of carbon dioxide in the primitive atmosphere shows how the authors linked the possibility of life on Earth to these primitive conditions.

French astronomer Alexandre Dauvillier (1882–1979) was the co-author, with Etienne Desguins, of a theory of the photochemical origin of life (Dauvillier and Desguins 1942), which focused on the effect of solar light on carbon dioxide. He did not believe that conditions on other planets in our solar system could produce the same results. However, he claimed that these conditions could exist elsewhere in the galaxy, and he tried to evaluate the proportion of this type of planet. He wrote that there would be 10^8 systems analogous to ours in the galaxy, and 10^7 planets with advanced life (Dauvillier 1958, 208). Thus, he distinguished suitable planets from those on which life actually appears. This distinction is very similar to the factor in the Drake Equation that is the focus of this chapter.

Finally, we note that Dauvillier did not provide a way to compute his estimate. In the last lines of his book, it appears that it was based on a philosophical conception:

> Life and thought are perpetual only in a statistical manner and play no role in the dynamic evolution of the Universe. Life appears, evolves and disappears on a myriad of planets in a cycle of a few billion years, limited in duration by cosmic accidents or because of his (its?) own evolution.[3]

An important point to underline is that Dauvillier also defined suitability by using analogy with terrestrial conditions.

In 1952, Harold Urey (1893–1991) suggested that some experiments could approximate the primitive conditions in which life appeared:

> It seems to me that experimentation on the production of organic compounds from water and methane in the presence of ultra-violet light of approximately the spectral distribution estimated for the sunlight would be most profitable. The investigation of possible effects of electric discharges on the reactions should also be tried since electric storms in the reducing atmosphere can be postulated reasonably. (Urey 1952, 362)

Urey's suggestion was based on combining data from planetology and geology. He included carbon in the form of methane in the primitive atmosphere, and according to him, these simplified conditions were accurate. On this point, he firmly contested Melvin Calvin's (1911–1997) experiments in which the biochemist introduced carbon dioxide. Indeed, one year earlier Calvin had obtained formaldehyde after exposing a carbon dioxide solution to gamma rays (Garrison et al. 1951).

Opposition between the two scientists focused on the presence or absence of carbon dioxide, and their claims depended on the planetological references that they used. Calvin used Abelson's references and Urey his own conclusions.

The problem of conditions for the synthesis of organic compounds was central to ideas about the origin of life. We see the emergence of a consensus around the idea of a reduced primitive atmosphere. Urey agreed with Oparin, and this point became a very important characteristic of Earth as a planet suitable for life.

At the middle of the twentieth century, discussions regarding the origins of life focused on conditions on the primitive Earth. For scientists, it was clear that any scenario that explained the origin of life depended on these conditions. The focus was on Earth, but this would become the basis of broader speculative thought about the conditions under which life could appear.

Studying the origin of life: Between understanding a past phenomenon and the development of a model

After the Second World War, research on the origin of life was dominated by the emergence of a new methodology: prebiotic chemistry. Research on the origin of life on Earth also led many to reflect on life in other places in the universe.

Prebiotic chemistry: Exploring methods of the past

Melvin Calvin's and Stanley Miller's (1930–2007) experiments, which occurred respectively in 1951 and 1953, were the basis of prebiotic chemistry. Miller's experiment was based on Urey's hypothesis regarding Earth's simplified primitive atmosphere (Miller 1953). Under prebiotic conditions, he exposed methane, ammonia, hydrogen, and water to electric discharges for one week and got amino acids. The idea behind this experiment was to test chemical reactions in the supposed conditions of primitive Earth. Prebiotic chemistry was a way to explore this phenomenon of initial conditions.

Prebiotic chemistry underwent a very important development after 1953 because it seemed to be a way to systematically explore processes behind the origin of life on Earth. Chemists were very active in this field and produced a broad panel of molecules found under prebiotic conditions, or in other words, under the supposed conditions of primitive Earth. In this synthesis, they were inspired by the results of molecular biology, which revealed the fundamental importance of proteins and nucleic acids.

Table 7.1 shows the chronological correspondence between these two scientific domains (Tirard 2000; 2002). The left column shows some of the most important results in prebiotic chemistry and the right three important results in molecular biology. The table demonstrates that, beginning in the 1960s, a three-phase model existed about the origin of life: the first was synthesis of simple organic compounds (amino acids, sugars), the second was synthesis of macromolecules, and the third came from formation of primitive cells.

The study of the history of life on Earth using prebiotic chemistry gave us a model for the origin of life here. If this model worked on Earth, it also seemed applicable to analogous situations. During the 1950s, the universality of life was accepted by a very large part of the community of specialists on life's origins.

It is clear that the model produced by the prebiotic chemistry did not give any absolute answer to questions about the existence of life elsewhere in the universe. Nevertheless this model gave us a plausible scenario for its emergence on Earth and allowed some scientists to easily imagine that among a very large number of planets, a fraction could present the same conditions. Therefore,

Table 7.1 *Developments in prebiotic chemistry and molecular biology*

Origin of life and prebiotic chemistry	Molecular biology
1953 – Miller: synthesis of amino acids.	1953 – Discovery of DNA structure
1958 – Fox and Harada: Thermolymerization of amino-acids	1958 – Crick: Central doctrine of molecular biology DNA –> RNA –> Protein
1960 – Beadle: The first replicative molecule could be a sequence of adenine and thymine.	
1960 – Synthesis of adenine	1960 – Discovery of messenger RNA
1961 – Synthesis of uracil	
1965 – Fox: Synthesis of microspheres	

an extrapolation of the Earth model could be very useful for calculating the f_l factor of the Drake Equation.

Life: Unique or universal?

As we have seen, on the one hand, comprehension of the molecular mechanisms of life highlighted the process of the origin of life on Earth because it gave some milestones to its prebiotic chemistry. On the other hand, molecular biology described very complex mechanisms in the cell, which were very difficult to link to a contingent process. So did progress in biology become both a help and an obstacle for conceiving the probability of life on other planets?

The French biologist Jacques Monod (1910–1976), with François Jacob (1920–2013) and André Lwoff (1902–1994), won the 1965 Nobel Prize for physiology and medicine for their work on the regulation of gene expression and for their participation in the discovery of messenger RNA. In 1970, in his book *Le Hasard et la Nécessité*, Monod reflected philosophically on the new biology that has emerged with molecular biology. He described the tension between chance, which is central to evolutionary phenomenon, and necessity, on which, according to him, molecular mechanisms seemed based. He called this aspect teleology. The particular point that interests us is that Monod denied the possibility of the universality of life because of its complexity.

In chapter 8 of his book, entitled "The boundaries," Monod insisted on the fact that two big problems persist in biology: the origin of the first living system and the complexity of the brain. He described the process of the origin of life in three steps, according to the model produced by prebiotic chemistry, and gave details about these mechanisms. His references are based on the results of prebiotic chemistry and on his own biological knowledge. According to him, the first step is formation of the fundamental compounds of living beings: nucleotides or amino acids. The second is the formation, from previous molecules, of replicating macromolecules. The last step is the evolution of a teleonomical apparatus around these replicative macromolecules to achieve the primitive cell.

He finished with this question about the probability of the origin of life:

> Life appeared on Earth: what was the probability of this event? Because of the current structure of the biosphere, the hypothesis that this decisive event did not just occur once could not be excluded. Which would mean that its a priori probability was near non-existent.[4]

He said very clearly that he was convinced that the origin of life was very rare at the scale of the universe, and he did not hesitate to say that it could be unique. Monod argued that the origin of life on Earth was the result of a succession of steps, the probability of each of which was infinitesimally small. Using a metaphor that became famous in France, he wrote that, on the Earth, life has won at "the game of Monte Carlo." He was very clear that it is almost impossible that life exists elsewhere.

Monod's claim about the origin of life was discussed by biologists. Three years later, the Belgian biologist Ernest Schoffeniels (1927–1992) responded to Monod in his 1973 book *L'Anti-hasard*. His argument was the complete opposite, and denied all conceptual notions: teleonomy, finality, function, and chance. To him, the process of the origin of life was the result of a chain of chemical reactions that was able to go only one way and was perfectly reproducible.

The opposition between these two scientists shows that opinions on the probability of life in the universe are associated not only with the notion of a suitable planet. In their opposition, Monod and Schoffeniels discussed the probability of the origin of life without any formulas, and their positions were more the results of their philosophical convictions about life.

The 1950s saw a very complex tension between opposing sides about the existence of life elsewhere in the universe. Prebiotic chemistry gave us a very useful way to test the supposed initial conditions on the Earth, and it was theoretically possible to transpose this method to anywhere in the universe. The discovery of the complexity in biological mechanisms, however, tended to show that the appearance of life is a succession of steps, and according to specialists, each one varies in its probability. The estimate of this probability will depend on the philosophy of each scientist.

Conclusion

Before 1961, thought on the origin of life was focused mainly on the Earth. Moreover, this focus did not preclude the possibility that analogous processes could occur on other planets. Earth, as the only example we have of a planet with life, was the case that had to be studied, and also served as our reference point for the development of a comprehensive model of the origin of life as well as for the difficulties that come with attempts to transpose this model.

The appearance of life on other suitable planets was not obvious to all scientists. There is a degree of subjectivity in thought about the probability of the origin of life. Indeed, each scientist introduces his or her own philosophical approach to life. Therefore, there is a large spectrum of conceptions, from the certainty that life is unique, to the contrary certainty that the mechanisms revealed by prebiotic chemistry could be universal.

References

Arrhenius, Svante. 1908. *Worlds in the Making: The Evolution of the Universe*. New York: Harper and Brothers Publishers.

Becquerel, Paul. 1910. "L'action abiotique de l'ultraviolet et l'hypothèse de l'origine cosmique de la vie." *Comptes rendus hebdomadaires des séances de l'Académie des sciences* 151: 86–88.

Bernal, John D. 1951. *The Physical Basis of Life*. London: Routledge and Kegan Paul.

Dauvillier, Alexandre. 1958. *L'Origine Photochimique de la Vie*. Paris: Masson.

Dauvillier, Alexandre, and Etienne Desguins. 1942. *La genèse de la vie, Phase de l'évolution chimique*. Paris: Hermann.

Flammarion, Camille. 1877. *La pluralité des mondes habités*. Paris: Gauthier-Villars.

Flammarion, Camille. 1884. *Les Terres du ciel*. Paris: C. Marpon et E. Flammarion.

Flammarion, Camille. 1892. *La Planète Mars et ses conditions d'habitabilité*. Paris: Gauthier-Villars.

Garrison, W. M., D. C. Morrison, J. G. Hamilton, A. A. Benson, and M. Calvin. 1951. "Reduction of Carbon Dioxide in Aqueous Solutions by Ionizing Radiation." *Science* 114: 416–18.

Haldane, John B. S. 1929. "The Origin of Life." In *Rationalist Annual, 1929. Repr. in J.B.S. Haldane On Being the Right Size and Other Essays*, ed. John Maynard Smith, 101–12. Oxford University Press, 1991.

Loison, Laurent. 2011. *Qu'est-ce que le néolamarckisme? Les biologistes français et la question de l'évolution des espèces*. Paris: Vuibert.

Miller, Stanley L. 1953. "A Production of Amino Acids under Possible Primitive Earth Conditions" *Science* 117: 528–29.

Monod, Jacques. 1970. *Le hasard et la nécessité: Essai sur la philosophie naturelle de la biologie moderne*, Paris: Le Seuil.

Oparin, Alexandre I. 1924. *Proiskhozhdenie zhizny [The Origin of Life]*. Translated by Ann Synge in *The Origin of Life*, ed. J. D. Bernal. London: Weidenfeld and Nicholson, 1967.

Oparin, Alexandre I. 1938. *The Origin of Life*. Translated by Sergius Morgulis. New York: Macmillan.

Raulin-Cerceau, Florence. 2006. *A l'écoute des planètes*. Paris: Ellipses.

Schoffeniels, Ernest. 1973. *L'anti-hasard*. Paris: Gautier-Villars.

Thomson, William. 1872. *"Address of Sir William Thomson (President)." British Association for the Advancement of Science, Edinburgh, Report – 1871*. London: John Murray.

Tirard, Stéphane. 2000. "Les origines de la vie: Un problème des disciplines." *Aster* 30: 105–22.
Tirard, Stéphane. 2002. "Les origines de la vie, un problème historique." In *Exobiologie: aspects historiques et épistémologiques*, ed. Florence Raulin-Cerceau and Stéphane Tirard. *Cahiers François Viète* 4: 35–48.
Tirard, Stéphane. 2006. "William Thomson (Kelvin): Histoire physique de la Terre et histoire de la vie." In *Pour comprendre le XIXe siècle Histoire et philosophie des sciences à la fin du siècle*, ed. Jean-Claude Pont, 297–306. Geneva: Olski.
Tirard, Stéphane. 2010. "Origin of Life and Definition of Life, from Buffon to Oparin." *Origins of Life and Evolution of Biospheres* 40(2): 215–20.
Urey, Harold. 1952. "On the Early Chemical History of the Earth and the Origin of Life." *Proceedings of the National Academy of Science USA* 38: 351–63.
van Tieghem, Philippe. 1884. *Traité de botanique*, Paris: Librairie Savy.

Notes

1 "LA Vie se développe san fin dans l'espace et dans le temps; elle est universelle et éternelle; elle emplit l'INFINI, des ses accords, et elle règnera à travers les siècles des siècles, durant l'interminable éternité" (Flammarion 1884, 769).

2 "La végétation de la Terre n'est qu'une très petite partie de la végétation de l'Univers. Une fois apte à la vie végétale, elle s'est peuplée de plantes comme se peuple encore aujourd'hui une île émergée ou un rocher éboulé par l'apport de germes venus de leur voisinage. La seule objection qu'on puisse faire est le prétendu isolement matériel de la Terre dans l'espace. Mais tout le monde n'admet pas cet isolement. La chute des météorites est là d'ailleurs pour le démentir. Il aurait suffi qu'une fois, ou un petit nombre de fois, quelques germes enfermés dans un météorite ou apportés par un tout autre moyen parvinsent au globe terrestre après son refroidissement. La Terre une fois ensemencée, tout se serait développé à partir des germes positifs. La végétation de la Terre a eu un commencement et aura une fin; mais la végétation de l'Univers est éternelle comme l'Univers lui-même" (van Tieghem 1884, 982).

3 "La vie et la pensée ne sont perpétuels que d'une manière statistique et ne jouent aucun rôle dans l'évolution dynamique de l'Univers. La vie apparaît, évolue, et disparaît sur une myriade de planètes selon un cycle de quelques milliards d'années, limité en durée par les accidents cosmiques ou par l'effet de sa propre evolution" (Dauvillier 1958, 208).

4 "La vie est apparue sur la terre: quel était *avant l'évènement* la probabilité qu'en fût ainsi? L'hypothèse n'est pas exclue, au contraire, par la structure actuelle de la biosphere, que l'événement decisif ne se soit produit qu'une seule fois. Ce qui signifierait que sa probabilité a priori était quasi nule" (Monod 1970, 183).

8

Fraction of suitable planets on which life actually appears, f_l, 1961 to the present

David J. Des Marais

Abstract

An assessment of f_l depends heavily upon perspectives gained from the single known example of our own biosphere. Extrapolating this single data point to a quantitative estimate of f_l amounts to sheer speculation. But recent discoveries have identified perspectives about our biosphere and early planetary environments that are quite relevant to f_l and that also benefit our search for a second example of life elsewhere. This chapter addresses these topics from the following perspectives: concepts of life, life's environmental requirements, early conditions on rocky planets, and the origins of life. The habitability of a planetary environment is defined by the intrinsic environmental requirements of life, which, in turn, arise from the most universal attributes of life itself. Key environmental requirements include a suitable solvent, the chemical building blocks of life, biologically useful sources of energy, and environmental conditions that favor the survival of key complex molecules and structures. The earliest evidence of our biosphere now extends back to more than 3.7 billion years ago, essentially as old as the oldest rocks that could have preserved recognizable evidence. The most ancestral characteristics of microbial metabolism are broadly consistent with the resources that were likely available as early as 4.4 billion years ago. Organic compounds relevant to prebiotic chemistry have been discovered in primitive bodies (e.g., meteorites, comets) and in interstellar space, and these might have been delivered intact to planetary surfaces. Thus, life might have begun very soon after habitable conditions were established. Regarding the origins of life, biochemical research is narrowing the knowledge gap between prebiotic chemicals and the first living systems. RNA molecules could have served as both self-replicators and enzymes. The earliest functional protein enzymes might have been much smaller in size and thus perhaps

The Drake Equation, ed. Douglas A. Vakoch and Matthew Dowd. Published by Cambridge University Press.

easier to develop than previously imagined. The prebiotic environment could have provided molecules that formed vesicles as precursors of cellular envelopes. Although the evidence certainly cannot prove the notion that another young Earth-like planet probably also sustained an origin of life, the evidence is at least quite consistent with the notion that life can arise early on Earth-like planets. Also, the evolution of stars and planets follows trajectories that allow reasonable estimations to be made of long-term changes in planetary climates and habitability. Thus, although recent discoveries have not yet reduced the enormous uncertainty in estimating f_l, they have substantially improved our strategies for seeking a second example of life that would, in turn, substantially reduce that uncertainty.

Introduction

Discoveries during the space age have broadened immensely our comprehension of the diversity of planetary systems, environments, and attributes of life that bear directly on the topic of this chapter. This progress has increased dramatically our recognition of the sheer number of factors that are relevant to f_l and therefore must be assessed before we can make even an educated guess of its magnitude.

This chapter addresses f_l by exploring the following perspectives: (1) a working concept of life, (2) characteristics of habitable environments, (3) the earliest environments on Earth, (4) the origins of life, and (5) improving our estimates of f_l.

But first, what do we mean by "life," and, therefore, what kind of planetary environment would qualify as being "suitable"?

A "working concept" of life

Any attempt to quantify f_l depends fundamentally upon a working concept of life that, in turn, specifies the kinds of planetary environments that are required for life's origins and survival. Rather than attempting to present a "universal definition of life," which some contend is currently impossible (Cleland and Chyba 2007), this chapter presents a working concept of life and its requirements as a starting point, with the understanding that future discoveries will inevitably necessitate revisions.

An assessment of f_l must start with perspectives gained from our own biosphere. Life-sciences research has illuminated many attributes that are shared by all known living systems. Although one is tempted to equate such attributes to those that are universal for all life elsewhere, at least some of these known attributes actually might be life's local solutions to the challenges of environmental changes that have been unique to Earth. Yet this notion of local solutions does imply one potentially

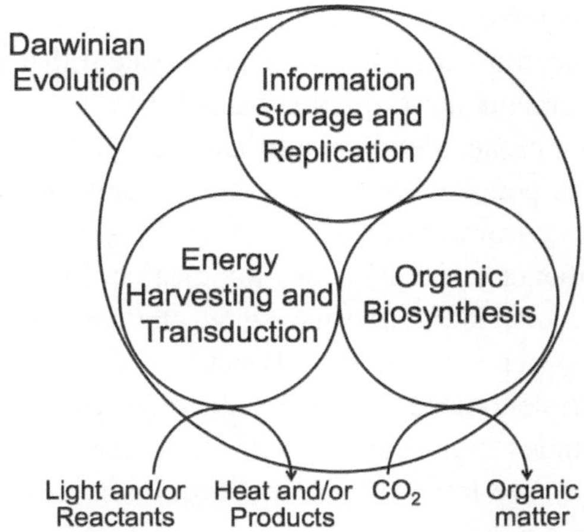

Figure 8.1 Life's basic functions.
Credits: David DesMarais

universal attribute, namely that the origin and evolution of any particular example of life are inextricably linked to the origin and evolution of its host planetary environment. The environment had to provide the essential resources and conditions, and environments inevitably changed in response to planetary evolution. Accordingly, our own biosphere provides a useful "working concept" that, of course, is subject to revision. And we must admit that our earthly example is likely only a small subset of the full diversity of life. Indeed, some truly exotic life forms might not conform to the familiar concepts articulated below. But still, the basic attributes of any "weird life" must first be postulated and its habitability requirements also specified before its corresponding f_l can be envisioned and quantified.

Recent studies (e.g., Baross 2007) have identified the following attributes shared by life as we know it: (1) It must exploit thermodynamic disequilibrium in the environment in order to perpetuate its own disequilibrium state; (2) it most probably consists of interacting sets of covalently bonded (i.e., bonds where the two adjacent atoms share electrons nearly equally) molecules that include a diversity of atoms in addition to carbon (C) and hydrogen (H) (e.g., nitrogen [N], oxygen [O], phosphorous [P], sulfur [S], etc., as in Earth-based life) that promote chemical reactivity; (3) it requires a liquid solvent that supports these molecular interactions; and (4) it employs a molecular system capable of Darwinian evolution.

The above list implies the following basic universal functions (Figure 8.1): (1) Life harvests energy from its environment and converts it to forms of chemical energy that directly sustain its other functions; (2) it sustains "metabolism," namely, a network of chemical reactions that synthesize all of the key chemical compounds that are required for maintenance, growth, and self-replication; and (3) life sustains an "automaton," a multi-component system that is essential for self-replication and self-perpetuation (von Neumann 1966).

Perhaps the most unique, defining attribute of life is its capacity for self-replication that creates populations for natural selection by Darwinian evolution. When von Neumann first presented his theory of the automaton, he predicted the basic functional components required for biological self-replication. He achieved this well before molecular biologists discovered the essential components of the DNA-RNA-protein system in the mid-twentieth century. The sophistication implied by these multiple molecular components requires an environment wherein several relatively complex molecules can self-replicate more rapidly than they are degraded. This perspective imposes essential requirements that any environment must meet in order to become habitable.

Characteristics of habitable planetary environments

The concepts and attributes of life articulated above imply the following key determinants of habitability (Des Marais et al. 2008): (1) A solvent must be available with chemical properties and persistence that maintain metabolic activity essential for long-term survival; (2) thermodynamic disequilibria must be available as biologically accessible energy sources; (3) physicochemical environmental parameters (e.g., temperature, pH, salinity, radiation) must allow covalent and hydrogen bonds in biomolecules to be stable, and these parameters must provide energy for living systems to maintain biological molecules and structures in a functional state; and (4) the environment must provide bioessential elements in the form of biologically accessible chemicals. On Earth, these elements are principally C, H, N, O, P, S, and a variety of metals. A habitable environment must provide all of the above key resources and conditions simultaneously and with a frequency over time that is sufficient to allow rates of self-maintenance and growth to prevail over rates of degradation (Figure 8.2).

Solvent: The case for water as the cosmic choice

The key functions of life require molecules that are relatively complex and stable and that these molecules assume specific configurations to become functional and interact with other molecules in specific ways. A solvent plays critical roles in promoting biological organization (e.g., Tanford 1978). Water is indeed the solvent for life on Earth, but how can we assess the extent to which it might assume this key role elsewhere?

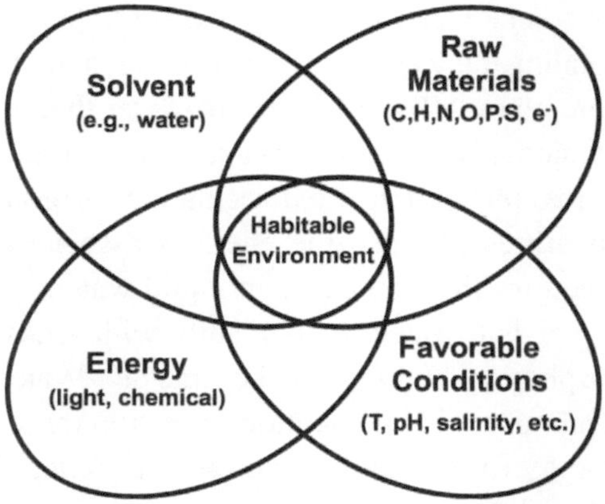

Figure 8.2 Resources and conditions that a habitable environment must provide simultaneously.
Adapted from Hoehler (2007)

Noncovalent interactions between molecules strongly modulate the structures and functions of living cells (Pohorille and Pratt 2012). (Noncovalent interactions are weaker than chemical bonds in stable molecules. Examples include interactions between water molecules and dissolved positive or negative ionic species, or interactions between individual hydrocarbons in oil films or in aqueous colloidal mixtures.) For example, noncovalent interactions govern protein folding, interactions between enzymes and ligands, the regulation of gene expression, the self-assembly of boundary structures, and ion transport across membranes. In order to be effective, these interactions must be in the appropriate energy range. Noncovalent attractions between molecules must be sufficiently strong so that inherent thermal processes do not dismantle functional configurations. The system should be sufficiently stable so that it functions properly over the range of environmental temperatures. However, if these interactions are too strong, a potentially prohibitive expenditure of energy might be required to regulate local thermodynamic equilibria that are essential for maintaining a functional metabolism. Essential biomolecular interactions would become practically irreversible.

Electrostatic interactions between biomolecules cannot be too strong; therefore, the solvent should have a high dielectric constant. This constant is a measure of how easily a material can be polarized by the imposition of an electric field on an insulating material. Because water has a relatively high dielectric constant, it lowers the strength of the electrostatic interactions between any positively and negatively charged ions in solution. But interactions between nonpolar molecules or groups should be sufficiently strong to favor their organization. Water satisfies all of these requirements and thus is an excellent solvent for life. It has a high dielectric constant and also facilitates hydrophobic interactions between nonpolar species. (A hydrophobe is a "water avoiding" molecule [e.g., a hydrocarbon] that disrupts hydrogen bonding between water molecules and therefore tends to be expelled into a separate phase.) The attributes of water that are most essential for supporting life are remarkably constant over a wide range of temperatures and pressures. The structures, interactions, and regulation of biomolecules all exploit the characteristic similarities between the energies associated with hydrophobic and electrostatic interactions that can occur in liquid water.

Being an excellent solvent, water reacts readily with rocks to release essential nutrients such as phosphorous, sulfur, and key metals. Water participates extensively with other crustal materials in oxidation-reduction (redox) reactions that can serve as sources of energy for metabolism and growth (e.g., Shock et al. 2005).

Water is not the only solvent having attributes that appear to be ideal for life (see also Chapter 6). Formamide and di-alcohols also have high dielectric constants. Vesicles can form in some of these other solvents due to solvophobic ("solvent-avoiding") effects (McDaniel, McIntosh, and Simon 1983). But very few solvents have all of the essential favorable properties. Formamide has several favorable properties but has not yet been sufficiently investigated. However, it seems inconceivable that any of these promising alternative solvents come close to matching the stability, abundance, and broad distribution of water in the known universe. Thus, even if alternative solvents actually do serve exotic examples of life in some cosmic niches, water is very likely the most pervasive solvent of choice for extraterrestrial life.

Raw materials: The case for carbon compounds

The notion that carbon plays prominent roles in the chemistry of life elsewhere in the universe also seems quite valid. This is true even if alternative life forms that are based on other elements also exist in some environments. A key requirement of life is that the molecular system involved in self-replication and catalytic enzymes must be at least moderately complex (e.g., von Neumann 1966). Accordingly, the unique ability of carbon to form highly stable chains, rings, and molecules that include N, O, P, S, and so forth and thereby create stable complex molecules is a crucially important attribute.

Silicon has been viewed as the most viable alternative to carbon because it is the lightest element having a half-filled outer shell with four unpaired valence electrons, analogous to the outer shell of the carbon atom. Carbon and silica form methane (CH_4) and silane (SiH_4), respectively, and these are gases under ambient conditions. Alternating carbon and oxygen atoms form chain-like compounds called polyacetals; silicon and oxygen can form silicone chains. But silicon and oxygen readily form silicon dioxide (SiO_2), the simplest oxide of silicon, which typically leads to a wide variety of relatively insoluble solids. In contrast, because carbon dioxide is a gas and is quite soluble in water, it can easily supply, participate in, and be excreted by biological processes.

Living systems must store energy from their environment, and carbohydrates are critically important storage media. These "polyformaldehyde" compounds function as "biochemical batteries" and their polymer lengths represent their "battery state of charge." Silicon cannot form stable compounds that are analogous to carbohydrates.

Longer silicon chains, especially those having six atoms or more, become increasingly susceptible to aqueous hydrolysis. Molecules with silicon-chain backbones that have the size and complexity required for biological functions would be highly unstable in the presence of water or other polar solvents. Thus, for example, silicon cannot form compounds analogous to enzymes, whose chirality and complexity impart the attributes of chemical selectivity and regulation that are so essential for metabolism and genetic functions. (A chiral molecule is one that has a nonsuperimposable mirror image. For example, a molecule having a carbon atom that is bonded to four different atoms is a chiral molecule [e.g., the amino acid alanine].)

Carbon is the fourth most abundant element in the universe. Nuclear reactions in aging stars produce carbon along with oxygen (third most abundant element) and nitrogen (fifth most abundant by atom fraction). Carbon compounds are observed in the envelopes of these stars and in cooler, diffuse dark clouds. Because carbon chemistry is so pervasive in the universe, complex organic compounds were probably incorporated into protoplanetary disks and planets. In contrast, only silicon dioxide and more complex silicates have been detected in the outer layers of cool stars. Although silicones, silanes, or other reduced (i.e., relatively H-rich) silicon compounds might be important precursors of silicon-based life, none of these compounds have been detected as of this writing.

The notion that *all* life in the universe is necessarily based on carbon compounds cannot be demonstrated, of course. However, the chemical versatility and widespread cosmic distribution of organic compounds argue that carbon plays prominent roles in extraterrestrial life. If life ever arose elsewhere in environments similar to those on early Earth, its biochemistry is probably based upon carbon.

Therefore, seeking carbon-based life is a high-priority, near-term search strategy. If living systems indeed exist elsewhere, probably most will consist of organic compounds. This by no means precludes future astrobiologists from pursuing truly exotic life forms based on alternative elements and compounds. But their searches for truly exotic life should be accompanied by laboratory and theoretical studies that explore the feasibility of such life and determine the specific appropriate attributes and environments to be targeted for exploration.

Energy for life

Living systems must consume and transform energy from their environment; this requirement is very likely a universal attribute of life (Schrödinger 1944). Considerations of the nature and availability of energy provide broad insights into the viability, activity, and distribution of organisms in nature (Hoehler 2007). This is particularly relevant to investigations of extreme environments, some of which are relevant analogs for potential abodes of life elsewhere in the solar system and beyond. Regarding life's origins, the availability of energy makes it possible to achieve specific outcomes of otherwise low probability. Examples of such outcomes that come to mind are the earliest functional proteins or RNA-like molecules. Energy is required to "pay for the cost" of increasing the information content of a molecular system that, at some stage, is capable of self-replication and evolution. The extent to which energy can be utilized probably places limits upon the probability and complexity of increasingly sophisticated prebiotic molecular systems that emerge along the pathway to life. Life on Earth utilizes either light or chemical disequilibria as sources of energy. Chemical energy can emerge from oxidation-reduction reactions; those further from thermodynamic equilibrium provide more energy. (A redox reaction changes the oxidation states of atoms in the reacting species, i.e., a net transfer of electrons occurs between chemical species.) A planet that can provide nutrients and liquid water at its surface creates the potential for light-harvesting mechanisms to develop. The crust of a geologically active planet might provide water, nutrients, and redox energy (i.e., energy released as a consequence of redox reactions) in the subsurface and perhaps also at the surface (e.g., Shock at al. 2005).

Environmental contexts for life through time

For the purposes of this discussion, a planet that is "suitable" for life is defined as one that provides the resources and environments that sustain life as described above, namely, life as we know it. Should truly exotic life exist, then an assessment of f_l that is based upon "life as we know it" would underestimate the true cosmic f_l value.

The frequency with which life appears on environmentally suitable planets depends upon the duration of the suitable environments. Thus a highly favorable planet establishes habitable environments soon after its formation and sustains them thereafter. Alternatively, if billions of years of favorable conditions pass before life can begin, f_l would be much smaller. A related example is the widely shared view that "highly complex life" analogous to our plants or animals is probably much less common in the universe. This view seems justified because, on Earth, billions of years of habitable conditions passed before these highly complex life forms developed.

Therefore, it is useful to assess when some early planetary environments apparently became suitable for the appearance of life, using Earth and Mars as examples.

Early environments on Earth and Mars

One of the crowning achievements of the earth sciences since 1961 has been the discovery of a substantial record of microbial life prior to the dawn of the Phanerozoic eon some 543 million years ago. The earliest evidence of our biosphere now extends back more than 3.7 billion years (Figure 8.3; Rosing 1999; Schopf 1983). But a rock record is always required to determine whether Earth was actually inhabited at some time in the past. An important observation about Earth's rock record is that evidence of life appears in the oldest rocks whose alteration has not eradicated the fossil evidence. Crustal processes such as tectonics, metamorphism, weathering, and erosion apparently have eradicated essentially all evidence of life prior to about 3.7 billion years ago. Therefore, life appeared sometime earlier than that, perhaps hundreds of millions of years earlier.

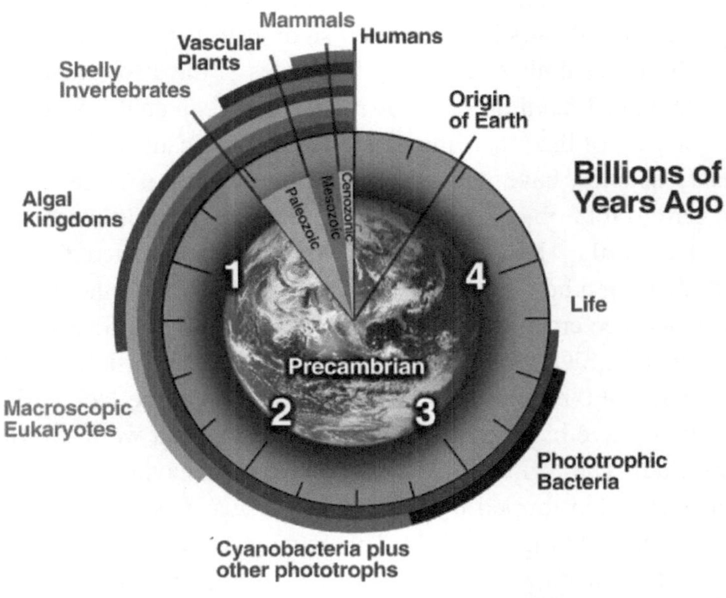

Figure 8.3 Evidence for key categories of life (adapted from Des Marais 2000). The white numbers provide a clock-like timescale (in billions of years) that indicates the antiquity of this evidence.

Earth's earliest environments

Despite the absence of a robust rock record, advances in planetary sciences since 1961 have provided key insights about Earth's earliest environments. Relatively soon after its accretion, Earth sustained an impact by a Mars-sized (0.1 Earth mass) object that led to the formation of the Moon. The energy of this impact melted Earth's mantle and, together with lunar tidal heating, maintained melted-rock temperatures near the surface for millions of years (Zahnle et al. 2007). Thereafter the steam atmosphere condensed and a ~100 bar carbon dioxide–rich atmosphere maintained near-surface temperatures around ~500 degrees Kelvin for another 10 million years or perhaps much longer (Zahnle et al. 2007). (One bar is approximately equal to the pressure of Earth's atmosphere at sea level.) As the crust cooled further, eventually carbonate minerals became stable, and both carbonates and hydrous minerals could be subducted (injected along with descending dense crustal plates) into the upper mantle at rates sufficient to lower atmospheric carbon dioxide levels substantially. The timing and effectiveness of this early subduction of volatiles are highly uncertain; perhaps several tens of millions of years were required to reduce atmospheric carbon dioxide to levels where the surface environment became habitable. In any case, by 4.4 billion years ago Earth's surface environment very likely hosted standing water and surface temperatures habitable by life as we know it.

Subsequent impacts were considerably smaller than the Moon-forming event, yet the largest among them still constituted significant threats to the continuity of habitable conditions (Zahnle et al. 2007). Estimates of the sizes and consequences of impacts older than 3.8 billion years are model-based, given the near-absence of a rock record. Studies of the lunar impact record, a significant achievement of post-1961 planetary science, have helped to calibrate the rate of large impacts. Anywhere from zero to four objects having diameters exceeding 300 kilometers might have struck the Earth. These would have partially evaporated the ocean, and perhaps only thermophilic ("heat-loving") life harbored within the coolest regions of the near subsurface crust might have survived (Zahnle and Sleep 2006).

Ancient zircon ($ZrSiO_4$) mineral grains offer additional insights about Earth's environment before 4.0 billion years ago. Many zircons in the Jack Hills region of Western Australia have been recycled from other rocks that were several hundred million years older. These zircons formed from magmas having major components of continental crust that formed in the presence of water near the Earth's surface (Mojzsis, Harrison, and Pidgeon 2001). This finding is consistent with the presence of a hydrosphere interacting with the crust by 4.3 billion years ago. Some zircons are as old as 4.4 billion years (Valley et al. 2014). The age distribution of zircon grains older than 4.0 billion years indicates that tectonic activity at that time recycled oceanic and continental crust at about ten times the present rate (Sleep

2007). Heat derived from Earth's accretion and the decay of more abundant radionuclides created a hotter mantle and a crustal heat flow that was approximately three times its present value.

The elevated tectonics and heat flow in Earth's ancient crust sustained elevated volcanism and hydrothermal activity. These processes delivered a continuous supply of carbon dioxide, hydrogen gas, and reduced species of sulfur and redox-sensitive metals such as iron and manganese to surface and near-surface environments. The utilization of these chemical species as sources of energy and nutrients is so universal among the bacterial, archaeal, and eukaryotic domains of microbial life that they probably played central roles in the earliest biosphere (e.g., Sleep 2010). The most ancestral characteristics and requirements of microbial metabolism are broadly consistent with the resources that were likely available at that time. Suitable ingredients, energy sources, and conditions that were required to support habitable environments existed on Earth perhaps as early as 4.4 billion years ago.

Mars

Mars provides an accessible second example of a rocky planet to explore for evidence of habitable environments and life. In the near term, Mars exploration allows us to strengthen our understanding of at least the planetary environmental aspects of f_l, namely, the potential of diverse rocky planets to provide conditions suitable for life early in their histories.

During its history, the climate of Mars has been more similar to that of Earth than the climate of any other planet in our solar system (Figure 8.4). Today, however, wherever the surface and shallow subsurface are at or close to being equilibrated thermodynamically with the atmosphere, any combination of temperature and water activity in the Martian shallow subsurface is considerably below the threshold conditions required for terrestrial life to propagate (MEPAG SR-SAG 2006). In warm soil, the chemical activity of water is several orders of magnitude too low to support life – there simply is insufficient water to dampen the soil by the required amount. Water activity can attain a value of unity at the frost point, but then the temperature is far too low to support propagation.

But orbital observations by the Mariner and Viking missions mapped numerous features indicating that water once flowed abundantly across the Martian surface (Carr 2007). These features, together with evidence of more vigorous geologic activity in the past, indicate that early climates were wetter and perhaps also somewhat warmer. A denser atmosphere was required for liquid water to be stable at the surface and also would have provided substantial protection from radiation, as it does on Earth today. Redox energy from volcanism, hydrothermal activity,

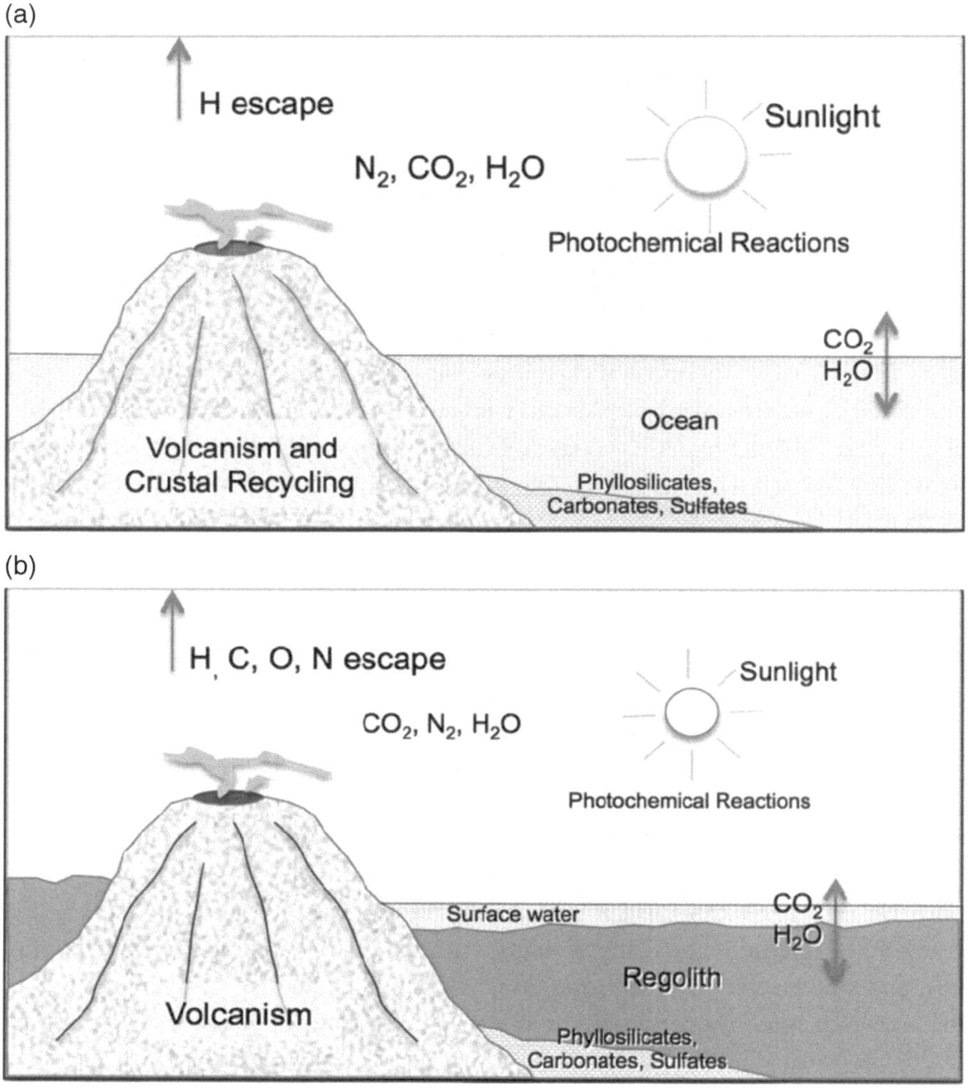

Figure 8.4 Schematic depicting key attributes of early environments on Earth (A) and Mars (B). Both planets sustained volcanism, liquid water, other key ingredients for life, aqueously formed minerals, and relatively clement conditions. Evidence found since 1961 is consistent with the hypothesis that, at least early in their history, rocky planets typically hosted environments that were suitable for life. *Credits*: David DesMarais

and weathering of crustal materials would have been more readily available (Des Marais, Jakosky, and Hynek 2008). Life might have persisted in long-lived hydrothermal systems, particularly those that had been associated with ascending magmas (molten rock) that were emplaced in near-subsurface crust or erupted by volcanos (Gulick 1998). The likelihood of any life at the surface would have been greatest during early wetter epochs. Scenarios envisioned for early Mars are strikingly similar to those envisioned for early Earth.

Beyond Mars?

Chapter 6 in this volume discusses the variety of potentially habitable environments that might have developed on moons in the outer solar system and on exoplanets.

Appearance (origins) of life

An estimate of the fraction of suitable planets where life appears must address the phenomena and constraints surrounding the origins of life. The need to comprehend our origins runs deep in human culture and has spawned a variety of approaches that span religion, mythology, and science (Morowitz 1992; Chapter 7). The religious view is that life began through the actions of a Creator. A second view is that life on Earth arose by natural causes but through a set of events that were so highly improbable that they might have been unique. A third view holds that life begins as a consequence of the laws of nature and that some environments lead inevitably to origins of life. This third approach is consistent with the religious view if one accepts that the Creator acts through the laws of nature. The second approach, in the extreme, implies an origin event so rare that it cannot practically be investigated via a systematic, scientific strategy. This discussion follows the science-based third approach.

Multiple origins of life might be an ongoing process in the universe; indeed, perhaps someday they will be ongoing in biotechnology research laboratories. Historically, origins-of-life research has focused on synthesizing organic monomers in reducing atmospheres and examining their assembly into structures (e.g., "proteinoids") and macromolecules (i.e., peptides and polynucleotides) whose roles embody life's most diagnostic properties, namely its ability to replicate and evolve (Miller and Orgel 1974). Following naturally from the traditional theory that life arose within a reducing organic-rich aqueous broth, the earliest cells were assumed to have been fermentative heterotrophs (Miller and Orgel 1974). (A fermentative heterotroph obtains its carbon from organic molecules and its energy from redox reactions involving organic molecules.) A long-standing paradox has been that the functions performed by the nucleic acid-protein translation apparatus were assumed to be essential for life to begin, yet the required apparatus seemed too complex to have arisen in the prebiotic milieu (see Chapter 7). The discovery of "ribozymes" (Cech 1989) offers one potential solution to this paradox because, early in life's history, a single class of compounds might have executed both information storage and enzymatic catalysis. But it still seems extraordinarily difficult for ribozymes to have arisen in the absence of some kind of "proto-life."

More recent research has explored the potential roles that environmental conditions might play in prebiotic evolution (i.e., evolution leading to origins of life). Some studies have recognized that prebiotic chemical processes required sustained energy sources that were actually available in the environment. For example, Wächtershäuser (1992) developed a model in which prebiotic "evolutionary biochemistry" was driven by oxidation and reduction reactions involving iron and sulfur species. Such energy-rich species occur abundantly in hydrothermal systems within planetary crusts. Amphiphilic compounds (e.g., lipids) assemble spontaneously into vesicles that resemble cellular membranes and that create chemical microenvironments favorable to the development of a "proto-metabolism" (Cody 2007). Similar amphiphilic compounds (molecules having both polar and nonpolar components, e.g., a fatty acid) have been identified in meteorites. The increased complexity of these molecular systems imparts emergent properties (e.g., structures, specific molecular interactions) that probably facilitated life's origin but that require specific attention by theorists and experimentalists. This relationship between theories of complexity and the origins of life continues to be explored (Kauffman 1995; Pohorille 2012).

Organic compounds relevant to prebiotic chemistry have been discovered in primitive bodies (e.g., meteorites, comets) and in interstellar space, and these might have been delivered intact to planetary surfaces. Biochemical research is narrowing the knowledge gap between prebiotic chemicals and the first living systems. Simpler "prebiotically plausible" RNA-like molecules might have functioned as both self-replicators and enzymes (Ricardo and Benner 2007). The earliest functional protein enzymes might have been much smaller in size and thus perhaps easier to develop than previously imagined (e.g., Golynskiy, Haugner, and Seelig 2013). The prebiotic environment could have provided molecules that formed vesicles as precursors of cellular envelopes (Deamer 2007).

Changing perceptions related to f_l

Since 1961, perceptions about the magnitude of f_l have reflected advances in our knowledge about several key factors that affect f_l. This chapter addressed the following factors: concepts of life, characteristics of habitable planetary environments, key roles played by liquid water and organic matter, the antiquity and attributes of Earth's early biosphere, early Martian environments, and origins of life.

Developments during the 1960s probably elevated the perceived value of f_l. Paleontological research uncovered a microbial fossil record of our biosphere that extends back billions of years (see Schopf 1983 for a summary), indicating that life could arise early in the history of a habitable planet. The onset of space exploration elevated interest in this subject. Preparations for the first US mission to land on Mars built expectations that we might find a second example of life there.

The 1970s engendered pessimism about f_l. The Viking missions did not detect life but documented instead that the present Martian surface environment is highly inhospitable to life as we know it. Also, earlier claims of biological organic matter preserved in Earth's earliest rock record were challenged by evidence of contamination by younger organics.

Perceived estimates of f_l stabilized during the 1980s and 1990s. Paleontologists confirmed that our early biosphere was at least as old as the sedimentary rocks that could preserve fossil evidence (Schopf 1983). Ongoing orbital observations of Mars documented additional evidence that liquid water existed on its surface long ago. Rising interest in the search for planets orbiting other stars created optimism about the prospects to address f_l by direct observations.

Since 2000, many of the factors affecting f_l that were amenable to investigation have heightened perceptions that f_l just might not be infinitesimally small. But these are perceptions, not actual reality. Much critical work remains to be done.

Improving our estimates of f_l

Advances in the planetary sciences indicate that environments that provide many of the key requirements for life as we know it might be widespread in the universe. Earth-like planets might typically be suitable for life soon after their accretion. Life on Earth began remarkably early, perhaps much earlier than the oldest rocks capable of preserving fossil evidence. These two observations are at least consistent with the possibility that f_l might be reasonably large.

However, our near total lack of understanding about how life begins severely constrains any quantitative estimate of f_l. It is critical to pursue both theoretical and experimental approaches in the context of planetary environments that favored the origins of life. For example, extraterrestrial organic matter from cosmic dust and ice rained down on the early Earth (Chyba and Sagan 1992). If this external supply was available only early in a solar system's history yet was essential for life to begin, it imposes a key constraint on life's distribution in the cosmos. Perhaps planetary processes also created prebiotic organic matter. A mildly reducing early atmosphere that was sustained by thermal processes acting upon a more reduced crust and upper mantle might have facilitated prebiotic organic synthesis (e.g., Cody and Scott 2007). A cold early surface environment would have allowed prebiotically important species to survive and accumulate (Bada, Bigham, and Miller 1994). Continued investigations of the indicators of early Earthly and Martian environments are essential for characterizing the conditions where life can begin. For example, assessing the relative importance of various organic sources should at least partially constrain the range of plausible prebiotic scenarios. The Afterword in this volume provides additional perspectives about how the deep mysteries surrounding the origins of life challenge our ability to quantify f_l. That chapter also articulates future investigations that might prove to be fruitful.

The evolution of stars and planets follows characteristic trajectories that allow reasonable estimates to be made of the origins and evolution of planetary climates and habitability. Recent discoveries have improved our prospects and strategies for seeking a second example of life. Such a seminal discovery would, in turn, substantially reduce the uncertainty in estimating f_l. Biochemists are gaining insights about how assemblages of relatively simple molecules might perform key cellular processes.

Ultimately, coordinated efforts in astronomy, planetary science, molecular biology, and the information sciences will be essential for reducing the large uncertainty that currently surrounds f_l.

References

Bada, Jeffrey L., C. Bigham, and Stanley L. Miller. 1994. "Impact Melting of Frozen Oceans on the Early Earth: Implications for the Origin of Life." *Proceedings of the National Academy of Sciences* 91: 1248–50.

Baross, John A. 2007. *The Limits of Organic Life in Planetary Systems*. Washington, DC: National Academies Press.

Carr, Michael H. 2007. *The Surface of Mars*. New York: Cambridge University Press.

Cech, Thomas R. 1989. "RNA as an Enzyme." *Biochemistry International* 18: 7–14.

Chyba, Christopher, and Carl Sagan. 1992. "Endogenous Production, Exogenous Delivery and Impact-Shock Synthesis of Organic Molecules: An Inventory for the Origins of Life." *Nature* 355: 125–32.

Cleland, Carol E., and Christopher F. Chyba. 2007. "Does 'Life' Have a Definition?" In *Astrobiology and Life*, ed. Woodruff T. Sullivan III and John A. Baross, 119–31. Cambridge University Press.

Cody, George D., and James H. Scott. 2007. "The Roots of Metabolism." In *Astrobiology and Life*, ed. Woodruff T. Sullivan III and John A. Baross, 174–86. Cambridge University Press.

Deamer, David W. 2007. "The Origin of Cellular Life." In *Astrobiology and Life*, ed. Woodruff T. Sullivan III and John A. Baross, 187–209. Cambridge University Press.

Des Marais, David J. 2000. "When Did Photosynthesis Emerge on Earth?" *Science* 289: 1703–5.

Des Marais, David J., Bruce M. Jakosky, and Brian M. Hynek. 2008. "Astrobiological Implications of Mars Surface Composition and Properties." In *The Martian Surface: Composition, Mineralogy and Physical Properties*, ed. James A. Bell, 599–623. Cambridge Planetary Science 9. Cambridge University Press.

Des Marais, David J., Joseph A. Nuth III, Louis J. Allamandola, Alan P. Boss, Jack D. Farmer, Tori M. Hoehler, Bruce M. Jakosky, Victoria S. Meadows, Andrew Pohorille, Bruce Runnegar, and Alfred M. Spormann. 2008. "The NASA Astrobiology Roadmap." *Astrobiology* 8: 715–30.

Golynskiy, Misha V., John C. Haugner III, and Burckhard B. Seelig. 2013. "Highly Diverse Protein Library Based on the Ubiquitous (β/α)8 Enzyme Fold Yields Well-Structured Proteins through In Vitro Folding Selection." *ChemBioChem* 2013(14): 1553–63.

Gulick, Virginia C. 1998. "Magmatic Intrusions and a Hydrothermal Origin for Fluvial Valleys on Mars." *Journal of Geophysical Research* 103: 19365–88.

Hoehler, Tori M. 2007. "An Energy Balance Concept for Habitability." *Astrobiology* 7: 824–38.

Kauffman, Stuart. 1995. *At Home in the Universe*. New York: Oxford University Press.

McDaniel, Robert V., Thomas J. McIntosh, and Sidney A. Simon. 1983. "Glycerol Substitutes for Water in Lecithin Bilayers." *Biophysical Journal* 41: A116.

MEPAG SR-SAG. 2006. "Findings of the Mars Special Regions Science Analysis Group." *Astrobiology* 6: 677–732.

Miller, Stanley L., and Leslie E. Orgel. 1974. *The Origins of Life on the Earth*. Englewood Cliffs, NJ: Prentice-Hall.

Mojzsis, Stephen J., T. Mark Harrison, and Robert T. Pidgeon. 2001. "Oxygen-Isotope Evidence from Ancient Zircons for Liquid Water at the Earth's Surface 4,3000 Myr ago." *Nature* 409: 178–81.

Morowitz, Harold J. 1992. *Beginning of Cellular Life*. New Haven, CT: Yale University Press.

Pohorille, Andrew 2012. "Processes that Drove the Transition from Chemistry to Biology: Concepts and Evidence." *Origins of Life and Evolution of Biospheres* 42: 429–32.

Pohorille, Andrew, and Lawrence R. Pratt. 2012. "Is Water the Universal Solvent for Life?" *Origins of Life and Evolution of Biospheres* 42: 405–9.

Ricardo, Alonso, and Steven A. Benner. 2007. "The Origin of Proteins and Nucleic Acids." In *Astrobiology and Life*, ed. Woodruff T. Sullivan III and John A. Baross, 154–73. Cambridge University Press.

Rosing, Minik T. 1999. "^{13}C-Depleted Carbon Microparticles in the >3700-Ma Sea-Floor Sedimentary Rocks from West Greenland." *Science* 283: 674–76.

Schopf, J. William. 1983. *Earth's Earliest Biosphere*. Princeton University Press.

Schrödinger, Erwin. 1944. *What is Life? The Physical Aspect of the Living Cell*. Cambridge University Press.

Shock, Everett L., Melanie Holland, D'Arcy R. Meyer-Dombard, and Jan P. Amend. 2005. "Geochemical Sources of Energy for Microbial Metabolism in Hydrothermal Ecosystems: Obsidian Pool, Yellowstone National Park, USA." In *Geothermal Biology and Geochemistry in Yellowstone National Park*, ed. William P. Inskeep and Timothy R. McDermott, 95–112. Bozeman: Montana State University Press.

Sleep, Norman H. 2007. "Plate Tectonics through Time." In *Treatise on Geophysics*, vol. 9, edited by G. Schubert, 101–17. Oxford: Elsevier.

Sleep, Norman H. 2010. "The Hadean-Archaean Environment." *Cold Spring Harbor Perspectives in Biology* 2: a002527.

Tanford, Charles. 1978. "The Hydrophobic Effect and the Organization of Living Matter." *Science* 200: 1012–18.

Valley, John W., Aaron J. Cavosie, Takayuki Ushikubo, David A. Reinhard, Daniel F. Lawrence, David J. Larson, Peter H. Clifton, Thomas F. Kelly, Simon A. Wilde, Desmond E. Moser, and Michael J. Spicuzza. 2014. "Hadean Age for a Post-Magma-Ocean Zircon Confirmed by Atom-Probe Tomography." *Nature Geoscience* 7: 219–23.

von Neumann, John 1966. *Theory of Self-Reproducing Automata*, ed. A. Burks. Urbana: University of Illinois Press.

Wächtershäuser, Günter. 1992. "Groundwork for an Evolutionary Biochemistry: The Iron-Sulphur World." *Progress in Biophysics and Molecular Biology* 58: 85–201.

Zahnle, Kevin J., and Norman H. Sleep. 2006. "Impacts and the Early Evolution of Life." In *Comets and the Origin and Evolution of life*, 2nd ed., ed. P. J. Thomas et al., 207–51. New York: Springer.

Zahnle, Kevin J., Nick Arndt, Charles Cockell, Alex Halliday, Euan Nisbet, Franck Selsis, and Norman H. Sleep. 2007. "Emergence of a Habitable Planet." *Space Science Reviews* 129: 35–78.

9

Fraction of life-bearing planets on which intelligent life emerges, f_i, pre-1961

Michael J. Crowe

Abstract

Among ancient authors, only the Epicureans believed in the existence of extraterrestrial intelligent beings (ETI). Astronomy was shaken at its roots after 1543, when Nicholas Copernicus presented his heliocentric system. By the end of the sixteenth century, Giordano Bruno speculated that stars are inhabited suns surrounded by inhabited planets. By 1750, growing acceptance of (1) the principle of plenitude (the idea that God or nature would populate most regions of the universe) and (2) versions of the so-called Copernican principle (the claim that all regions of space must be more or less identical to the region in our solar system) led to widespread belief in a large value for f_i. An influential advocate for a high value of f_i was Christiaan Huygens, who in his *Cosmotheoros* (1698) relied heavily on the latter principle. Belief in ETI was widespread in the eighteenth century, including such pioneers of stellar astronomy as Johann Lambert, Immanuel Kant, and William Herschel. Belief in ETI remained widespread from 1750 to 1870, despite counterevidence from such scientific laws as the inverse-square laws for gravitation, light, and radiant heat.

The first modern author who seriously challenged claims for a high value for f_i was William Whewell, who in 1853 questioned whether higher forms of life could survive elsewhere in our solar system, doing this partly in terms of the inverse-square laws. Whewell's *Of the Plurality of Worlds: An Essay* outraged many of his contemporaries but gradually found limited acceptance among such astronomers as Richard Proctor, who recognized the need for a much lower value for f_i. Roughly simultaneously, the Darwin-Wallace theory of

The Drake Equation, ed. Douglas A. Vakoch and Matthew Dowd. Published by Cambridge University Press.

evolution by natural selection led scientists to a deeper and more naturalistic understanding of the development of higher forms of life. Moreover, Wallace in 1904 stressed the improbability of beings as complex as humans developing elsewhere.

During the first half of the twentieth century, belief in ETI was at a relatively low level, partly because various scientific factors, such as encounter theories of planet formation, were taken seriously. Around 1950, belief in ETI became more widespread after the Miller-Urey experiment (1953) and the development of radio telescopes raised hopes that ETI might be contacted.[1]

Extraterrestrial life debate before 1800

Among ancient authors, only the Epicureans believed in the existence of many other systems comparable to our Earth-centered system and thus inhabited by ETI. These systems centered on an Earth-like body, orbited by a Sun-like source of light. The Epicureans did not claim that we could observe these Earths; rather, they based their position on their belief that the universe is infinite and that such bodies form from the concourse of an infinite number of atoms. Platonists, Aristotelians, and most ancient and medieval thinkers rejected such claims, a partial exception being Cardinal Nicholas Cusa (1401–1464), whose *De docta ignorantia* (1440) favored inhabited planets and stars.[2]

The publication in 1543 by Nicholas Copernicus (1473–1543) of a book advocating the heliocentric theory gradually led some authors to believe that our Earth orbits the Sun, that the stars are suns, and that these suns are surrounded by planets on which ETI dwell. Although Copernicus never mentioned extraterrestrials (even in his unpublished writings), his claim that Earth orbits the Sun led some thinkers to infer that if the Earth is a planet, then the planets must be earths and possibly stars are suns. Very gradually, this idea promoted the claim that there is nothing special about our Earth and that all regions of space are comparable. This principle, later labeled the Copernican principle even though Copernicus would almost certainly have opposed it, seemed to some a sensible conclusion to draw from heliocentricity. Nevertheless, by 1600, only twelve astronomers had adopted heliocentricity; the most speculative of these was Giordano Bruno (1548–1600), who championed a universe in which stars are suns surrounded by inhabited planets.

Scientific advances made in the first half of the seventeenth century by such natural philosophers as Galileo, Kepler, and Descartes added plausibility to heliocentricity, although none of these scientists adopted the extreme views of Bruno nor did they typically champion belief in ETI. For example, Kepler's universe included ETI on the Moon and at least some planets, but he insisted that the Earth is a special place – the best observation point in the universe. By 1650, a minority of intellectuals advocated ETI.

By 1750, belief in ETI had become widespread and remained so until well into the nineteenth century. Three late seventeenth-century authors played a role in this transition. One of the most engaging authors to advocate ETI was Bernard le Bovier de Fontenelle (1657–1757), who successfully reached a wide audience – including women – with his *Entretiens sur la pluralité des mondes* (*Conversations on the Plurality of Worlds*, 1686). Quickly translated into ten languages, the book presents a dialogue between a philosopher and a marquise, in which the philosopher presents Copernican and Cartesian ideas to this intellectually able woman. Fontenelle incorporates a great deal of discussion about ETI. Not only does he assert their existence, he also discusses their characteristics. His Mercurians, for example, "are so full of Fire, that they are absolutely mad." But the philosopher assures the marquise that ETIs would be perfectly at home in their own environment, even one we would find unbearable (Crowe 2008, 79–80). Fontenelle based his claims not only on the Copernican principle but also the principle of plenitude: this is the idea that God or nature has acted in such a manner that ETI exist throughout the universe; otherwise the multitude of planets would be wasted.

Another highly influential late seventeenth-century advocate of ETI had impeccable scientific credentials; this was Christiaan Huygens (1629–1695), whose *Cosmotheoros*, or, as the English version has it, *The Celestial Worlds Discover'd: or, Conjectures Concerning the Inhabitants, Plants and Productions of the Worlds in the Planets*, appeared posthumously in 1698. Huygens sprinkled his book with scientific information compatible with belief in ETI, but he repeatedly returned to his key argument: the Copernican principle. For example, he states that "the other Planets are not inferior in dignity to ours" and asserts that he believes creatures similar to humans must inhabit these planets:

> otherwise our Earth would have too much the advantage of them, in being the only part of the Universe that could boast of such a Creature so far above, not only Plants and Trees, but all Animals whatsoever: a Creature that has a Divine somewhat within him, that knows, and understands, and remembers such an innumerable number of things. (Crowe 2008, 96)

A curious feature of Huygens's reliance on the Copernican principle is that it seems in tension with a scientific result derived from the fact that the Copernican system leads to the conclusion that there are major differences among the planets. In particular, the Copernican system, unlike the geocentric system, made it possible to calculate each planet's relative distance from the Sun. From knowing these distances and observationally measuring each planet's apparent diameter, astronomers could assign relative diameters to the planets, at least as compared to the Earth and Sun. When this is done, it reveals that the planets differ greatly in size. We are certain that Huygens recognized this diversity in size because in his 1698 book he provides a diagram comparing the sizes of the objects in our system. This diagram reveals that although Mercury, Venus, Earth, and Mars are similar in size, they are miniatures compared to giant Jupiter and Saturn, which are dwarfed by our Sun. This hardly fits with the Copernican principle.

A third influential seventeenth-century author was Isaac Newton (1642–1727). Scattered throughout his writings are passages supporting ETI; one of the most important appeared in the "General Scholium," which he inserted into the 1713 edition of his *Principia mathemtica* (1687). He wrote, "And if the fixed stars be centers of similar systems, all these, constructed by a similar plan, will be under the rule of One" (Crowe 2008, 112). In his *Principia*, Newton presented an important result derived from his famous law of gravitation. He showed how by using this law he could weigh the Earth, Sun, Jupiter, and Saturn, and also determine their densities and compare the gravitational forces on the surface of each. In particular,

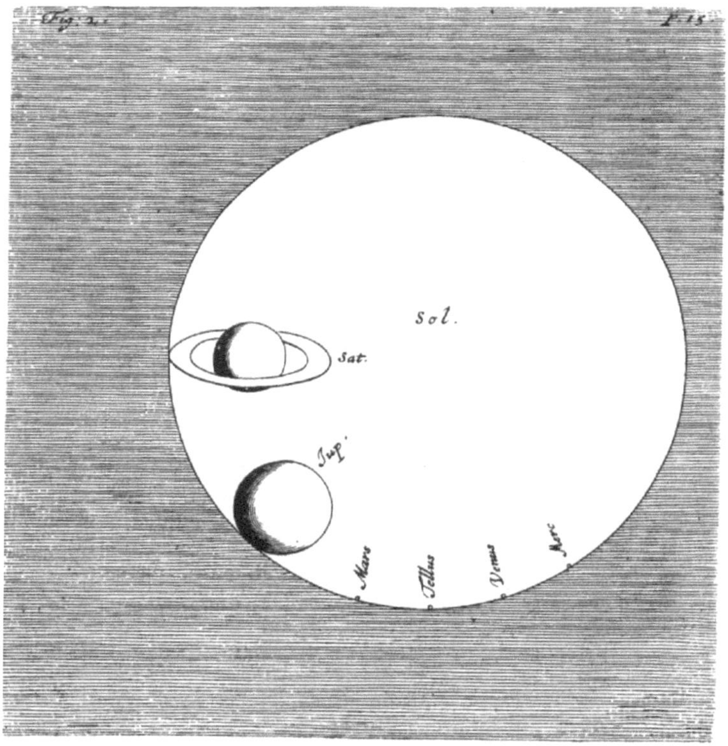

Figure 9.1 Huygens's diagram of sizes of solar system objects from his *Cosmotheoros*.

Table 9.1 *Planetary data from Newton's* Principia *(Crowe 2008, 112)*

	Sun	Jupiter	Saturn	Earth
Mass	1	1/1076	1/3021	1/169282
Density	100	94.5	67	400
Weight of person on	10000	943	529	435

he derived the striking results presented in Table 9.1. Newton did not comment on what these results indicate regarding ETI or the diversity within the solar system, nor for the most part did other authors over the next 150 years. We can now see that these results pointed to the problematic nature of the Copernican principle and to the perils in inferring that the other planets are analogous to the Earth. Put differently, this information was bad news for Jupiterarians, Saturnians, and Solarians.

In the early eighteenth century, Rev. William Derham joined the many authors who believed in ETI, publishing in 1714 his *Astro-Theology, or A Demonstration of the Being and Attributes of God from a Survey of the Heavens*. This popular work based partly on Huygens linked ETI to Newtonianism and also to Natural Theology, which sought to demonstrate God's power and goodness by examining the natural world. One striking feature of the book is that Derham divides the history of astronomy into three periods: the Ptolemaic, Copernican, and "New Systeme," which goes beyond the Copernican by assuming that every star is surrounded by inhabited planets, urging that this system "is far the most magnificent of any; and worthy of an infinite CREATOR" (Crowe 2008, 123).

At Cambridge University, William Whiston, who in 1702 succeeded Newton as Lucasian professor of mathematics, repeatedly advocated ETI. As early as his *New Theory of the Earth* (1696), Whiston urged that planets and planetary systems have inhabitants subject to moral trials. Two decades later in his *Astronomical Principles of Religion*, Whiston proposed denizens dwelling in the interiors of the Sun, planets, and comets; moreover, he posited "Not wholly Incorporeal, but Invisible Beings" living in planetary atmospheres (Crowe 1986, 31).

By 1750, belief in ETI was winning international acceptance. Nevertheless, from the perspective of the present, its foundation was frail, resting on analogical arguments of dubious force, on metaphysical principles such as the principle of plenitude and the Copernican principle, and on a scattering of astronomical observations.

The most important development in astronomy between 1750 and 1800 was the creation, chiefly by four authors, of stellar astronomy. Moreover, these four authors, who believed deeply in ETI, brought this commitment to their interest in stellar astronomy.

The three claims central to the founding of stellar astronomy were (1) that the Milky Way is an optical effect arising from the scattering of millions of stars over a planar area, (2) that these stars form a giant disk-shaped structure many light years in diameter, and (3) that numerous nebulous patches seen in the heavens are galaxies comparable in magnitude to our own Milky Way.

In 1750, Thomas Wright published his *An Original Theory or New Hypothesis of the Universe*, a volume rich in illustrations from Wright's hand, speculations from his active imagination, and claims about ETI. His goal was not only to delineate the structure of the universe but also to integrate it with the spiritual. In attempting this, he came close to suggesting that the Milky Way can be explained as consisting of a vast planar array of stars. Moreover, some historians suggest that he attained the idea that various patches of stars constitute other universes (Crowe 1986, 45–47). His pious purposes appear in his calculation that we can see possibly 170 million inhabited globes.

In 1755, Immanuel Kant published his *Allgemeine Naturgeschichte und Theorie des Himmels*, the first book to set out all three claims central to sidereal astronomy, which he does in the first two parts of this volume and which throughout draws on the principle of plenitude. In the third part, centered on the solar system, Kant's commitment to ETI is especially evident. Relying on various assumptions, including the Great Chain of Being, and the idea that one can infer the nature of a planet's inhabitants from the type of matter dominant on the planet, Kant asserts that planetarians "become more excellent and perfect in proportion to the distance of their habitats from the sun" (Crowe 2008, 145). Kant's Mercurians and Venusians are consequently dullards, whereas his earthlings occupy "exactly the middle rung ... on the ladder of beings" and his Jovians and Saturnians are greatly superior beings.

The third pioneer of sidereal astronomy was Johann Lambert, a brilliant mathematician, physicist, and philosopher, who was the first person to discern the structure of the Milky Way. Although Lambert did not publish the disk theory until 1761 when he brought out his *Cosmologische Briefe uber die Einrichtung des Weltbaues*, he had attained that idea in 1749. It is a complex question whether Lambert attained the notion of nebulae as other universes, but it is indisputable that this book was as richly stocked with ETI as those of Wright and Kant, and went beyond them in his heavy reliance on the Copernican principle and in his stress on comets as carrying ETI.

The substantial level of interest in Lambert's book among his contemporaries may have derived more from its ideas about ETI than from its pioneering views in stellar astronomy. Ironically, these three pioneers may have had little more concern about stars than their contemporaries; it was rather inhabited planetary systems that interested them and that they sought to arrange into systems. The Milky Way for them was not primarily a giant array of glowing globes, but rather a visible collection of sources of heat and light serving countless ETI on the admittedly unobservable planets of the stellar systems. Although Kant's dull Mercurians and super Saturnians, and Lambert's itinerant cometarians, are now seen as bizarre companions to their more durable doctrines, eighteenth-century readers may have held the opposite view.

Sir William Herschel (1738–1822) was the chief pioneer of stellar astronomy. If, however, his life is looked at broadly, it emerges that he was also deeply involved with ETI. He turned to astronomy only after two decades centered on music, partly by reading *Astronomy Explained upon Sir Isaac Newton's Principles* by James Ferguson. Passages in Ferguson led Herschel to believe not only that ETI dwell on the Moon but also that telescopic observations might confirm their presence. Such hopes inspired Herschel to build telescopes, which by 1781 had made him famous for discovering the planet Uranus. Ironically, Herschel may have believed that he had made a more important discovery in 1776 when, as we learn from his unpublished manuscripts, he believed he had discovered "growing substances," in particular, a forest on our Moon (Crowe 2008, 177–78). He built ever-larger reflecting telescopes, possibly hoping to confirm this discovery, but eventually used them to observe stars and nebulous patches. The genius of Herschel is evident in the fact that around 1780, astronomers knew of about one hundred such nebulae – a number Herschel's observations increased to 2,500. Part of his passion for these objects was that he viewed them as island universes, as structures comparable to the Milky Way and correspondingly filled with ETI. Moreover, Herschel developed (and observationally supported) a theory that our Sun (and therefore all stars) consist of a cool solid interior covered by a glowing layer of gas. He further suggested that ETI live on the surface of the Sun's solid interior. For some decades, astronomers took Herschel's model of the Sun quite seriously (Crowe 2011).

By 1800, belief in ETI was more widespread than ever before or since. If, however, we examine this belief from the present perspective, it is clear that its conjectural component was large. Broad analogy more than detailed astronomy, physicotheology more than physics, and teleology more than telescopes had been used to erect a vast edifice on what nineteenth-century scientists gradually found to be a frail foundation. Moreover, although some authors analyzed the question of other worlds in largely scientific terms, the majority, explicitly or implicitly, invoked religious or metaphysical considerations.

In the case of science, so also in regard to religion: serious difficulties lay just beneath the surface. The success of the natural theological enterprise, whether practiced by Christians or deists, tended to emphasize "Nature's God" while downplaying the idea of an incarnated redeemer. In short, whereas by the 1790s pluralism had reached a rapprochement with theism, tensions with Christianity had not as yet been fully faced. This is at least the conclusion suggested by the sensation that resulted when, in the 1790s, Thomas Paine launched a vigorous attack on Christianity based on belief in ETI.

Entitling his book *The Age of Reason*, Paine argued that what astronomy teaches about ETI makes it impossible for thinking persons to accept the central Christian notions of a divine incarnation and redeemer. Paine argues that although the existence of intelligent life only on the Earth is not a specific Christian doctrine, it is nonetheless "so worked up therewith from ... the story of Eve and the apple, and the counterpart of that story – the death of the Son of God, that to believe otherwise ... renders the Christian system of faith at once little and ridiculous" (Crowe 2008, 224). Paine's *Age of Reason* attracted an immense readership both in Britain, where 60,000 copies of it were printed, and in America, where a single Philadelphia bookshop sold over 15,000 copies.

The extraterrestrial life debate, 1800–50

On one level, the confidence in ETI common around 1800 is surprising. As noted earlier, Newton had used his law of gravitation to determine the relative masses and densities of the Sun, Earth, Jupiter, and Saturn and also the relative force of gravitation on the surface of each of these objects. His results revealed immense diversity among these objects. For example, he found that it would take 157 Earths to match the mass of Jupiter and 169,282 Earths to match the Sun's mass. This suggests the problematic character of claims that these giant bodies are analogous to the Earth. Even more striking was Newton's calculations that the density of the Earth is more than four times greater than the density of the Sun, Jupiter, or Saturn and that an earthling, if transported to Jupiter, would weigh twice as much and more that twenty-three times more if on the Sun.

If one turns from gravitational to optical and thermal considerations, the situation becomes more complex historically, but no less striking. Consider first of all light, especially the fundamental photometric result that the amount of light radiating from a point source decreases according to the inverse-square law. This law for light propagation was well known at least from 1720, when Pierre Bouguer published experiments demonstrating it (Mach n.d., 13–17). This makes it easy to determine that Saturn, being about 9.5 times farther from the Sun than Earth, must receive about ninety times less light per unit surface area than Earth.

Regarding the amount of radiant heat each planet receives from the Sun, the situation was still more complicated, partly because the nature of radiant heat remained under discussion in 1800. By 1850, however, scientists recognized that the amount of heat from our Sun reaching any planet would, like the amount of light, drop off at a rate governed by the inverse-square law.

Severe as these problems may seem, proponents of ETI rarely took them seriously, or at most (say in the case of light) viewed them as making clear why God had provided Saturn with a ring. One author who used these considerations against extraterrestrials was Thomas Young, who, commenting in 1807 on Herschel's arguments for solar ETI, suggested that the Sun's mass would make human-sized solarians weigh over two tons (Crowe 1986, 168).

In the 1830s, arguably the most prominent British astronomer was John Herschel (1792–1871), and his *Treatise on Astronomy* (1833) was the most respected presentation of astronomy available in English. Regarding the heat–light problem, he states: "The intensity of solar radiation is nearly seven times greater on Mercury than on the earth, and on Uranus 330 times less; the proportion between these two extremes being that of upwards of 2000 to one" (Herschel 1833, 277–78). Regarding the gravity issue, he declared, "the intensity of gravity, or its efficacy in ... repressing animal activity on Jupiter is nearly three times that on Earth, on Mars not more than one third, and on the four smaller planets probably not more than one twentieth; giving a scale of which the extremes are in the proportion of sixty to one" (1833, 278). Regarding density, he states that Saturn's is about one-eighth that of Earth's, "so that [Saturn] must consist of materials not much heavier than cork." All this does not lead him away from ETI but to remark,

> Now, under the various combinations of elements so important to life as these, what immense diversity must we not admit in the conditions of that great problem, the maintenance of animal and intellectual existence and happiness, which seems, so far as we can judge by what we see around us in our own planet, and by the way in which every corner of it is crowded with living beings, to form an unceasing and worthy object for the exercise of the Benevolence and Wisdom which presides over all! (1833, 278)

It is an interesting and important fact that the quotations from Herschel cited in this paragraph, including the last, reappeared with no significant changes not only in his famous *Outlines of Astronomy* (1849) but also in the 1869 edition of *Outlines*. As we shall see, in 1853 one of Herschel's closest friends analyzed these matters quite differently.

Although it is certainly true that in early-modern times, Christian authors were hesitant about embracing ETI, and although it is true that authors such as Thomas Paine in the 1790s used ETI to attack fundamental Christian beliefs, it would be seriously mistaken to assume that early nineteenth-century religious thinkers opposed claims for extraterrestrials. For example, two major religious groups founded in the first half of the nineteenth century – the Church of Jesus Christ of Latter-day Saints and the Seventh-day Adventists – incorporated ETIs into their scriptures. Numerous religious figures sought to accommodate ETIs to their Christian beliefs. Examples are such prominent theologians as America's Timothy

Dwight and Scotland's Thomas Chalmers. To illustrate the degree to which such accommodation extended, let us look at a disciple of Chalmers, Thomas Dick (1774–1857), who from his observatory near Dundee published books blending astronomy and religion. In his *Christian Philosopher* (1823), Dick asserts that the wisdom of God is shown by our Sun being placed at just such a distance as best to benefit us. Dick hastens, however, to add that the Sun's position does not prevent other planets from being happily inhabited by beings appropriately formed for their varying distances from the Sun. We learn from this book that ETI dwell not only on all the planets but also on the Moon and Sun. For example, Dick states that God placed within the immense body of the Sun "a number of worlds . . . and peopled them with intelligent beings" (Crowe 1986, 196). Turning to the Moon, he predicts that "*direct proofs*" of the Moon's habitability will be forthcoming, supplementing this by appendices in which he discusses whether various German astronomers who claimed to have observed evidence of extraterrestrials were correct. Dick, moreover, boldly claims that the existence of ETI "is more than once asserted in Scripture" (Crowe 1986, 197).

Dick presents similar ideas in his *Philosophy of Religion* (1826) and his *Philosophy of a Future State* (1828). In the former book, he asserts that "the grand principles of morality . . . are not to be viewed as confined merely to the inhabitants of our globe, but extend to all intelligent beings . . . throughout the vast universe [in which] *there is but one religion*" (Crowe 1986, 197, emphasis in original). In the latter book, he calculates that 2,400,000,000 inhabited worlds exist in the visible creation. In his *Celestial Scenery* (1836), he provides a table of the population of each planet, including even the ring, and the edge of the ring, of Saturn.

Extraterrestrial life debate, 1850–1910

Soon after 1850, astronomers found themselves entering the ETI debate equipped with telescopes of enhanced quality and powerful spectroscopic methods, which expanded astronomy to include astrophysics and astrochemistry. Not only was geology becoming increasingly relevant and available, but also the formulation by the Darwin-Wallace theory of evolution by natural selection provided a more powerful analysis of how organisms develop.

As of 1850, the belief that ETI populate a very high percentage of planets (not to mention stars and moons) in the universe remained widespread. By 1915, however, estimates of the number of planets with ETI had plummeted nearly to zero for our solar system and probably for other solar systems, if such exist. Who drove these ETI from the universe? Who first discerned the nearly desolate solar system beheld in the early twentieth century? This section offers an answer.

Table 9.2 *Thomas Dick's population table for the solar system from his* Celestial Scenery

	Square Miles	Population	Solid Contents
Mercury	32,000,000	8,960,000,000	17,157,324,800
Venus	191,134,944	53,500,000,000	248,475,427,200
Mars	55,417,824	15,500,000,000	38,792,000,000
Vesta	229,000	64,000,000	10,035,000
Juno	6,380,000	1,786,000,000	1,515,250,000
Ceres	8,285,580	2,319,962,400	2,242,630,320
Pallas	14,000,000	4,000,000,000	4,900,000,000
Jupiter	24,884,000,000	6,967,520,000,000	368,283,200,000,000
Saturn	19,600,000,000	5,488,000,000,000	261,326,800,000,000
Saturn's outer ring	9,058,803,600		
Inner ring	19,791,561,636	8,141,963,826,080	1,442,518,261,800
Edges of the rings	228,077,000		
Uranus	3,848,460,000	1,077,568,800,000	22,437,804,620,000
The Moon	15,000,000	4,200,000,000	5,455,000,000
Jupiter's satellites	95,000,000	26,673,000,000	45,693,970,126
Saturn's satellites	197,920,800	55,417,824,000	98,960,400,000
Uranus's satellites	169,646,400	47,500,992,000	84,823,200,000
Amount	78,196,916,784	21,894,974,404,480	654,038,348,119,246

In 1853, Rev. William Whewell (1794–1866), Master of Trinity College, Cambridge, launched a major campaign against ETI by publishing anonymously *Of the Plurality of Worlds: An Essay.* As one might expect, both scientists and religious authors rallied around ETI. They won this *battle*, but by 1915 had lost the *war.* Good evidence indicates that what first led Whewell to adopt his anti-ETI position was that he was unsure that ETI could satisfactorily be reconciled with the Christian notion of a divine redeemer (Crowe 1986, chapter 6). Nonetheless, the reasons he offered against ETI were scientific, not religious. Not only does a careful reading of his book support this claim, but Whewell also made it explicit. According to Whewell, Critic Y had chastised him for building "the philosophy of your Essay on a religious basis [and taking] for granted the truths of Revealed Religion, and reason[ing] from them." In response, Whewell stated, "I do not reason in the way which you ascribe to me. I obtain my views of the physical universe from the acknowledged genuine sources: observation and calculation" (Whewell 2001, 486).

In particular, Whewell frequently takes up astronomical information long available but largely neglected. He argues that solar radiation, given the inverse-square law of light propagation, must make the inner planets excessively hot and the outer planets excessively cold (Crowe 2008, 345). (Recall, for example, the passages from John Herschel previously cited.) Whewell adopts Herschel's information, but dismisses Herschel's attempt to save ETI by drawing on divine powers. Having taken the inverse square laws seriously in regard to solar radiation, Whewell introduces a concept that is now central to discussions of life beyond the Earth. Whewell states, "*The Earth's Orbit is the Temperate Zone of the Solar System*" (Whewell 2001, 196, emphasis in original). Thus, Whewell developed the concept of a Habitable Zone, which since the 1950s has come to play a central role in debates about ETI. Similarly, Whewell took Newton's determination that the density of Jupiter is about one-quarter that of Earth to indicate that Jupiter may well be a planet without a solid surface (Crowe 2008, 345–49). Also, Whewell used Herschel's observations of the Magellanic Clouds to suggest that the nebular patches that some claimed to be other universes comparable to our Milky Way are not composed of vast numbers of stars. This and other arguments created serious difficulties for the island-universe theory (Crowe 2008, 344). The breadth of Whewell's analysis is evident from the fact that he used the best available information on the age of the Earth to critique those who maintained that God's efforts would have been wasted had the vast universe not contained plentiful intelligent life. Against this, he maintained that the best geological determinations of the Earth's age were such as to reveal that throughout most of its vast age, no intelligent beings were present. In other words, he argued that though the Earth may occupy an extraordinarily small region of space, we cannot infer from this that intelligent life must be widespread unless we are also willing to accept the idea that God's efforts were wasted because through vast periods of the past, Earth lacked intelligent beings (Crowe 2008, 343–44).

What is crucial to understand is that Whewell argued that the question of the existence of ETI should be determined on scientific, not religious grounds. In fact, Whewell played a key role in *opposing* and *freeing* the ETI debate from the religiously based arguments so prominent in the eighteenth and nineteenth centuries.

In the preface to his book, Whewell correctly predicted, "It will be a curious, but not a very wonderful event, if it should now be deemed as blameable to doubt the existence of inhabitants of the Planets and Stars as, three centuries ago, it was held heretical to teach that doctrine" (Crowe 2008, 335). In fact, his book generated at least twenty other books and more than fifty journal publications – some scientific, some religious, and most a blend of the two. Over two-thirds of these publications opposed Whewell (Crowe 1986, chapter 7).

The most interesting and influential response to Whewell's claims came (gradually) from Richard Anthony Proctor (1837–1888), a prolific British expositor of astronomy. Proctor first won a widespread audience for his *Other Worlds than Ours* (1870), which showed much enthusiasm for ETI, although he had dispatched Jupiterians for Whewellite reasons and labeled life on the Sun as "too *bizarre* [for] consideration" (Crowe 1986, 370). Although mentioning

Darwin only once in his book, he adopted an evolutionary approach, seeing the solar system and the celestial realm in general as evolving. Moreover, around 1870, Proctor and others were becoming skeptical of the island-universe theory. By 1875, Proctor had moved much further in a "Whewellite" direction. In an essay he published that year, he withdrew ETI from most planets of our solar system and also of other systems, but added an evolutionary caveat by claiming "Each planet, according to its dimensions, has a certain length of planetary life, the youth and age of which include the following eras: – a sunlike state; a state like that of Jupiter or Saturn, when much heat but little light is evolved; a condition like that of our earth; and lastly, the stage through which our moon is passing, which may be regarded as planetary decrepitude" (Crowe 2008, 402). Within this perspective, he admits that not only most planets, but also most solar systems lack ETI at any given time. Nonetheless, he saves ETI by arguing that rare as presently inhabited planets may be, there are "millions on millions, nay, the millions of millions of suns which people space, millions have orbs circling round them which are at this present time the abode of living creatures" (Crowe 2008, 403–4).

A very dramatic situation regarding ETI arose around 1877. By then, it appeared that Mars was the last best hope for ETI in our solar system. In that year, the prominent Italian astronomer Giovanni Schiaparelli announced that he had observed "canali" on Mars. Within a few years he was drawing Martian maps showing hundreds of "canals," as these structures were called in the English-speaking world. Some astronomers failed to see them, but others such as Percival Lowell supported Schiaparelli's results and indeed detected dozens more canals and was quicker to label these as features constructed by Martians. For nearly forty years, there was intense debate involving thousands of publications and many prominent astronomers, some accepting Martian ETI, others labeling the canals an optical illusion. The key figures in the latter group were Edward W. Maunder and Eugène Antoniadi, who eventually discredited the Martians by showing that no trustworthy observations (telescopic or spectroscopic) supported their existence (Crowe 1986, 480–546).

It is important to understand the full significance of the recognition that in our solar system, higher life exists only on Earth. This implied that the area around other stars may be similarly desolate.

Did the formulation in 1858 by Charles Darwin and Alfred Russel Wallace of the theory of evolution by natural selection encourage belief in ETI? Looked at historically, the situation is complex. Certainly, evolutionary theory provided a mechanism by which organisms could develop, possibly even attain intelligence. Historically, however, almost no authors writing before the late nineteenth century about extraterrestrials were concerned with lower organisms or their development.

Their concern centered on whether ETI exist. The number of authors who wrote about whether cabbages or caribou exist on Mars or on another planet seems close to zero. Once the Darwin-Wallace theory had gained respectability, such questions could be meaningfully discussed. By giving primacy to developmental questions and a naturalistic approach, evolutionary theory eventually affected the extraterrestrial debate in significant ways, but people seem surprised that many experts on evolution were hesitant to assume the existence of ETI. In fact, in 1904, Alfred Russel Wallace himself spoke out strongly against the probability of ETI precisely on evolutionary grounds. His argument was that the number of chance variations needed to produce beings as complex as humans is so vast that intelligent life elsewhere seems improbable (Crowe 2008, 427–37). In particular, Wallace stated, "the evolutionary improbabilities now urged cannot be considered to be less than perhaps a hundred millions to one; and the total chances against the evolution of man, or an equivalent moral and intellectual being, in any other planet, through the known laws of evolution, will be represented by a hundred millions of millions to one" (Crowe 2008, 435). Moreover, over the last fifty years probably the majority of leading evolutionists have been skeptical of ETI. One thinks of such examples as George Gaylord Simpson, Ernst Mayr, Jared Diamond, and Simon Conway Morris.

Extraterrestrial life debate, 1910–61

Overall, ETI fared rather poorly in the first half of the twentieth century. In biology, Wallace's arguments against ETI were picked up in 1921 by the American paleontologist William Diller Matthew (Matthew 1921). By the 1950s, biologists had become more open to the possibility of ETI, although continuing to share Wallace's view that they would be very different from humans. Examples of such scientists are the anthropologist Loren Eiseley and the geneticist Herman M. J. Muller. In 1957, the former asserted, "Of men elsewhere, and beyond, there will be none forever," and the latter stated in 1961 that ETI "may be expected to have followed radically different courses in regard to many of the features" of their evolution, although adding that they "would certainly be capable of achieving much mutual understanding" with us (Dick 1996, 391–92).

Interest in ETI certainly picked up after 1953, when Stanley Miller, working under the direction of Harold Urey, carried out experiments testing whether biological materials can form from prebiotic materials in a supportive atmosphere comparable to Earth's primitive atmosphere. To this end, Miller mixed methane, ammonia, and hydrogen with water vapor and passed an electric discharge through the mixture. After about a week, several amino acids and other organic compounds were detected. (For more on prebiotic chemistry, see Chapters 7 and 8.)

As noted previously, astronomers gradually recognized the large dispersion in the properties (mass, temperature, atmosphere, etc.) of objects orbiting our Sun. Similarly, in the late nineteenth and the early twentieth century, astronomers realized that the dispersion among stars is no less extraordinary. The recognition of binary stars (around 1800) and the discovery of stellar parallax (1838) showed that stars do not necessarily have identical properties. Nevertheless, the first decade of the twentieth century was the first time that astronomers talked in terms of dwarf and giant stars. This distinction became clearer in 1913 after Henry Norris Russell published his spectrum-luminosity diagram, which showed not only the broad range of spectral types among the stars, but also that stars differ greatly in luminosity. Another step forward came in 1920 when Albert Michelson with his stellar interferometer made the first accurate measurement of the angular diameter of a star (DeVorkin 1984, 103).

Astronomers realized that stars can differ in mass, luminosity, and lifetimes by factors sometimes measured in the millions. It was also evident that the development of intelligent life on a planet takes longer than for lower forms of life, which in turn suggests that searches for ETI should be directed to areas around long-lived stars. There were, however, severe challenges in measuring, for example, even such fundamental factors as the age of the universe and of our solar system. The creation of the expanding universe theory and its Hubble constant (1929) provided a method for estimating the age of the universe, whereas the age of our solar system and of Earth could be determined by geological studies of radioactive elements in rocks. The complexity of the situation is evident from the fact that around 1942, geological studies of radioactive elements in rocks indicated about 3.5 billion years for the age of the Earth, whereas estimates of the age of the universe based on the Hubble constant, were about 2 billion years. This created an obvious problem because it made the age of the universe less than the age of the Earth (Bartusiak 2010, 258).

In the decade before the formulation of the Drake Equation, the situation significantly improved. In 1956, Clair Patterson used radiometric methods to set the age of the Earth as 4.55 billion years, a value close to present estimates (Kragh 1996, 274). In the same year, Allan Sandage and others revised Hubble's constant, which was key to estimating the age of the universe. This set the age of the universe at 5.5 billion years, from which point estimates continued to rise to the present value of 13.8 billion years (Kragh 1996, 273). These figures were very important for discussions of ETI, especially as studies of stars produced values for the lifetimes of the different types of stars.

As Steven Dick has written, "all the problems associated with the possibilities of extraterrestrial intelligences . . . could be leapfrogged if only a method were found for direct communication with such intelligence" (Dick 1998, 200). The development of radio telescope technology provided hope for exactly such a result, which increased interest in the ETI question after 1961; indeed, it was the major factor in the creation of the Drake Equation in 1961.

The Drake Equation was first presented by Frank Drake at a meeting in November 1961 at the National Radio Astronomy Observatory at Green Bank, West Virginia. The Space Science Board of the National Academy of Sciences convened the meeting with ten scientists invited, including Drake, Carl Sagan, Su-Shu Huang, J. C. Lilly, and J. P. T. Pearman. The participants discussed the likelihood of success of a radio telescope search, specifically a search (Project Ozma) carried out by Drake. Two curious aspects of this event may be noted. First, it appears that there is no Drake Equation paper. There are papers from the 1961 period by Drake, Sagan, Pearman, and others that discuss the Drake Equation, but no paper by Drake designed to present it. This is not surprising because Drake created his equation as a way of organizing this meeting. The second curious feature is that it is unclear whether the group designed the Drake Equation to estimate (1) the number of communicative civilizations existing in 1961 or (2) the number at some generalized time in the future. The point of this distinction is that, given the relatively long period required for intelligent life to develop, the value of f_i must be heavily time dependent. We know that on Earth, intelligent life took over three billion years to develop. It appears that the scientists at Green Bank were looking for an estimate of the number of communicative civilizations existing in 1961. This is indicated by Pearman's statement in his conclusion that the members sought an estimate that specifies "the *present number* of communicative societies in our galaxy" (Pearman 1963, 292, emphasis added). The second point is significant particularly in regard to the value for the f_i factor in the equation. Perhaps the closest publication to what might be described as a Drake Equation paper is the account of the meeting written by Pearman for a collection of papers titled *Interstellar Communication* (Cameron 1963, 287–93).

Pearman's account of the f_i factor reveals a surprising result. The value the group adopted for f_i was 1. This implies they had concluded that essentially all planets that had attained life must also have ETI. David Darling has explained this as due to the presence of J. C. Lilly at the meeting. Lilly, who had just published a book on dolphins and maintained that dolphins have attained intelligence, convinced the group that intelligence had originated twice on Earth, which made those present comfortable adopting the highest possible value for f_i (Darling 2014). In fact, the group labeled itself the "Order of the Dolphin." To this historian, such a high value, indeed a 100 percent value, seems unjustified.

One other author in the Cameron volume deserves attention in relation to the f_i factor. This is astrophysicist Su-Shu Huang, who contributed five papers. In these papers, Huang gives extensive attention to the issue of the type of stars that can support higher forms of life. Present-day discussions of the possibility of ETI stress, for example, that giant and super-giant stars have lifetimes too short for intelligent life to develop in their vicinity, whereas stars more like our Sun are far better candidates for supporting higher forms of life. It is interesting that a concept central to Huang's analysis is "habitable zone," the zone around a star that permits the development of life. Wikipedia attributes creation of this concept, which is now extensively used in discussions of ETI, to either Huang or to Hubertus Strughold in his *The Green and Red Planet: A Physiological Study of the Possibility of Life on Mars* (1953). In the article, the claim is made that Strughold came up with the idea, but that Huang in 1959 was the first to use the term "habitable zone" ("Circumstellar Habitable Zone" 2014). In fact, William Whewell (as is shown earlier in this essay) had introduced this concept fully a century earlier, in 1853, in his *Of the Plurality of Worlds: An Essay*. Whewell's term was "Temperate Zone" (Whewell 2001, 196).

References

Bartusiak, Marcia. 2010. *The Day We Found the Universe*. New York: Vintage.

Cameron, A. G. W. (ed.) 1963. *Interstellar Communication: A Collection of Reprints and Original Contributions*. New York: W. A. Benjamin.

"Circumstellar Habitable Zone." 2014. Available at http://en.wikipedia.org/wiki/Circumstellar_habitable_zone/.

Crowe, Michael J. 1986. *The Extraterrestrial Life Debate 1750–1900: The Idea of a Plurality of Worlds from Kant to Lowell*. Cambridge University Press. Repr. as *The Extraterrestrial Life Debate 1750–1900*. Mineola: Dover, 1999.

Crowe, Michael J. (ed.) 2008. *The Extraterrestrial Life Debate, Antiquity to 1900: A Source Book*. University of Notre Dame Press.

Crowe, Michael J. 2011. "The Surprising History of Claims for Life on the Sun." *Journal of Astronomical History and Heritage* 14(3): 169–79.

Crowe, Michael J., and Matthew F. Dowd. 2013. "The Extraterrestrial Life Debate from Antiquity to 1900." In *Astrobiology, History, and Society: Life Beyond Earth and the Impact of Discovery*, ed. Douglas A. Vakoch, 3–56. Heidelberg: Springer.

Darling, David. 2014. "Drake Equation." Available at www.daviddarling.info/encyclopedia/D/DrakeEq.html.

DeVorkin, David. 1984. "Stellar Evolution and the Origin of the Hertzsprung-Russell Diagram." In *Astrophysics and Twentieth-century Astronomy to 1950*, ed. Owen Gingerich, 90–108. Cambridge University Press.

Dick, Steven J. 1982. *Plurality of Worlds: The Origins of the Extraterrestrial Life Debate from Democritus to Kant*. Cambridge University Press.

Dick, Steven J. 1996. *The Biological Universe: The Twentieth-Century Extraterrestrial Life Debate and the Limits of Science*. Cambridge University Press.

Dick, Steven J. 1998. *Life on Other Worlds: The Twentieth-Century Extraterrestrial Life Debate*. Cambridge University Press.

Guthke, Karl S. 1990. *The Last Frontier: Imagining Other Worlds from the Copernican Revolution to Modern Science Fiction*. Translated by Helen Atkins. Ithaca, NY: Cornell University Press.

Herschel, John. 1833. *Treatise on Astronomy*. London: Longman.

Jaki, Stanley. 1978. *Planets and Planetarians*. Edinburgh: Scottish Academic Press.

Kragh, Helge. 1996. *Cosmology and Controversy: The Historical Development of Two Theories of the Universe*. Princeton University Press.

Mach, Ernst. n.d. *The Principles of Physical Optics: An Historical and Philosophical Treatment*. New York: Dover.

Matthew, William Diller. 1921. "Life in Other Worlds." *Science* 54: 239–41.

Pearman, J. P. T. 1963. "Extraterrestrial Intelligent Life and Interstellar Communication: An Informal Discussion." In *Cameron* 1963: 287–93.

Tipler, Frank. 1981. "A Brief History of the Extraterrestrial Intelligence Concept." *Quarterly Journal of the Royal Astronomical Society* 22: 133–45.

Whewell, William. 1859. *Of the Plurality of Worlds: An Essay*, 5th ed. London: John W. Parker and Son.

Whewell, William. 2001. *A Dialogue on the Plurality of Worlds*. In *Of the Plurality of Worlds,* facsimile edition, ed. Michael Ruse. University of Chicago Press.

Notes

1 I am very indebted to my colleague, Dr. Matthew F. Dowd, for insightful comments on the drafts of this paper.

2 Many references in this paper are to writings by two authors, Steven Dick and myself, who have written extensively on the history of the extraterrestrial life debate. These references allow the reader to locate additional material on the matter referenced. See also Dick 1982; Guthke 1990; Jaki 1978; and Tipler 1981.

10

Fraction of life-bearing planets on which intelligent life emerges, f_i, 1961 to the present

Lori Marino

Abstract

The question of how intelligence evolves on different planets is a central factor in the Drake Equation and informs the fields of bioastronomy, astrobiology, and SETI (the search for extraterrestrial intelligence). In this chapter, I trace the history of our conceptions of intelligence through changes and growth in our understanding of brain evolution, genetics, and animal behavior, and present a modern view of intelligence that places human intelligence in an evolutionary context and linked to the multiple intelligences inhabiting this planet. Much of our current understanding of intelligence as an astrobiological question and, specifically, the nature and much-vaunted uniqueness of human intelligence, should be updated by modern knowledge and divested of outdated ideas such as the *scala naturae*, progressive evolution and teleology, and the anthropic principle. These notions continue to fuel a fundamental misconception of intelligence as a uniquely human phenomenon with little or no evolutionary or comparative context and, therefore, no way to understand its true biological nature. In this chapter, I will discuss these issues in detail and replace these outmoded notions with new information and insights about how and why intelligence evolves and the levels and distribution of intelligence across species on this planet. Modern understanding of intelligence shows that it is continuous across all animal life on Earth and that the human brain is embedded in the evolutionary web of primate brain evolution and contains the hallmarks of nervous-system evolution traced back to the first life forms on this planet. These updated ideas provide a biological context for understanding the mechanisms and range of intelligence on this planet and should therefore serve the critical purpose of revising notions of f_i, leading to more productive outcomes for the study of the evolution of intelligence on this and other planets.

The Drake Equation, ed. Douglas A. Vakoch and Matthew Dowd. Published by Cambridge University Press.

Starting at the endpoint: The beginnings of SETI

Radio waves were known to exist and could be detected in the late 1800s, and by the early 1900s, physicists began to think about whether they could be observed from astronomical sources. The first such discovery, radio waves from the galactic center, was made by Karl Jansky in the 1930s, and with that came the development of the radio telescope and the first realistic hope of searching for extraterrestrial intelligence (ETI) empirically (although no one was yet doing this).

But by the 1950s, physicists, astronomers, and many others were pondering the use of radio technology for such a search. In a watershed paper in *Nature* in 1959, physicists Giuseppe Cocconi and Philip Morrison argued that radio telescopes had become sensitive enough to detect any transmissions that might be deliberately broadcast into space by civilizations orbiting other stars. Such messages, they specified, might be transmitted at a wavelength of 21 centimeters (or 1,420.4 megahertz). This is the wavelength of radio emission by neutral hydrogen, the most common element in the universe, and they reasoned that other intelligences might see this as a logical starting point for communications in the radio spectrum (Cocconi and Morrison 1959).

The search for Princess Ozma

Seven months after Cocconi and Morrison's paper, radio astronomer Frank Drake, who was already thinking about using radio for detecting ETI, became the first person in history to start a serious data-based search for signals from intelligent extraterrestrials. Drake used the 25 meter telescope at the National Radio Astronomy Observatory (NRAO) in Green Bank, West Virginia, to listen to two Sun-like stars near ours. Drake named his search Project Ozma after the princess of the land of Oz, which was, according to L. Frank Baum, reachable by radio. Drake followed Cocconi and Morrison's suggestion by systematically scanning frequencies close to the 21 centimeter wavelength for six hours per day from April to July 1960. There was little pretense that we would be able to decipher an alien radio message if found, but that didn't matter. If we could determine that a radio signal was artificial (by ruling out all possible natural sources) then we would have the answer to the question: Are there other intelligences out there?

Obviously, the chances of finding a signal on the first attempt were spectacularly remote, but Drake and others realized the potential of radio telescope technology to make the search for ETI an empirical endeavor, fitting squarely within the domain of observational science. No longer was discourse about ETI limited to mere speculation; now it could be probed systematically as any other scientific endeavor. SETI (the search for extraterrestrial intelligence) was born and came to refer to the search for artificial electromagnetic signals from extraterrestrial civilizations.

The Order of the Dolphin

One year later, in November 1961, Drake created and presented the famous Drake Equation at an historic meeting held at the NRAO. The purpose of the meeting, convened by the National Academy of Sciences Space Science Board, was to quantify whether SETI had any reasonable chance of successfully detecting civilizations around other stars. Drake developed the famous equation as a way to organize thinking and discussion at the meeting about the possible number of intelligent extraterrestrial civilizations and not as a way to make definitive calculations. Organization was most probably needed because the invitees were a small group of brilliant men who had deeply thought about these issues and were eager to share their ideas. They were organizer Peter Pearman of the National Academy of Sciences, Frank Drake, physicist Philip Morrison, businessman and radio amateur Dana Atchley, chemist Melvin Calvin, astronomer Su-Shu Huang, neuroscientist John C. Lilly, inventor Barney Oliver, astronomer Carl Sagan (who was a young postdoc at the time), and radio-astronomer Otto Struve (Dick 1998).

The first few variables in the Drake Equation focused on probabilities in research areas in which the conference participants were engaged. That is, they dealt with star formation, planet formation, habitable zones, and the emergence of life. The probability of life on other planets had been given support from pioneering experiments designed by Stanley Miller and Harold Urey (Miller 1953; Miller and Urey 1959) and the work of Aleksandr I. Oparin (Oparin 1924). The only chemist in the group, Nobel laureate Melvin Calvin, was enthusiastic about the relevance of such work for the probabilities attached to the "life term," f_l, in the Drake equation. Other participants, particularly Sagan, who worked on origin-of-life experiments, were conversant in the science of biology, and thus the group was able to come to a reasonable (at the time) estimate for the term.

But the next term, the probability of life becoming intelligent, or f_i, clearly challenged the group, which, despite its formidable expertise in areas relevant to the early terms in the Drake Equation, was dealing with a topic that only one member – John C. Lilly – was studying scientifically. Although Lilly's later work was controversial both scientifically and ethically, at the time he knew more than anyone in the world about a complex nonhuman brain on this planet – that of the dolphin – and he reported on his studies, suggesting that "intelligence" had evolved twice on this planet. Lilly's accounts so impressed the group that they dubbed themselves "The Order of the Dolphin." Dolphins were viewed as the "other intelligence" because of their obvious smarts and elaborated brains, which possess functional parallels with our own; the relationship between dolphins and SETI was to become a long one. Conference presentations and discussions about complexity, the role of convergence in the evolution of intelligence, and the analysis of dolphin signal repertoires as pilot studies for deciphering ETI radio signals notwithstanding, the real importance of dolphin intelligence for SETI has not yet been fully realized because we have not yet examined the dolphin brain and intelligence in the fuller comparative context of brains and intelligence on Earth to inform thinking about ETI. Suffice it to say, many astronomers and other scientists found dolphins interesting and perhaps somewhat relevant to SETI, but, generally, most could not overcome the concern that dolphins do not have manual technology and are never going to be able to build radio telescopes no matter how smart they are. So, although dolphin research and SETI have always been closely linked, dolphins were not considered serious candidates for the kind of intelligence SETI was seeking because they are a "non-technological" intelligence. It has been argued that the inclusion of John Lilly in the Green Bank discussions was one factor in the group's optimistic conclusion that close to 100 percent of planets with life also have intelligent life (Darling 2014), and while such a high value for this term seems unjustified to some (see Chapter 9) the Green Bank group came closer, arguably, to a scientifically valid approach to estimating f_i than even they might have realized. Paradoxically, despite the group's enthusiasm for the probability of ETI, SETI necessarily depended on an anthropocentric approach. But this focus on the search for a specific outcome – radio-telescope-using civilizations – became a mindset that unnecessarily infiltrated the broader domain of theorization about ETI. Even today, many, if not most, scientists embrace this mindset and view the question of ETI as one referring to our own kind of intelligence. Basing their estimates for f_i on the highly specific historical contingencies required to produce human intelligence, they view complex ETI as exceedingly unlikely (Burger 2002; Conway Morris 2003; Ward and Brownlee 2000). Others point to the range of nonhuman intelligence on Earth and other indicators as the basis for less anthropocentric estimates (Schwartzman and Rickard 1988), or make more optimistic estimates based on the sheer number of possibilities in the universe.

The "unique human intelligence" premise

Frank Drake's early radio searches and the Green Bank meeting were immense steps forward in developing a roadmap for answering the question of whether there are intelligences on other planets and, even more importantly, ushered in the scientific age of SETI. But, there was an unintended consequence of the Green Bank meeting that became, in this scientist's view, an impediment to thinking about ETI in a broader sense. SETI focuses upon the search for "human-like intelligence" because – although we are not the only technological species on the planet – we are the only one capable of sending electromagnetic signals into space. Thus, SETI is anthropocentric for purely pragmatic reasons. But at some point this workable approach became something else: a hardcore assumption that even Earthly intelligence is restricted to our own species. That is, the practicalities of SETI were transmogrified into the belief that human intelligence is entirely unique, thus biasing our thinking about intelligence on other planets.

Human intelligence is but one outcome within an interdependent network of intelligence that necessarily includes all of the other life forms on Earth. Put simply, one cannot understand human intelligence (or the chances of a radio telescope–using civilization of humanoids emerging) without understanding it within an evolutionary framework. The idea that there is one intelligence on this planet is a scientific impossibility, as human intelligence did not arise *de novo* but is the product of billions of years of evolution starting with the first organisms on the planet. Beyond radio searches, the real question is not who can send an electromagnetic signal, it's how does intelligence evolve in general? The answer to this question is grounded in evolutionary biology and should therefore be addressed by astrobiology.

The astrobiology paradigm

Astrobiology is one of several terms used to refer to the study of life throughout the universe; others are xenobiology, exobiology, and bioastronomy (the last of which was coined in 1982 by Michael D. Papagiannis of Boston University). Xenobiology is typically reserved for the study of life forms based on a chemistry and genetic code different from life on Earth. Exobiology and astrobiology came to be the more familiar terms for the study of the origin, evolution, distribution, and future of life in the universe, both extraterrestrial and on Earth. They are multidisciplinary fields that encompass a broad spectrum of studies ranging from habitable zones, prebiotic chemistry, and the origins and evolution of life, including the ability of life to adapt to various challenges on Earth and in space. More to the point, astrobiology deals with the question of whether life exists beyond Earth and how we can detect it if it does.

At the time that radio telescopes were searching for extraterrestrial technological intelligence, people were thinking of other approaches as the possibility of sending experiments to detect life in space and other planets became feasible. In 1960, Joshua Lederberg, Nobel laureate and one of the leaders in the field, published a seminal paper in *Science* titled "Exobiology: Approaches to Life Beyond Earth" (Lederberg 1960). In doing so, he laid the groundwork for NASA's first exobiology program established in 1960. Over the years, the program grew into the Exobiology and Evolutionary Biology Program and expanded to include studies of evolutionary biology, the origin and evolution of prebiotic elements and compounds in the universe, the search for extrasolar planets, and the future of life in the universe. One of the best-known efforts was NASA's Viking missions to Mars, launched in 1976, which included three exobiology experiments designed to look for evidence of life on that planet. Although none was found, the mission was "proof of concept" that exobiology was a genuine field of scientific experimentation. And, by the mid-1990s, the Exobiology Program had evolved into a thriving multidisciplinary field of science and astrobiology, and had become one of NASA's signature programs in the Planetary Science Division.

Since then there have been numerous lines of research addressing astrobiological questions. These include searching for extrasolar planets and looking for biomarkers in the atmospheres of those planets, studying extremophiles on Earth, which broaden our understanding of the range of conditions in which extraterrestrial life could exist, and searching for liquid water in our solar system, among numerous other endeavors. The scientific community organizes biennial astrobiology conferences (called "AbSciCon") to report on the latest research findings and discuss the next wave of studies. Today, astrobiology is a well-established field of scientific study and attracts enormous attention from the public, which is naturally interested in the prospect of extraterrestrial life.

Astrobiology necessarily encompasses the study of biological evolution and how life responds to various environments. The basic astrobiological paradigm is elegant: see how life arose and evolved on this planet and apply that knowledge to detecting and understanding extraterrestrial life. This "earth as a natural laboratory" paradigm is a logical starting point for asking complex questions about extraterrestrial life. Therefore, the study of the evolution of intelligence would seemingly have found a natural "home" in that field. From a sensible and very straightforward point of view, one should study the evolution of complexity and intelligence on this planet to inform our estimates of the range of possibilities on other planets. But in its fifty-year history, there has been little work within astrobiology on intelligence. Despite intense public and scientific interest, the study of biological evolution in astrobiology has been largely limited to studying biochemistry and single-celled organisms and the connection to more complex intelligences has not been made. As Linda Billings describes:

> One aspect of the study of life in the universe that scientists have not yet found a way to adequately explain to non-expert audiences is the vast knowledge void that remains to be filled between scientific understanding of the emergence of life and the emergence of intelligence in life. Non-experts have far less trouble than scientists do in condensing and simplifying the immense "spaces" of time and complexity that lead to prebiotic chemistry and then to life, to molecules and then to cells, and to microbial life and then to intelligent life. (Billings 2011)

SETI and astrobiology have, historically, originated from two opposite points on a spectrum articulated by the Drake Equation. SETI explicitly relies for success upon the existence of intelligent technological life on other planets. Astrobiology has focused on the earlier terms in the Drake Equation, such as the distribution of habitable zones, prebiotic chemistry, and the origin and early evolution of life. Ironically, few if any people saw the direct relevance of astrobiology (focused on early, simpler life forms) to SETI (the search for very complex technical intelligence). The reasons for this may have much to do with the complexity of intelligence but also the presence of ancient assumptions about intelligence that hinder bringing the study of ETI to fruition.

The landscape of intelligence on Earth

Intelligence is, by its nature, a fuzzy concept. That is, there are no strict boundaries on it and there is no scientific consensus on its definition. The study of intelligence, therefore, is complex and multidisciplinary, and the absence of an exact definition necessitates a strong allegiance to empirical description of a range of phenomena rather than a hunt for a precise exemplar. Whereas it might be somewhat more straightforward to determine if something is or is not a prebiotic element or even if something is alive or not (although this too can be complex) intelligence is not an all-or-nothing phenomenon. Rather, it is a multidimensional phenomenon, which expresses itself in varying phenotypes and levels of complexity.

By intelligence, we are referring to something akin to a level of cognitive complexity: in other words, how an individual acquires, processes, stores, analyzes, and acts upon information and circumstances (Shettleworth 2010). This bare-bones definition will suffice as a "place-holder" for this complex adaptation. But, despite its complexities, intelligence is amenable to scientific investigation in the same way any biological adaptation is and, therefore, should be treated no differently from any of the other scientifically accessible terms in the Drake Equation. The fact that informed study of ETI has lagged behind most of the other terms in the equation is, in part, due to the influence of an ancient model of nature called the *scala naturae*.

Scala naturae

Despite all evidence to the contrary, notions of intelligence on this planet are still largely entrenched in an Aristotelian model of nature from the third century BC (but which probably goes back much further). This notion, known variously as the *scala naturae*, Great Chain of Being, or ladder of progress, continues to influence conceptualizations of intelligence even in the post-Darwinian era.

According to this model, nature is arranged on a linear scale of progression with inorganic substances and objects, such as rocks, depicted on the lowest rung, then plants, then up from "lower" animals (invertebrates), to vertebrates to "higher" mammals such as primates, and, finally, humans, who occupy a separate, superior "step" of the ladder above all other animals. Humans are considered the most "advanced" life form on the planet, possessing unique biological and, in some religious traditions, metaphysical (or spiritual) characteristics (Marino 2007). It is easy to see how this notion, flattering though it may be for our species, is fraught with premises that lead to misconceptions about intelligence.

Notions of ETI in the post-Darwinian era

Charles Darwin established that all species of life have descended over time from common ancestors and, in 1859 in *The Origin of Species*, introduced (along with his contemporary Alfred Russel Wallace) a scientific theory on the major mechanism that drives biological evolution: natural selection (Darwin 1859). In 1871, in *The Descent of Man*, Darwin made the logical argument that humans had descended with other animals from a common ancestor and that we were part of, and not separate from, the animal kingdom (Darwin 1871). By the 1870s, the scientific community (but only a subset of the general public) had accepted evolution as a fact (though accepting humans as part of that process remains a struggle for some people to this day). It was not until the emergence of the modern evolutionary synthesis from the 1930s to the 1950s, however, that a consensus developed in which natural selection was the basic (but not exclusive) mechanism of evolution. In modernized form, Darwin's work unified the field of biology and provided a fertile and substantive foundation for theorizing about ETI.

From the big bang to the big brain?

Although Darwin and the modern synthesis laid the groundwork for immunity against notions of human superiority and separation from the rest of nature, *scala naturae* notions persist and shape misconceptions about intelligence to this day. Interestingly, even Darwin's establishment of a branching (nonlinear) pattern of species relationships over time (called the phylogenetic tree) is still referred to as a "phylogenetic or phyletic scale," preserving the linear notions of *scala naturae*.

One of the conceptual errors engendered by this thinking is known as *sui generis*, meaning "one of its own kind" in Latin and referring to the idea that humans are qualitatively different from other forms of life on Earth, particularly as it relates to the phenomenon of mind. The implication of this separation of humans from the rest of the animal kingdom is a fundamentally mistaken assumption that human intelligence is the only kind of intelligence we are (or should be) searching for in astrobiology. This error leads to anthropocentric notions (typically expressed by the anthropic principle) that are directly antithetical to the astrobiology paradigm. The historical result has been an unending argument about contingency versus convergence between those who view ETI as highly improbable and those who do not. As far back as 1904, Alfred Russell Wallace expressed strong reservations about ETI on the basis that the number of steps it would take to create an intelligence as complex as that of a human is too improbable (Crowe 2008, 429–36). And over the last several decades, many leading scientists, such as Gaylord Simpson, Ernst Mayr, and Simon Conway Morris, have vociferously argued that the emergence of a human-like intelligence is based on a highly improbable set of events (contingencies) that cannot be repeated elsewhere (see Chapter 9). But these interminable debates, arguably, miss the point by centering on an anthropocentric notion of intelligence, leaving little, if any, room for considering the relevance of the other animals on the "steps of the *scala naturae*," that is, the myriad of other intelligences with which we share the planet.

Scala naturae thinking is also closely related to teleological assumptions about intelligence. Teleology is the appeal to a goal-directed, purposeful progression toward an end point as a means of explaining phenomena. Teleological thinking is quite compatible with the *scala naturae* and, from the point of view of intelligence, would suggest that all life on Earth was moving toward the eventual emergence of human intelligence. It is entirely understandable that one would propose this after examining average relative brain size (a biological proxy for intelligence) across species over the last 200 million years. The result is a graph depicting an upwardly sloping line as time moves from early Earth to the present, which seems to make the case for a progressive increase in relative brain size and intelligence over time.

But this is an illusion in the fossil record. When understood properly, this graph represents the fact that brains simply need time to get large and, thus, larger brains are found in more recent species. When proper phylogenetic analyses are conducted, there is no directional trend. There is no evidence for any progressive linear trends in biological evolution that lead to humans. Life on this planet resembles a branching tree, not a ladder. There is no "lower," no "higher," no "advanced," and no "primitive" among extant species. Even human evolution itself is anything but linear. Human evolution shows the same branching pattern of speciation and extinction found in all animal groups on Earth. The fact that we are the only hominid species on the planet should not fool us into thinking that we are a product of a directional progressive process. To summarize, teleological and anthropocentric notions about intelligence on Earth, or in ETI for that matter, are not supported by any scientific evidence.

Major milestones in the evolution of intelligence on Earth

Questions about the origin and evolution of intelligence on Earth are more tractable than ever before because of our sophisticated methods of collecting, storing, and analyzing large quantities of data. Thus, by setting old ideas and assumptions aside, we can empirically explore the evolution of intelligence on Earth. Although it would be impossible to document each step in one chapter, we can ask these questions: What are the major milestones and characteristics of intelligence on this planet, and what do these milestones tell us about the nature of intelligence that may be relevant to ETI?

Milestone 1: Brainlike functions in unicellular organisms

Before beginning, it is important to acknowledge a caveat. If one uses extant species as "stand-ins" for ancestral species, employing the comparative phylogenetic method, one must be aware of the limitations of this approach. But this kind of analysis has allowed us – when based on genetic, fossil, and other kinds of information – to reconstruct the traits of extinct species with some degree of certainty. Therefore, although modern-day bacteria, for instance, are not the same as the unicellular organisms that comprised early life on Earth, they are, roughly, "working" surrogates for early life.

All organisms need to find food and avoid hazards in a changing environment. Brains exist because of the varied distribution of resources in the environment that affect survival. An immobile organism or one living in a predictable, stable environment would not have much need for a brain. But if one moves around, then there must be a way to detect and analyze all of the variable information in the environment and make adaptive behavioral responses to it. Some of the most basic features of nervous systems and brains are found in single-celled organisms, such as bacteria. They sense their environment, briefly store the information, integrate it in different channels of input and then act upon it, producing basic adaptive behavior. These brain-like processes work at a molecular level, but they reveal a common ancestry with all modern brains.

One example of a bacteria with brain-like capacities is *Escherichia coli,* which senses its environment through a dozen different types of protein receptors embedded in its cell wall that allow chemicals outside the cell to communicate with mechanisms inside. Each receptor specializes in a specific kind of information, such as toxins, sugars, amino acids, and so forth. The input is integrated and the result is the "decision" to behave in a certain way, for example, to use its flagellar motors to swim toward or away from the environmental input (Allman 1999, 6–8). The fundamental functional features of brains, such as sensory integration, memory, decision making, and behavior, are found in these single-celled organisms and was likely present in the earliest organisms on Earth.

The mechanism by which unicellular organisms detect their environment is known as "membrane excitability" and involves sensitivity to electrochemical gradients typically accomplished through the flow of ions into and out of the cell. This basic mechanism was built upon and adapted to specialized cells in metazoans. These specialized cells with sensitive or excitable membranes became the first neurons. Evidence for continuity derives from the fact that the membranes of modern neurons operate by the same electrochemical principles (voltage-gated ion movement across a membrane) as in single-celled organisms (Allman 1999, 2–12).

The fact that all mobile unicellular organisms possess the fundamental characteristics of nervous systems suggests that the two critical factors for the early emergence of intelligence on any planet are environmental variability and mobility. With environmental variation, a way to sense and act upon that environment to optimize one's survival will inevitably evolve. Thus, environmental variation is key to the emergence of nervous system functions and the potential for complex intelligence.

Milestone 2: Multicellularity and neurons

Nervous system–like functions might have remained at the level of complexity found in unicellular organisms had it not been for the evolution of metazoans, that is, multicellular animals with functionally specialized cells. (And the emergence of metazoans was related to increased oxygen levels in the atmosphere). In metazoans, the excitable membrane of single-celled organisms was co-opted for the evolution of cells known as neurons that are specialized for information processing. The neuron is the basic unit of information processing in all nervous systems on Earth and has remained, fundamentally, the same for the past 600 million years with only some smaller modifications along the way, such as myelinization to increase efficiency.

A neuron is an electrochemically polarized cell with a characteristic architecture consisting of a body with branching processes and dendrites containing chemical receptors. When an electrochemical threshold is reached, a long axon transmits an electrical pulse to terminal branches, which then release chemicals that are picked up by the excitable membranes of the dendrites of another neuron, and so forth. This highly simplified description reveals some important steps in the evolution of nervous systems. First, neurons form fiber tracts, which allow information to travel along functional pathways. In the earliest animals with nervous systems, such as jellyfish, transmission is bidirectional. In bilaterians, neurons are unidirectional, so information flows only one way (from the dendrites to the terminal branches), thus increasing the specificity of neural transmission. Second, although the dendrites sum and integrate graded potentials initiated by the input of chemicals (called neurotransmitters), a neural pulse down an axon, called an action potential, is an all-or-nothing electrical phenomenon. Thus, the earliest neurons provided a way for a digital set of signals to travel along specific paths, creating neural circuits.

The first organisms to possess neurons were the earliest metazoans – those with radially symmetric bodies such as cnidarians, ctenophores, hydrozoa, and so on – that appeared in the seas approximately 600 million years ago. They possess a decentralized nerve net (a loose arrangement of sensory neurons connected to motor neurons by interneurons). In many of these animals, there are the very beginnings of the next step in the evolution of nervous systems, centralization, in the form of nerve rings and clusters of neurons called ganglia, which allow these animals to exhibit complex behaviors requiring coordination of parts of the body. Despite possessing no brain, many animals with nerve nets evince simple learning capacities, such as habituation, and can navigate complex environments.

The shared characteristics of these early nervous systems and modern brains are striking. Voltage-gated sodium channels are present in jellyfish and all modern brains. The neurotransmitters coursing through the brains of all modern species have precursors in single-celled organisms and were present in the nerve nets of the earliest metazoans. This is not to say that there are not some exceptions that prove the rule. These include sponges and slime molds, which do not have the characteristic nervous systems with neurons but possess cells capable of communicating with one another to produce behavior (Jacobs et al. 2007; Sakarya et al. 2007). Slime molds process information and learn and anticipate temporal events (Saigusa et al. 2008). But they did not become a template for more complex intelligence. Bilaterians did.

Milestone 3: Bilateralization, centralization, and cephalization

Bilaterians, that is, bilaterally symmetrical animals with a hollow tube running from mouth to anus, emerged in the fossil record about 550–600 million years ago. Early bilaterians were wormlike creatures with a nerve cord running down the body and an enlarged ganglion at the front or head region – an early brain (or central nervous system). The process of enlarging the ganglionic cluster at the head region along with the sensory organs is called cephalization. An anterior brain connected to a nerve cord became the bauplan of the nervous system for all organisms with a central nervous system (Arendt et al. 2008; Striedter 2005).

The conservative plan of all central nervous systems runs deep. All vertebrate brains are segmented into four parts: telencephalon (part of the forebrain, including the neocortex), diencephalon (also forebrain), mesencephalon (midbrain), and rhombencephalon (hindbrain). And all vertebrates possess the same brain structures and functions within those segments. For instance, the hindbrain structure known as the locus coeruleus is involved in responses to stress across the animal kingdom, and also uses the same neurotransmitter in all major vertebrate groups (Adrio, Anadón, and Rodríguez-Morales 2002). And all mammals possess a layered neocortex (Allman 1999, 28), although the way the neocortex is organized or mapped in some species, such as cetaceans, is quite different in many respects from other species. The neocortex in all mammals, including cetaceans, is the most evolutionarily recent and pliable part of the brain, and thus, arguably, the vanguard of intelligence – in other words, the source of self-awareness, problem solving, sensory-motor integration, mental representation, and so forth. And despite differences in neocortical architecture, even cetaceans and primates display a striking degree of convergence in cognitive function (Marino 2002). Likewise, while there are differences, invertebrate and nonmammalian vertebrate brains rely upon the same basic principles of information processing as mammalian brains.

All of the major vertebrate neurotransmitters, which are ultimately derived from single amino acids, are also found in invertebrate brains (Messenger 1996), and most of them also served as signaling molecules before the first central nervous system evolved (Turlejski 1996). And embryonic development of nervous systems across species also reveals a highly conserved plan (although one obviously modified to create species differences) (Striedter 2005).

Advances in our understanding of the genetic mechanisms underlying nervous system formation reveal a striking level of conservation across all species. For instance, homeotic genes that segment insect bodies are used to segment parts of the vertebrate central nervous system. Other genes controlling the formation of part of the brain in fruit flies is replicated and regulates the newest part of the mammal brain, the neocortex. So even the human brain is organized by genes with antecedents going back half a billion years (Allman 1999, 57). The same basic genetic mechanisms are applied in different ways to create the gut of a fruit fly and the brain of a human. This is not to say that a fruit fly is the same as a human, or vice versa, but rather that ancient mechanisms underlie many of the most complex features of intelligence on earth.

Brain size issues

One issue that has received substantial attention in discourse about intelligence on this planet and ETI is brain size. Brain size has to do with the amount of neural tissue available for information processing, under the assumption that "more is better." Measures of brain size range from whole brain mass to the mass of specific structures in the brain to various ratios of neuron types. One measure of brain size or cephalization is the encephalization quotient (EQ), which is a measure of average brain size relative to body size taking into account brain–body allometric relations. EQ is calculated by linearly regressing average brain size on average body size across species. So the EQ metric allows one to compare directly the relative brain size of a squirrel and an elephant. Species with average brain size have an EQ of 1, falling right on the line of best fit. Modern humans have an EQ of 7 – that is, our brains are seven times larger than one would expect for an animal of our body size. Many cetaceans have EQs in the 4–5 range and great apes have brains 2.5–3 times larger than expected. We place a tremendous amount of emphasis on our high EQ as a way to explain our complex intelligence. We like to think that our large relative brain size and other features, such as our highly convoluted neocortex (wrinkled grey matter), have allowed our species to achieve a qualitatively different intelligence from other mammals. But brain size is a continuous variable, and the human brain, though large, is a product of changes in brain anatomy that are well predicted by scaling expectations for any nonhuman anthropoid primate (Sherwood et al. 2006); our degree of cortical gyrification is exactly as expected for a primate of our brain size (Zilles et al. 1989). Other aspects of our brain, such as relative size of our frontal lobes and the presence of specialized cells for information processing, are all either shared with other species or are just as expected for a primate of our body size. In summary, decades of neurobiological studies have failed to turn up a single property of the human brain that is qualitatively different from that of other species, that is, that is not explainable within a common framework of comparative evolution with the rest of the life forms on this planet.

Continuity of mind

The history of research on animal behavior, comparative psychology, and ethology is a history of decentration, that is, finding more and more shared cognitive characteristics across animals and fewer unique ones. The more we learn about other animals the more we come to realize that even "high-level" cognitive abilities, such as foreplanning, basic mathematics, self-recognition, and even the possession of a technological culture, are not unique to humans. It seems that there are some fundamental capacities that nervous systems, and in particular central nervous systems, allow that are omnipresent in nature. Basic cognitive processes, such as memory and learning, are found in all animals. While an exhaustive review of this rich literature is beyond the scope of this paper, there is abundant and ever-growing evidence for *continuity of mind* across all species on this planet.

Being brainy and boneless: Intelligence in invertebrates

The capacity to learn has been confirmed in many bilaterian invertebrate groups including the roundworms (Rankin 2004), molluscs (Kandel 2001), annelid worms (Friesen and Kristan 2008), arthropods (Morse 2000), and even the echinoderms that lack a central nervous system (Shulgina 2006). Moreover, Eric Kandel and his colleagues showed that the basic neural mechanisms of learning are the same "across the board" from molluscs to vertebrate brains (Kandel 2009).

Special attention has been given to the cephalopods (phylum Mollusca), particularly the octopus *Octopus vulgaris*, because of their complex cognitive and behavioral characteristics. Although octopi brains bear molluscan features, such as numerous interconnected ganglia dispersed throughout the body, they also have a highly elaborated three-lobed architecture, which gives these creatures a vertebrate-like intelligence (Hochner, Shomrat, and Fiorito 2006). There is evidence for observational learning in octopi (Fiorito and Scotto 1992), which is a high-level cognitive capacity involving learning by watching others. Perhaps most germane to the issue of ETI is the fact that octopi are flexible and creative tool users. Veined octopi (*Amphioctopus marginatus*) retrieve discarded coconut shells, manipulate them, transport them some distance, and then reassemble them to use as a shelter (Finn et al. 2009).

Honey bee and ant colonies possess a collective memory with features shared with the individual memory system of vertebrate brains (Couzin et al. 2002). These colonies perform very similarly on a range of psychological tests (Langridge et al. 2008; Passino 2010).

In addition to their collective intelligence, individual honey bees have complex cognitive abilities (Giurfa et al. 2003) and learning capabilities on a par with those of vertebrates (Bitterman 1996). Individuals understand the concepts of "sameness" and "difference" (Giurfa et al. 2001), can count from one to four (Dacke and Srinivasan 2008), and are able to accurately group visual stimuli into categories (Benard et al. 2006), to name just a few of their abilities.

Many other invertebrates, such as fruit flies and jumping spiders, possess complex cognitive abilities as well (Greenspan and van Swinderen 2004). Learning and memory are utilized by many arthropod species across a wide variety of contexts, including feeding, predator avoidance, and mating (Dukas 2008). And in addition to learning from the surrounding environment, some social insects, like octopi, are able to learn from one another through the use of social information (Leadbeater and Chittka 2009).

These findings of cognitive complexity in invertebrates reveal a striking insight about the evolution of intelligence on earth. The basic capacities of the human mind are demonstrable in the minds of beings who we commonly view as very different from us. These capacities – learning concepts, counting, and even tool use – are ubiquitous in the animal kingdom and point to the continuity of mind among humans and all other animals at a deep evolutionary level.

Vertebrate intelligence: Variations on a theme

It is fair to say that, even if one sets aside the intelligence of invertebrates, then the question of complex intelligence on earth was surely answered when vertebrate brains arose approximately 500 million years ago, because all vertebrate brains are variations on the same theme. Likewise, complex cognitive abilities – some found in invertebrates – are shared by all vertebrates. Most differences lie on a continuum of complexity and are shaped by the adaptive needs of each species from preexisting characteristics. Therefore, there is no cognitive ability that does not have roots in shared capacities. The striking implication of these facts, given the unusually wide scope of astrobiology, is that if we were to find an ETI on another planet whose behavior paralleled the complexity of a gobie using long-term memory and visual geometry to navigate back to a home tidal pool, or a scrub jay caching and remembering seeds in thousands of sites for the next season, or an elephant matriarch using her knowledge and wisdom to lead her family across the desert to water, we would know, for all intents and purposes, that a human level of ETI is very likely to exist.

The shared cognitive and social capacities across many vertebrate species are too numerous to list here. Suffice to say, learning and memory principles are nearly identical across species. And this is the case even in "noncognitive" domains. The highly evolutionarily conserved forebrain region called the limbic system (Striedter 2005) ensures shared basic emotions across all species. Likewise, personality structure is shared across numerous vertebrate (and invertebrate) species (Gosling 2008). These findings and others like them show that, in addition to cognitive abilities, there is basic continuity across general *psychology* in all vertebrates.

A nonexhaustive list of high-level capacities found in many nonhuman vertebrate species includes foreplanning by chimpanzees and orangutans (Osvath and Osvath 2008) and bottlenose dolphins (McCowan et al. 2000); mathematics by chimpanzees (Boysen et al. 1993; Tomonaga and Matsuzawa 2000); taking the visual perspective of another by domestic pigs (Nawroth et al. 2013), jays (Clayton et al. 2007), and dogs (Kaminski et al. 2013); empathy in elephants (Plotnik and de Waal, 2014); understanding physical causation by chimpanzees and orangutans (Mulcahy and Call 2006); and even intentional deception involving modeling of others' mental states by pigs (Held et al. 2000) and chimpanzees (Melis et al. 2006).

In many areas, members of other species either do as well (for example, spatial learning in fishes [Brown 2014]) or exceed the capacities of humans, such as working memory for sequences of numerals by chimpanzees (Inoue and

Matsuzawa 2007). And in many perceptual domains, there are no parallels in cognition in our own species, such as echolocation and its use in cross-modal mental representation of objects in dolphins (Pack and Herman 1995). All of this is not to deny the prodigious cognitive capacities of humans but, rather, to place them in context.

The big four

There are four cognitive domains in which we have, over the decades, argued quite energetically for human exclusivity. These are self-awareness, tool making and use, culture, and symbolic communication. But even these capacities are not entirely unique to our species.

Self-awareness

Self-awareness is the cognizance of oneself as an ongoing independent entity: in other words, the possession of an autobiographical sense or "I." Although all animals must have this capacity at some level, several species have demonstrated a complex human level of self-awareness in experimental studies of mirror self-recognition (MSR), metacognition (reporting on one's own thoughts), and other tests. MSR is the ability to recognize oneself in a mirror and is evinced by using a mirror to investigate parts of one's body. It has been convincingly displayed in all of the great apes (Anderson and Gallup 2011), bottlenose dolphins (Reiss and Marino 2001), Asian elephants (Plotnik et al. 2006), and magpies (Prior et al. 2008). Bottlenose dolphins have also demonstrated an awareness of their own body parts and behaviors in other paradigms (Herman 2012). Monkeys do not pass the mirror test but demonstrate abstract forms of self-awareness, such as metacognition, as do dolphins (Smith and Washburn, 2005). Hence, a complex sense of self is shared with many other species, including birds. At this point, it would be fair to say that our understanding of the distribution of self-awareness across species on this planet is incomplete and limited by our own species' ingenuity in developing methods to test it.

Tool making and use

Since the first time Jane Goodall observed wild chimpanzee David Greybeard make and use a tool (a "termite stick") in 1961, we have documented tool making and use in many primates, elephants, birds, dolphins, octopi, and a host of other species. Wild gorillas use branches to gauge the depth of water when crossing streams (Breuer et al. 2005). New Caledonian crows create and use stick tools with their beaks to extract insects from logs (Hunt and Gray 2003) and, in the laboratory, use analogical reasoning to use tools for accessing yet more tools (Taylor et al. 2007). A group of bottlenose dolphins off the coast of Western Australia uses pieces of sponge wrapped around their rostrum to prevent abrasions when searching for food on the sea floor (Krutzen et al. 2005).

Tool making and use is learned socially and through experimentation and the creation of novel solutions to various problems posed by the environment. This basic aspect of nonhuman tool making and use places it on a continuum with human technology and is, thus, highly relevant to hypotheses about the evolution of technological ETI. Chimpanzees will never be able to conduct SETI searches with termite sticks, yet, from a psychological point of view, termite sticks and radio telescopes are born of the same basic capacity to create something new from the environment for the purpose of achieving a goal. Moreover, the fact that tools are used by nonhanded individuals – for example, dolphin sponge carrying – is another relevant point in the astrobiological context.

Culture

The basic definition of culture is that of distinctive behavior originating in local populations and passed on through learning from one generation to the next. Culture is the main process by which behavioral innovation manifests itself. Tool making and use is often referred to as material culture, and it exists in a multitude of other species. We now know that the technology of chimpanzees (Boesch 2012), dolphins (Krutzen et al. 2005), and New Caledonian crows (Hunt and Gray 2003), for instance, is culturally transmitted. But culture in other behavioral domains also exists. For example, cultural transmission of specific dialects has been documented in orcas (*Orcinus orca*) (Rendell and Whitehead 2001).

Our sophisticated linguistic abilities have allowed human cultures to become extremely complex. But the fact that cultural transmission is shared with many other species on this planet means that we might expect the same in social ETI.

Symbolic communication

Most natural communication systems on this planet, including human language, derive from common principles of communication and information theory, allowing all organisms to accomplish the same basic communicative feats. But the comprehension and use of symbols is thought to be the defining characteristic of human language, allowing us to express abstract ideas, discuss objects and events that are not immediate in space and time, and give us the ability to create an unlimited set of utterances. This quality facilitates the cultural transmission of ideas, as mentioned above.

There have been numerous studies showing that members of other species, who do not appear to possess a human-like language, still can acquire a symbolic artificial language, including dolphins, who comprehend grammatical sentence structure (Herman et al. 1993), chimpanzees (Rumbaugh et al. 2003), African Grey Parrots (Pepperberg 2002), and many others. Importantly, chimpanzees in captivity use the symbolic elements of language in everyday settings (Lyn et al. 2011; Rumbaugh et al. 2003). Moreover, the symbolic element that is key to human culture is also found, though to a much more limited degree, in wild chimpanzees. For instance, in one chimpanzee group, arbitrary symbolic gestures are used to communicate desire to have sex, whereas in another group an entirely different symbolic gesture is used to express the same sentiment (McGrew 2011). The presence of symbolic culture in chimpanzees demonstrates that abstract concepts can be present without human language.

There are several points about communication capacities in other species relevant to ETI. First, the natural communication systems of many other species share features of human language. Second, we have vastly incomplete knowledge of the nature of communication in some groups of animals, such as cetaceans. Third, members of other species can comprehend and use symbols and human language in appropriate everyday settings and, thus, do possess the capacity to think in symbolic terms. Fourth, chimpanzees use symbolic gestures in cultural settings. Fifth, while symbolic communication has clearly become the forte of the human species, it is not entirely outside the range of mental capacities of many other species. Sixth, and most important, is that the human ability to communicate with a symbolic language could only have come from shared characteristics with other species.

Bringing the evolution of intelligence to the table

The existing body of scientific knowledge on the evolution of intelligence makes clear several points:

There are no scientifically valid reasons for treating the human brain and intelligence as a unique characteristic on this planet. Rather, human brains can only be understood in a comparative-evolutionary context.

Once the basic plan for brains evolved, everything that came afterward is a variation on a highly conserved theme.

There is a surprising degree of shared cognition across invertebrates and vertebrates.

There are, of course, differences in intelligence and mind across species, but they all can be understood in a common framework.

Finally, the realization that the human brain is one of many variations on a theme opens the door to a world of data waiting to be mined to gain insights about ETI.

References

Adrio, Fatima, Ramón Anadón, and Isabel Rodríguez-Moldes. 2002. "Distribution of Tyrosine Hydroxylase (TH) and Dopamine Beta-hydrozylase (DBH) Immunoreactivity in the Central Nervous System of Two Shondrostean Fishes (*Acipenser bneri* and *Huso huso*)." *Journal of Comparative Neurology* 448: 280–97.

Allman, John M. 1999. *Evolving Brains*. New York: Scientific American Library.

Anderson, James R., and Gordon G. Gallup Jr. 2011. "Which Primates Recognize Themselves in Mirrors?" *PLoS Biology* 9(3): e1001024.

Arendt, Detlev, Alexandru S. Denes, Gáspár Jékely, and Kristin Tessmar-Raible. 2008. "The Evolution of Nervous System Centralization." *Philosophical Transactions of the Royal Society B: Biological Sciences* 363(1496): 1523–28.

Benard, Julie, Silke Stach, and Martin Giurfa.2006. "Categorization of Visual Stimuli in the Honey Bee, *Apis mellifera*." *Animal Cognition* 9: 257–70.

Billings, Linda. 2011."Fifty Years of Exo/astrobiology." Available at http://doctorlinda.wordpress.com/2011/05/31/50-years-of-exoastrobiology/.

Bitterman, Morton E. 1996. "Comparative Analysis of Learning in Honeybees." *Learning and Behavior* 24(2): 123–41.

Boesch, Cristof. 2012. *Wild Cultures*. Cambridge University Press.

Boysen, Sarah T., Gary G. Berntson, Traci A. Shreyer, and Karen S. Quigley. 1993. "Processing of Ordinality and Transitivity by Chimpanzees (*Pan troglodytes*)." *Journal of Comparative Psychology* 107: 208–16.

Breuer, Thomas, Mireille Ndoundou-Hockemba, and Vicki Fishlock. 2005. "First Observation of Tool Use in Wild Gorillas." *PLoS Biology* 3(11): e380.

Brown, Culum. 2014. "Fish Intelligence, Sentience and Ethics." *Animal Cognition* doi 10.1007/s10071-014-0761–0.

Burger, William C. 2002. *Perfect Planet, Clever Species: How Unique Are We?* New York: Prometheus Books.

Clayton, Nicola S., Joanna M. Dally, and Nathan J. Emery. 2007. "Social Cognition by Food-Caching Corvids: The Western Scrub-Jay as a Natural Psychologist." *Philosophical Transaction of the Royal Society B* 362(1480): 507–22.

Cocconi, Guiseppe, and Philip Morrison. 1959. "Searching for Interstellar Communications." *Nature* 184: 844–84.

Conway Morris, Simon. 2003. *Life's Solution: Inevitable Humans in a Lonely Universe*. Cambridge University Press.

Couzin, Iain D., Jens Krause, Richard James, Graeme D. Ruxton, and Nigel R. Franks. 2002. "Collective Memory and Spatial Sorting in Animal Groups." *Journal of Theoretical Biology* 218(1): 1–11.

Crowe, Michael J. 2008. *The Extraterrestrial Life Debate, Antiquity to 1900: A Source Book*. University of Notre Dame Press.

Dacke, Marie, and Madyam Srinivasan. 2008. "Evidence for Counting in Insects." *Animal Cognition* 4: 683–89.

Darling, David. 2014. "Drake Equation." Available at www.daviddarling.info/encyclopedia/D/DrakeEq.html.

Darwin, Charles. 1859. *On the Origin of Species by Means of Natural Selection, or the Preservation of Favoured Races in the Struggle for Life,* 1st ed. London: John Murray.

Darwin, Charles. 1871. *The Descent of Man, and Selection in Relation to Sex,* 1st ed. London: John Murray.

Dick, Steven J. 1998. *Life on Other Worlds: The Twentieth-Century Extraterrestrial Life Debate*. Cambridge University Press.

Dukas, Reuven. 2008. "Evolutionary Biology of Insect Learning." *Annual Review of Entomology* 53: 145–60.

Finn, Julian, Tom Tregenza, and Mark D. Norman. 2009. "Defensive Tool Use in a Coconut-Carrying Octopus." *Current Biology* 19(23): R1069–70.

Fiorito, Graziano, and Pietro Scotto. 1992. "Observational Learning in *Octopus vulgaris.*" *Science* 256: 545–46.

Friesen, W. Otto, and William B. Kristan. 2008. "Leech Locomotion: Swimming, Crawling, and Decisions." *Current Opinion in Neurobiology* 17(6): 704–11.

Giurfa, Martin, et al. 2003. "The Effect of Cumulative Experience on the Use of Elemental and Configural Visual Discrimination Strategies in Honeybees." *Behavioral Brain Research* 145: 161–69.

Giurfa, Martin, S. Zhang, A. Jenett, R. Menzel, and M. V. Srinivasan. 2001. "The Concepts of 'Sameness' and 'Difference' in an Insect." *Nature* 410: 930–33.

Gosling, Samuel. 2008. "Personality in Nonhuman Animals." *Social and Personality Psychology Compass* 2(2): 985–1001.

Greenspan, Ralph J., and Bruno van Swinderen. 2004. "Cognitive Consonance: Complex Brain Functions in the Fruit Fly and its Relatives." *Trends in Neurosciences* 27(12): 707–11.

Held, Suzanne, Michael Mendl, Claire Devereux, and Richard W. Byrne. 2000. "Social Tactics of Pigs in a Competitive Foraging Task: The 'Informed Forager' Paradigm." *Animal Behaviour* 59: 569–76.

Herman, Louis M. 2012. "Body and Self in Dolphins." *Consciousness and Cognition* 21: 526–45.

Herman, L. M., Adam A. Pack, and Palmer Morrel-Samuels. 1993. "Representational and Conceptual Skills of Dolphins." In *Language and Communication: A Comparative Perspective,* ed. H. R. Roitblat, L. M. Herman, and P. Nachtigall, 403–42. Hillsdale, NJ: Lawrence Erlbaum.

Hochner, Binyamin, Tal Shomrat, and Graziano Fiorito. 2006. "The Octopus: A Model for a Comparative Analysis of the Evolution of Learning and Memory Mechanisms." *Biological Bulletin* 210(3): 308–17.

Hunt, Gavin R., and Russell D. Gray. 2003. "Diversification and Cumulative Evolution in New Caledonian Crow Tool Manufacture." *Proceedings of the Royal Society of London B* 270(1517): 867–74.

Inoue, Sana, and Tetsuro Matsuzawa. 2007. "Working Memory of Numerals in Chimpanzees." *Current Biology* 17(23): R1004–5.

Jacobs, Dave K., Nagayasu Nakanishi, David Yuan, Anthony Camara, Scott A. Nichols, and Volker Hartenstein. 2007. "Evolution of Sensory Structures in Basal Metazoa." *Integrative and Comparative Biology* 47: 712–23.

Kaminski, Julianne, Andrea Pitsch, and Michael Tomasello. 2013. "Dogs Steal in the Dark." *Animal Cognition* 16: 385–94.

Kandel, Eric R. 2001. "The Molecular Biology of Memory Storage: A Dialogue Between Genes and Synapses." *Science* 294: 1030–38.

2009. "The Biology of Memory: A Forty Year Perspective." *The Journal of Neuroscience* 29(41): 12748–56.

Krutzen, Michael, Janet Mann, Michael R. Heithaus, Richard C. Connor, Lars Bejder, and William B. Sherwin. 2005. "Cultural Transmission of Tool Use in Bottlenose Dolphins." *Proceedings of the National Academy of Sciences* 102(25): 8939–43.

Langridge, Elizabeth A., Ana B. Sendova-Franks, and Nigel R. Franks. 2008. "How Experienced Individuals Contribute to an Improvement in Collective Performance in Ants." *Behavioral Ecology and Sociobiology* 62: 447–56.

Leadbeater, Ellouise, and Lars Chittka. 2009. "Bumble Bees Learn the Value of Social Cues through Experience." *Biology Letters* 5: 310–12.

Lederberg, Joshua. 1960. "Exobiology: Approaches to Life beyond the Earth." *Science* 132: 393–400.

Lyn, Heidi, Patricia M. Greenfield, Sue Savage-Rumbaugh, Kristen Gillespie-Lynch, and William D. Hopkins. 2011. "Nonhuman Primates Do Declare! A Comparison of Declarative Symbol and Gesture Use in Two Children, Two Bonobos, and a Chimpanzee." *Language and Communication* 31: 63–74.

Marino, Lori. 2002. "Convergence in Complex Cognitive Abilities in Cetaceans and Primates." *Brain, Behavior and Evolution* 59: 21–32.

2007. "Scala Natura." In *The Encyclopedia of Human-Animal Relationships*, ed. Marc Bekoff, 220–24. Westport, CT: Greenwood Publishing Group.

McCowan, Brenda, Lori Marino, Erik Vance, Leah Walke, and Diana Reiss. 2000. "Bubble Ring Play of Bottlenose Dolphins: Implications for Cognition." *Journal of Comparative Psychology* 114: 98–106.

McGrew, W. C. 2011. "Pan symbolicus: A Cultural Primatologist's Viewpoint." In *Homo Symbolicus: The Dawn of Language, Imagination and Spirituality.* Ed. C. S. Henshilwood and F. d'Errico, 1–12, Amsterdam: John Benjamins.

Melis, Alicia P., Josep Call, and Michael Tomasello. 2006. "Chimpanzees (*Pan troglodytes*) Conceal Visual and Auditory Information from Others." *Journal of Comparative Psychology* 120(2): 154–62.

Messenger, John B. 1996. "Neurotransmitters of Cephalopods." *Invertebrate Neuroscience* 2: 95–114.

Miller, Stanley L. 1953. "A Production of Amino Acids under Possible Primitive Earth Conditions." *Science* 117: 528–29.

Miller, Stanley L., and Harold C. Urey. 1959. "Organic Compound Synthesis on the Primitive Earth." *Science* 130: 245–51.

Morse, Douglass H. 2000. "The Effect of Experience on the Hunting Success of Newly Emerged Spiderlings." *Animal Behavior* 60(6): 827–35.

Mulcahy, Nicholas, and Josep Call. 2006. "How Great Apes Perform on a Modified Trap Tube Task." *Animal Cognition* 9: 193–99.

Nawroth, C., M. Ebersbach, and E. von Borell. 2013. "Juvenile Domestic Pigs (*Sus scrofa domesticus*) use Human-Given Cues in an Object Choice Task." *Animal Cognition* 17: 701–13.

Oparin, Alexander I. 1924. *The Origin of Life*. Moscow: Moscow Worker.

Osvath, Mathias, and Helena Osvath. 2008. "Chimpanzee (*Pan troglodytes*) and Orangutan (*Pongo abelii*) Forethought: Self-Control and Pre-Experience in the Face of Future Tool-Use." *Animal Cognition* 11: 661–74.

Pack, Adam A., and Louis M. Herman. 1995. "Sensory Integration in the Bottlenose Dolphin: Immediate Recognition of Complex Shapes Across the Senses of Echolocation and Vision." *Journal of the Acoustical Society of America* 98: 722–33.

Passino, Kevin M. 2010. "Honey Bee Swarm Cognition: Decision-Making Performance and Adaptation." *International Journal of Swarm Intelligence* 1(2): 80–97.

Pepperberg, Irene M. 2002. *The Alex Studies: Cognitive and Communicative Abilities of Grey Parrots*. Cambridge, MA: Harvard University Press.

Plotnik, Joshua M., and Frans B. M. de Waal. 2014. "Asian Elephants (*Elephas maximus*) Reassure Others in Distress. *PeerJ* 2: e278.

Plotnik, Joshua M., Frans B. M. de Waal, and Diana Reiss. 2006. "Self-Recognition in an Asian Elephant." *Proceedings of the National Academy of Sciences* 103(45): 17052–57.

Prior, Helmut, Ariane Schwarz, and Onur Güntürkün. 2008. "Mirror-Induced Behavior in the Magpie (*Pica pica*): Evidence of Self-Recognition." *PLoS Biology* 6(8): e202.

Rankin, Catherine H. 2004. "Invertebrate Learning: What Can't a Worm Learn?" *Current Biology* 14: R617–18.

Reiss, Diana, and Lori Marino. 2001. "Self-Recognition in the Bottlenose Dolphin: A Case of Cognitive Convergence." *Proceedings of the National Academy of Sciences USA* 98(10): 5937–42.

Rendell, Luke E., and Hal Whitehead. 2001. "Culture in Whales and Dolphins." *Behavioural and Brain Sciences* 24: 309–24.

Rumbaugh, Duane M., Michael J. Beran, and E. Sue Savage-Rumbaugh. 2003. "Language." In *Primate Psychology,* ed. D. Maestripieri, 395–423. Cambridge, MA: Harvard University Press.

Saigusa, Tetsu, Atsushi Tero, Toshiyuki Nakagaki, and Yoshiki Kuramoto. 2008. "Ameobae Anticipate Periodic Events." *Physical Review Letters* 100: 18101-1–4.

Sakarya, Onur, Kathryn A. Armstrong, Maja Adamska, Marcin Adamski, I-Fan Wang, Bruce Tidor, Bernard M. Degnan, Todd H. Oakley, and Kenneth S. Kosik. 2007. "A Post-Synaptic Scaffold at the Origin of the Animal Kingdom." *PLoS One* 2(6): e506.

Schwartzman, David, and Lee J. Rickard. 1988. "Being Optimistic about the Search for Extraterrestrial Intelligence." *American Scientist* 76(4): 364–69.

Sherwood, Chet C., Cheryl D. Stimpson, Mary Ann Raghanti, Derek E. Wildman, Monica Uddin, Lawrence I. Grossman, Morris Goodman, John C. Redmond, Christopher J. Bonar, Joseph M. Erwin, and Patrick R. Hof. 2006. "Evolution of Increased Glia–Neuron Ratios in the Human Frontal Cortex." *Proceedings of the National Academy of Sciences USA* 103: 13606–11.

Shettleworth, S. J. (2010). *Cognition, Evolution and Behavior*. New York: Oxford University Press.

Shulgina, Galina I. 2006. "Learning of Inhibition of Behavior in the Sea Star, *Asterias rubens*." *Comparative and Ontogenic Physiology* 42(2): 161–65.

Smith, J. David, and David A. Washburn. 2005. "Uncertainty Monitoring and Metacognition by Animals." *Current Directions in Psychological Science* 14: 19–24.

Striedter, Georg F. 2005. *Principles of Brain Evolution*. Sunderland, MA: Sinauer Associates.

Taylor, Alex H., Gavin R. Hunt, Jennifer C. Holzhaider, and Russell D. Gray. 2007. "Spontaneous Metatool Use by New Caledonian Crows." *Current Biology* 17(17): 1504–7.

Tomonaga, Masaki, and Tetsuro Matsuzawa. 2000. "Sequential Responding to Arabic Numerals with Wild Cards by the Chimpanzee." *Animal Cognition* 3: 111.

Turlejski, Krzysztof. 1996. "Evolutionary Ancient Roles of Serotonin: Long-Lasting Regulation of Activity and Development." *Acta Neurobiologiae Experimentalis* 56: 619–36.

Ward, Peter, and David Brownlee. 2000. *Rare Earth: Why Complex Life is Uncommon in the Universe*. New York: Springer-Verlag.

Zilles, K., E. Armstrong, K. H. Moser, A. Schleicher, and H. Stephan. 1989. "Gyrification in the Cerebral Cortex of Primates." *Brain, Behavior and Evolution* 34(3): 143–50.

11

Fraction of civilizations that develop a technology that releases detectable signs of their existence into space, f_c, pre-1961

Florence Raulin Cerceau

Abstract

This chapter examines the prehistory of the search for extraterrestrial intelligence (SETI) prior to 1961. It reviews the first attempts to contact other planets at the scale of the solar system – that is, interplanetary communication (premodern SETI era). We emphasize the latter half of the nineteenth century because many efforts were made at that time to contact our neighboring planets through interplanetary telegraphy. Such a technique became conceivable in the 1860s thanks to many advances in the field of terrestrial communication (electrical telegraphy, telephone). Generally, the pioneers in interplanetary telegraphy proposed to send flashes using powerful lamps and reflectors to reach our neighboring planets, Mars and Venus. Considering their methodology, the early proposals using light flashes could be compared to modern Active SETI, or METI (messaging to extraterrestrial intelligence). The intellectual approach is similar to that of CETI (communication with extraterrestrial intelligence), since the first attempts were expected to be a two-way exchange of information. Even though they remained only theoretical, these attempts demonstrated that basic thought about a universal language had begun as early as the 1860s. At the turn of the twentieth century, new possibilities emerged with the birth of wireless telegraphy and the development of radio techniques used for telecommunications on Earth. Listening to Mars by means of radio waves was sporadically attempted in the 1920s. By the mid-twentieth century, developments in radio astronomy had a decisive influence on the birth of SETI because they allowed astronomers to contemplate the possibility of contact with extraterrestrials at a much larger scale, that of interstellar distances.

The Drake Equation, ed. Douglas A. Vakoch and Matthew Dowd. Published by Cambridge University Press.

Introduction

Today, the SETI focuses on the planetary systems that abound in our galaxy. Many exoplanets have been discovered since 1995, including terrestrial-like planets that could be habitable. First deployed only about fifty years ago, the current science of extraterrestrial communication in our galaxy is based on the detection of artificial signals using radio frequencies, of which Cocconi and Morrison demonstrated the possibility in 1959 (Cocconi and Morrison 1959). Optical-laser pulses are an alternative technique, as proposed by Schwartz and Townes as early as 1961 (Schwartz and Townes 1961). This science is new because it uses techniques that did not exist at the beginning of the twentieth century. However, SETI's "modern era" was preceded by a "premodern era," which began during the nineteenth century with sporadic schemes proposed by audacious pioneers. Speculations about life on the Moon or Mars encouraged discussions on the possibility of communicating with their hypothetical habitants (Crowe 1986, 393).

Nevertheless there are major differences between these two periods, despite their common interest in communication with other planets. One focuses on space and time: in the first case, signals are expected from stars located at distances beyond a few light-years, whereas the second case focuses on interplanetary communication only – especially targeting the closest celestial bodies, such as Mars. This point is important because it affects potential "dialogues" between Earthlings and hypothetical inhabitants of other planets, as well as message content. The approach used to communicate with other planets is the second main difference: SETI's current technique consists mainly of listening to the stars (even if active SETI – sending messages – is also part of SETI), while the approach used during SETI's premodern era was essentially sending signals. The difference is also seen in the implementation of research. The American astronomer Frank Drake initiated SETI experiments in 1960 at the National Radio Astronomy Observatory (Green Bank, West Virginia). Project Ozma, as Drake named it, aimed at detecting radio signals from two nearby Sun-like stars, Tau Ceti and Epsilon Eridani. It was the first step in a long series of SETI experiments, especially in the microwave region. Conversely, nineteenth-century projects, which were supposed for the most part to use light signals, remained only theoretical schemes and were never put into practice.

Despite these points of difference, the first attempts at contacting planets overlap in a number of interesting ways. First, they show that thought about techniques necessary to send interplanetary messages was already present in the West during the nineteenth century. Second, they provide pioneering proposals on a universal language involving a "code," that is, a correspondence between a series of signs and some basic transmitted information.

Premodern SETI, which runs from 1820 to 1960, seems to have no connection with the modern one (from 1960 on). It mainly calls for a very basic technology that uses light flashes to communicate with Mars or Venus, which was consistent with the state of thinking at that time about techniques required to contact neighboring planets. The period between 1920 and 1960 contains the (very few) first attempts to detect signals from Mars with the help of radio communication, a new field under development in the 1920s. However, even this period is still focused on interplanetary communication and could be considered an offshoot of the birth of radio broadcasting. Modern SETI started with an in-depth analysis of what could be the best region in which to explore for extraterrestrial signals (Cocconi and Morrison 1959). As is well known, modern SETI denotes interstellar communication and advocates searching for signals most efficiently in the microwave region, especially at the 21 centimeter wavelength of the hydrogen hyperfine transition. Other wavelengths such as optical have been proposed since then.

Premodern SETI prior to 1860: Construction of figures huge enough to be seen from the Moon or Mars

The very first schemes to communicate with other planets date back to the beginning of the nineteenth century, when many discussions were being held about life on the Moon. They are identified in several second-hand citations and very few texts. The few proposals from this period advocate huge visual schemes built on Earth that could be seen from the Moon or Mars. Changes in color or form could be a sign of an "intelligent" presence on Earth. Such schemes were not as simple as first thought because they required colossal construction works.

The first scheme of the premodern SETI era is generally attributed to the German mathematician Karl Gauss (1777–1855), although the sources of this proposal are unclear and still debated.[1] It seems credible that Gauss accepted life on the Moon, but no source reference is supplied for this proposal (Crowe 1986, 205). Based on the belief that mathematics would be the same on the Earth and Moon, Gauss is supposed in 1826 to have suggested using the Pythagorean theorem to arouse the curiosity of lunar inhabitants. A giant right-angle triangle and three squares (Figure 11.1) could be drawn by clearing trees in the Siberian forest. Differences in color would have given the design greater visibility to lunar residents. The outlines of the shapes would have been ten-mile-wide strips of pine forest, while the interiors could be rye or wheat. Different colors of vegetation, combined with seasonal changes, were supposed to be seen from the Moon. Because of the particular geometrical shape of the figure, Moon dwellers would have recognized our planet as inhabited by people who knew mathematics.

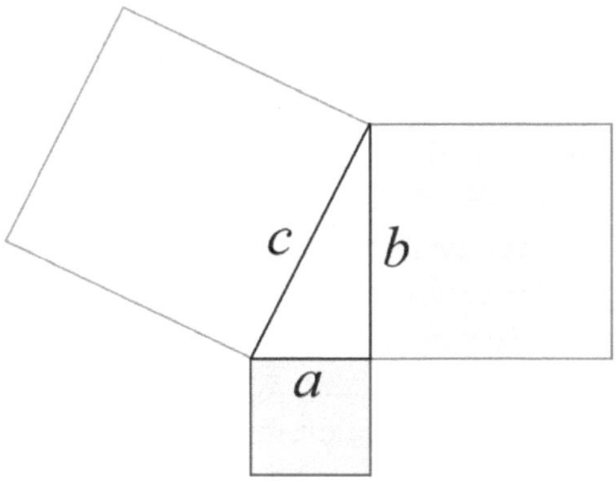

Figure 11.1 The symbolic representation of the Pythagorean theorem proposed by Gauss for an interplanetary communication: a giant right triangle and three squares built in the Siberian tundra and supposed to be seen from the Moon.

A discussion about this proposal between Gauss and the German astronomer Franz von Gruithuisen (1774–1852) is reported in an anonymous article in 1826 in the *New Philosophical Journal*. The discussion deals with the possibility of a correspondence with the inhabitants of the Moon:

> Gauss answered (to Gruithuisen), that the plan of erecting a geometrical figure on the plains of Siberia corresponded with his opinion, because, according to his view, a correspondence with the inhabitants of the moon could only be begun by means of such mathematical contemplations and ideas, which we and they must have in common. (Anonymous 1826, 390)

Another method to contact intelligent beings on nearby planets was proposed by Gauss and is better documented. It was reported in 1822 in a letter to Gauss's friend, the German astronomer Heinrich Wilhelm Matthäus Olbers (1758–1840) (Crowe 1986, 207). In 1818 Gauss had invented the heliotrope, a device that could be used to send light to the Moon. This instrument was believed able to reflect the Sun's rays in a focused beam viewable from very long distances. Gauss theorized: "With 100 separate mirrors, each of 16 square feet, used conjointly, one would be able to send good heliotrope-light to the moon." It would be, Gauss added, "a discovery even greater than that of America, if we could get in touch with our neighbors on the moon" (Gauss, in Crowe 1986, 207).

The Austrian physicist Joseph von Littrow (1781–1840), director of the Vienna Observatory, is generally associated with a second scheme to communicate with extraterrestrials. In 1840, he is believed to have suggested digging giant geometric shapes in the Sahara Desert as signaling devices. Circular excavations about thirty kilometers wide would be filled with water and kerosene, and set on fire at night to signal Earthlings' existence.

These early proposals have an important common point (beyond the fact that they are of uncertain origin): the use of huge geometrical shapes likely to prove "intelligent" intent. Gauss's suggestion bears an additional meaning related to math since his figures were intended to express the Pythagorean theorem (a triangle and three squares). Douglas Vakoch has explained that this last graphic representation can be considered iconic because there is a physical resemblance between the sign (the figure) and the signified (the theorem) (Vakoch 1998b).

However, these experiments were never undertaken, and it is easy to understand why. They were not very realistic (except the heliotrope): such gigantic structures had more to do with fiction than science. Even if gigantic signals could be produced so that they could be distinctly seen by the inhabitants of the Moon, their inefficiency as intentional messages seemed to make them impractical. However, they remain examples of the different ways imagination could be inspired to find means to communicate with aliens. It was a first step toward interplanetary communication.

Premodern SETI from 1860 to 1920: Interplanetary telegraphy

The context

In the West, the second half of the nineteenth century was characterized by a spectacular growth in technology, including electrical telegraphy, a new means of communication that led to a new era of information exchange. Such an exciting period seemed to offer limitless possibilities. If communication was suddenly easier on Earth, why would it not be possible over planetary distances? This hypothesis came naturally, as if no clear distinction existed between terrestrial distances and interplanetary ones.

In addition, the period was very prolific in Martian studies, especially about its surface. Thanks to the development of better telescopes and the existence of data accumulated over many years, the German astronomers Beer (1797–1850) and Mädler (1794–1874) in 1840 provided the first map of Mars. After 1877, more detailed maps were drawn by the British astronomer R. A. Proctor (1837–1888). Mars was presumed to be like Earth, with continents and seas, atmosphere and seasons, and even surface life. A strong interest emerged about this neighboring world that seemed to be "the miniature of our Earth," as asserted by Proctor in his book *Other Worlds than Ours* (Proctor 1870, 90). The red planet was thought to be older than Earth, with an existence long enough to allow the presence of a highly developed civilization. It was then easy to extrapolate and conclude that "intelligent" Martians would be able to communicate with Earthlings. The famous controversy about canals[2] strengthened the discussion about life on Mars, even if only very few astronomers finally glimpsed them and agreed with their existence. In the 1890s, the

American businessman Percival Lowell (1855–1916) published a series of books reporting his observations of the canals and strongly promoted the idea of life on Mars (see, for instance, Lowell 1895). Nonetheless, Lowell's work was largely discredited during his lifetime. In 1909, Franco-Greek astronomer Eugène M. Antoniadi (1870–1944) refuted the canals' existence (Antoniadi 1909; 1930), after using the 83 centimeter telescope at the Meudon Observatory in France, which allowed him to report that "the canal illusion was a product of poor seeing" (Antoniadi, in McKim 1993, 220).

Despite this, important figures in this period promoted interplanetary communication through celestial telegraphy. We review in the following sections the major contributors to celestial telegraphy from 1869 to 1896: the French poet and inventor Charles Cros (1842–1888); the French astronomer Camille Flammarion (1842–1925); the British multi-disciplinary scientist Francis Galton (1822–1911); the French amateur astronomer A. Mercier[3]; the Finnish military officer and mathematician Edvard E. Neovius (1823–1888); and the American astronomer William H. Pickering (1858–1938). The most frequently mentioned technical method was based on the global use of light beams: powerful lamplights focused on parabolic mirrors. Following this simple technique, flashes could be sent toward neighboring planets, especially Mars. All these early efforts in communication were then based on optical wavelengths. However, details of each scheme in this first period varied. Nevertheless, it should be noted that interplanetary communication was severely criticized by some who considered the idea foolish.

Still, an elaborate paper dealt with flashes of light: In 1920, the brothers H. W. Nieman and C. Wells Nieman (Nieman and Nieman 1920) proposed using Morse code to deliver detailed messages, arguing that the key to communication was the timing in the duration of the signal to produce dots and dashes, and in the gap between signals to produce pauses. Their proposal was the basis for the encoding system used by Frank Drake and his colleagues in the first 1974 interstellar message (Lemarchand 1998).

Charles Cros

Charles Cros, one of Flammarion's friends, was famous for his photographic techniques, including an early method of color photography. He also made improvements in telegraphs and invented processes capable of recording and reproducing sounds (the paleophone, a phonograph ancestor, in 1877). Cros believed that interplanetary distances could be overcome by new techniques and that a dialogue could then be established between people from different worlds. On July 5, 1869, the French geologist Stanislas Meunier presented to the French Academy of Sciences a study explaining possible ways to communicate with other planets (Meunier 1869). Meunier reported Cros's proposals related in a sixteen-page booklet titled *Etude sur les moyens de communication avec les planètes* (*Studies on the Means of Communication with the Planets*, Cros 1869). Cros's suggestion was to communicate with the closest planets, Venus and Mars, with the help of a powerful beam of light focused on a parabolic mirror. The targeted planet would be wrapped in the beam and the hypothetical creatures living on its surface could see where the beam originated (i.e., the Earth).

Cros thought that two main points were absolutely required to communicate through interplanetary distances with techniques available at that time (Cros 1869): (1) Using light beams, that is, one or more powerful electric lights focused on parabolic mirrors so as to be visible by the inhabitants of Mars or Venus; and (2) exchanging repetitive signals using periodic flashes as proof of an intentional message.

Cros paid particular attention to the search for an interplanetary language using numerical notations, pictures, and rhythms:

- Numerical notations: The message would consist of a numerical series translatable in patterns written as dots. The numerical series would start with a very simple message. The subsequent ones could be more and more sophisticated.

Figure 11.2 Light flashes to Mars: a powerful beam of light sent from the Earth could envelop the targeted planet.
Courtesy: http://www.smithsonianmag.com/history/hello-mars-this-is-the-earth-10699440/?no-ist.

- Pictures: Pictures could be reconstructed from the series of numbers. This series would be converted into lines or points of two different colors.
- Rhythms: The "artificial" (and thus "intelligent") nature of the signal could be proved by sporadic intervals between each optical flash. The only way to be understood (or not to be mistaken for a "natural" signal) would be to send a repetitive message.

Cros proposed to send periodic pulses of single, double, and triple flashes. He underlined the importance of rhythms: sporadic intervals, equivalent to zero, alternating with optical flashes, equivalent to one (a binary language that was ahead of its time).

As with his contemporaries, Cros held that Martians might have already tried to contact humans but that their messages had not been seen or understood. His fervent belief in interplanetary communication led him to regard the bright spots that have been noticed on Mars and Venus as extraterrestrial signals intentionally sent to the Earth but unfortunately misinterpreted. None of Cros's contemporaries, however, seemed to have been sufficiently enthused or adequately skilled to attempt his project (Crowe 1986, 394).

Edvard E. Neovius

In 1875, Edvard Engelbert Neovius published a small book in Swedish (the translated title is *The Most Important Task of our Time*) suggesting a way to contact the inhabitants of Mars. His work has been reported and analyzed by Lehti (1998). Neovius drew inspiration from beliefs of the Danish physicist Hans Christian Orsted (1777–1851) and the French astronomer Camille Flammarion about the theory of habitability and the principle of analogy applied to planets. He was convinced that intelligence and science were universal and believed that the Martians, who he thought represented a civilization more advanced than ours, might have (at least) developed techniques similar to ours, a view adopted by others. According to Neovius, an "optical channel of communication" between the two planets would be possible only if both participants cooperated in the undertaking. Moreover, we would have to try both to catch extraterrestrials' attention and to watch for possible return signals. In technical terms, Neovius proposed sending messages to Mars with the help of 22,500 electric lamps and a large array of simple generators, such as galvanic pairs. Referring to the most recent photometric literature of his time, Neovius calculated how a light source on Earth had to be powerful enough to be detectable on Mars (as a star of sufficient brightness would be).

Like Cros, Neovius seemed very optimistic about future celestial telegraphy. He was convinced that "it will not take many years before vigorous preparations have been carried out for a telegraphic contact with planets of our Solar System" (Neovius, in Lehti 1998, 732).

One can find the same feeling in Nikola Tesla's writings at the turn of the twentieth century (discussed later in this chapter). In Neovius's scheme, as well as Cros's, the message consisted of a series of signals that would gradually increase in complexity. Neovius's proposal started from simple arithmetical concepts to logic and the physics of the solar system, using first numbers and then geometry. The message, which was mainly based on scientific information, was conceived to be both precise and as simple as possible.

Francis Galton

Interest in Martian signals sharply increased during the 1890s following a series of reports of bright spots observed on the Martian surface. As previously suggested by Cros, William H. Pickering and Francis Galton connected the so-called bright spots with signals sent by a Martian civilization to Earth. Such an interpretation encouraged them to propose, separately, schemes to contact Mars by celestial telegraphy.

Francis Galton, Charles Darwin's cousin, was an English statistician and meteorologist. In 1892, he published a letter in the London *Times*[4] suggesting that a combination of giant mirrors could reflect sufficient sunlight to communicate with Martians (Crowe 1986, 395). In 1896, he presented an essay in the *Fortnightly Review* devoted to developing a language suitable for extraterrestrial communication based on mathematics and pictorial messages (Galton 1896). His method was quite elaborate and included constructing pictures from a series of flashes produced by an assemblage of large heliographs. Before this, a pedagogic method was supposed to be conveyed, involving three lengths of signal that he called "dot," "dash," and "line" to transmit elementary arithmetical concepts (Forrest 1974, 238). Simple operations such as addition and multiplication would be among the first items transmitted. In the next step, Galton suggested "picture-writing" to represent the shape of many objects. A "celestial syntax" (radius, area, perimeter, triangle, etc.) would then facilitate the transmission of mathematical representations and of a list of objects that could be indefinitely extended (Forrest 1974, 238).

Comparisons between Cros's, Neovius's, and Galton's proposals

The most interesting part of Cros's, Neovius's, and Galton's proposals, published respectively in 1869 (Cros 1869), 1875 (examined in Lehti 1998), and 1896 (Galton 1896), was the mention of a code based on mathematics to convey information, step by step. The final shape of the message could take different forms, such as pictorial representations (Cros, Galton) or signs translatable into "words" (Neovius). In every scheme, however, mathematics was used as a universal language.

All of their proposals were aimed at conveying human knowledge or universal facts of science. All of them used light pulses and tried to demonstrate an intentional meaning in several ways to accomplish this. Neovius used variations in duration and intensity of pulses. Galton proposed a series of flashes comparable to Morse code: three lengths of signal, dot, dash, and line (Raulin Cerceau and Bilodeau 2012). In each scheme, celestial telegraphy required solving what was regarded as important problems, namely, "being seen," then "being distinguished from natural signals," and finally "proving our human intelligence."

Camille Flammarion

During this stimulating time in celestial telegraphy, Camille Flammarion, who was well known for his *Pluralité des Mondes Habités* (1862), translated into several languages, had a strong influence on French views of an inhabited solar system. In a special chapter of his book *Rêves étoilés* devoted to interplanetary communication (Flammarion 1888, 167–84; this proposal was also published in Flammarion 1891), he proposed another method, which was closer to the very early schemes described by Gauss and von Littrow. Curiously, Flammarion was not a strong supporter of the "popular" method of celestial telegraphy. Instead, he proposed using geometry because it could be the basis for a possible dialogue since geometry is the same for the inhabitants of every world. He suggested drawing huge figures on Earth's surface by means of electrical lamplights or any other method that could build geometrical features likely to be seen from Mars.

In *Rêves étoilés*, Flammarion was probably more interested in the choice of a universal language than in perfecting high-performance techniques to reach neighboring planets. The central point of Flammarion's scheme is the same as Gauss's and von Littrow's: visual changes are necessary to demonstrate an "intelligent" intent behind the features. The signals that could be seen from Mars could be very large-sized figures of about 100 kilometers or more: triangles, squares, and circles. Triangles moving in squares would demonstrate an intelligent cause, a message distinguishable from a natural phenomenon. Each geometrical shape could move successively to a different one: a circle could be changed in a square, which could be changed again in a triangle, and so on. This constituted a geometrical language, easily understandable and impossible to confuse with a "natural" signal (Raulin Cerceau and Bilodeau 2012).

In an 1891 paper published in the journal *L'Astronomie,* Flammarion announced that Clara Goguet Guzman, a Frenchwoman from Pau-France who had read

Flammarion's proposal in *Rêves étoilés*, had offered a prize of 100,000 francs to anybody able to communicate with any planet within the next ten years (Flammarion 1891). Administered by the French Academy of Sciences, the *Pierre Guzman Prize* (named after Mrs. Guzman's son) was described as follows by Flammarion:

> A prize of 100,000 francs is bequeathed to the Institute of France (Science Section) for the person of whatever nation who will find the means within the next ten years of communicating with a star (planet or otherwise) and of receiving a response.
>
> The testatrix designates the planet Mars, on which the attention and the investigations of all scientists are already directed. If the Institute of France does not accept the legacy, it will pass to the Institute of Milan, and in the case of a new refusal, to that of New York. (Flammarion 1891, translated in Crowe 1986, 395)

Flammarion contributed to the popularity of this prize, and Tesla would later try to claim it (Tesla 1937). Nobody won, but it can be regarded as a symbol of the new view of possible interplanetary communication.

Mercier

Another proposal should be considered, even though it came from an obscure Frenchman, the amateur astronomer A. Mercier.[5] In 1899, he wrote a booklet titled *Communications avec Mars* (Mercier 1899), where he questioned Mars's habitability and made suggestions about interplanetary communication by celestial telegraphy.

Mercier believed that various signaling methods were possible and explained two of them (Mercier 1899, 13–27). His first suggestion was to place one or several reflectors on the Eiffel Tower in Paris, France, in such a way that these reflectors could receive bright sunbeams during sunset. They would then be pointed at the targeted planet (Mars). A movable screen could interrupt the signals from time to time to give a "non-natural" meaning to the message. The second suggestion was an ambitious arrangement involving huge mirrors: solar rays would be reflected at sunset from the sunlit side of a mountain to a mirror atop the mountain, then to a mirror on the dark side of the mountain, and finally to Mars. The advantage of this method would be that the solar rays would attain increased visibility by being seen against a dark background (Crowe 1986, 397).

Mercier asked for subscriptions to put at least one of his projects into practice. Although he received a few, the results were apparently insufficient to carry out his plan (Raulin Cerceau 2006).

As was the case with many of his contemporaries involved in interplanetary communication, Mercier's aim was also philosophical with optimistic perspectives. He believed that attracting extraterrestrials' attention and making the

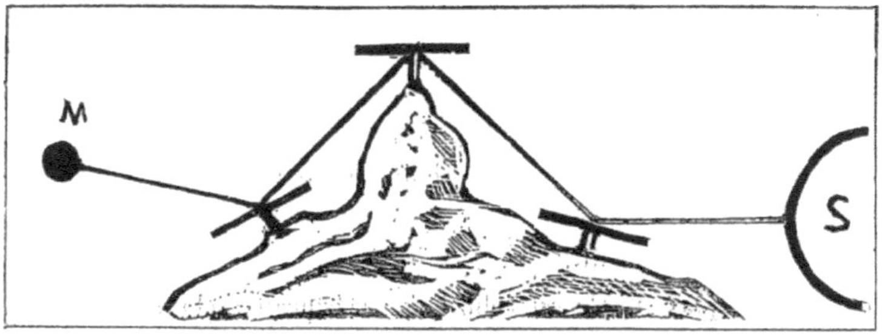

Figure 11.3 One of Mercier's proposals to communicate with Mars: a succession of mirrors used to transmit the solar rays toward Mars at sunset.
Credit: Mercier (1899, 16)

message understood would be of the highest importance for humanity. His contribution to that question, however, remained relatively unrecognized.

William H. Pickering

In the United States, William H. Pickering played an important part in the Martian canal controversy. He also proposed an example of celestial telegraphy aimed at Mars and saw the 1909 opposition of Mars as a good opportunity to contact the inhabitants of the red planet. With the help of a huge system of mirrors combined with sunlight, signals could be sent rhythmically in the hope that Martian astronomers working with their telescopes would notice them.

Pickering proposed mirrors so large that they would need to be moved by machinery:

> My plan of communication would require the use of a series of mirrors so arranged as to present a single-reflecting surface toward the planet. As the surface necessary for reflecting the sunlight 35,000,000 miles would have to be more than a quarter of a mile long, a single mirror would not be practicable. We should have to use many of them. These mirrors would all have to be attached to one great axis parallel to the axis of the earth, run by motors and so timed as to make a complete revolution every twenty-four hours, thus carrying the reflecting surface around with the axis once a day and obviating the necessity of continually readjusting it to allow for the movement of the planets. As far as the people of Mars are concerned this reflector would not, of course, be apparent to the naked eye, but through lenses of such magnitude as we have to-day the reflection would be easily discernible and would undoubtedly attract attention at once. The best time for transmitting such a reflection would be in the morning, a little after sunrise. The cost of such an undertaking would be about $10,000,000. (Pickering, in Anonymous 1909b)

This proposal revived the debate in the United Sates about the idea of life on Mars, despite the fact that Antoniadi, as previously mentioned, defused the canals controversy the same year. Pickering's idea of signaling to Mars by means of a huge system of mirrors seemed to have called forth schemes from other scientists. Many articles were published in 1909 around the topic "Signaling to Mars" in the *Correspondence* of *Scientific American* (vol. 100). However, Pickering was aware that a lot of money was needed to build his experiment for communicating with Mars and asked for more certitude about life on the red planet. He wanted to take advantage of Mars's exceptional and very favorable position before beginning to set up huge projects of interplanetary communication (Pickering 1909).

H. W. Nieman and C. Wells Nieman

About ten years later, celestial telegraphy faded. Yet, it reappeared in a paper entitled "What Shall We Say to Mars? A System for Opening Communication Despite the Absence of Any Common Basis of Language," in which the authors H. W. Nieman and C. Wells Nieman proposed using Morse code to establish contact with Martians (Nieman and Nieman 1920). In this paper, the authors are clearly more focused on language than techniques. They did not suggest any particular form of apparatus for signaling. They proposed a plan of building up a common language with the Martians, who they supposed had a civilization far older than ours. They recommended sending regularly spaced dots and dashes, using either wireless telegraphy (as early as 1920) or flashes of light.

At the end of each block of about twenty signals, there would be a pause, and after a series of blocks, a long delay, which would mean the end of the message. The basic idea was that a string of radio or light pulses, coded as on-off, black-white, or 0-1, could be arranged in a rectangular raster pattern to form a two-dimensional pictorial image (Basalla 2006, 140–42). It was inspired by Native American schemes of stringing black and white beads that were arranged to form patterns.[6] The Martians were expected to convert the flashes into black and white squares. Different sequences of dots and dashes yielded different visual messages.

The Niemans' technique of communication allowed senders to convey information from simple mathematical figures to more complex pictures, such as the picture of a man (see Figure 11.4). As the Martian gallery of pictures increased, our interplanetary interlocutors would have gained a better idea of our activities and human characteristics. In addition, the pedagogic code was supposed to introduce the meaning of words and sentences, and then the understanding of whole pages of Morse code: "At this stage our language will develop very rapidly" (Nieman and Nieman 1920, 312).

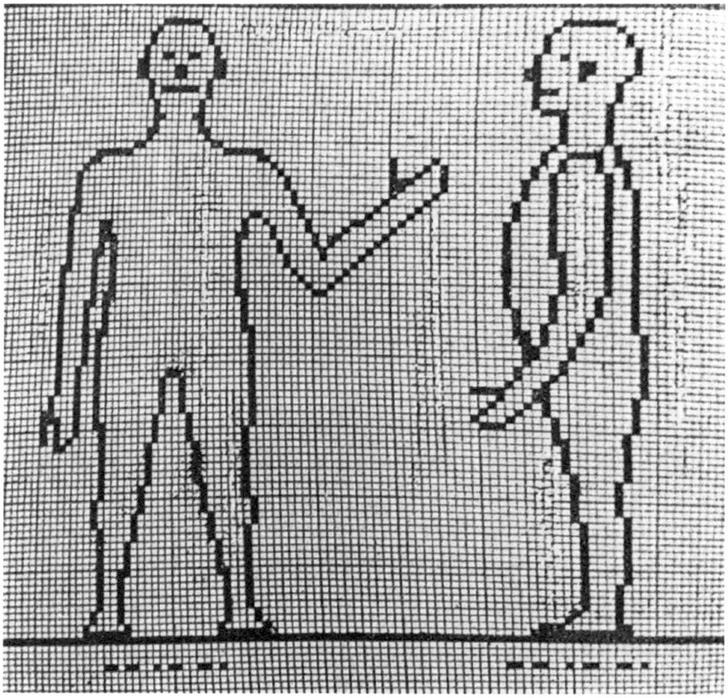

Figure 11.4 Nieman and Nieman's proposal to send increasing information to Mars: "A style more complicated message, depicting a man."
Credit: Nieman and Nieman (1920, 298)

Premodern SETI from 1920 to 1960: First attempts in wireless interplanetary telegraphy

In 1887, the German physicist Heinrich Hertz (1857–1894) discovered the existence of another type of electromagnetic waves much longer than visible light: radio waves (Drake and Sobel 1994). This was extraordinary because radio waves had such long wavelengths that they used extremely low amounts of energy. They were easy to transmit as well as receive, but technologies had to be invented for such a program to succeed. Wireless Hertz telegraphy started as early as the 1890s thanks to pioneers such as Branly, Lodge, and Popov. However, a few decades were necessary to develop remote techniques and explore all the extraordinary possibilities of these new wavelengths. In 1901, the establishment of the first trans-Atlantic radio by Marconi was of utmost importance (Shklovskiy 1963, 13). This cheaper and easier way to communicate, applied at first to communication on Earth, spread around the world in the 1920s thanks to the Italian physicist and inventor Guglielmo Marconi (1874–1937), one of the developers of commercial radio equipment. However, we now know that before this date the Serbian (and

Figure 11.5 Physicist and inventor Nikola Tesla (1856–1943), portrait by Sarony.

naturalized American) physicist and inventor Nikola Tesla (1856–1943) developed innovations in radio technology as well as electric power, another one of his chosen fields.

A physicist and electrical engineer, Tesla is famous for his alternating-current motor. Overall, his work deals with the generation and application of electric power. He believed that substantial amounts of power could be used for wireless communication, and at the end of the nineteenth century he proposed wireless communication by conducting electricity through natural media. He also believed it was possible to send signals to Mars by means of extremely powerful electrical discharges, carried out through the ether that was believed to exist between the planets: "The question is, can we transmit electrical energy to that immense distance? This I think myself competent to answer in the affirmative" (Tesla 1907, 121).

His suggestion to contact Mars came in 1899 after an unexpected event that occurred one night when he was in his laboratory at Colorado Springs. In "Talking with the Planets," published in *Collier's Weekly* (Tesla 1901), Tesla explained that he was working on improving his machines for the production of intense electrical actions when he discovered strange electrical disturbances he attributed (two years later) to signals coming from Mars: "The changes I noted were taking place periodically, and with such a clear suggestion of number and order that they were not traceable to any cause then known to me" (Tesla 1901, 5). He was convinced he had intercepted Martian signals: "The feeling is constantly growing on me that

I had been the first to hear the greeting of one planet to another. A purpose was behind these electrical signals" (5).[7]

Tesla was so excited by his discovery that, after this date, he spent a lot of time trying to develop new devices capable of conveying more powerful electrical messages to Mars (Tesla 1921). The aim was to establish a dialogue between Mars and Earth:

> Communication once established, even in the simplest way, as by a mere interchange of numbers, the progress toward more intelligible communication would be rapid. Absolute certitude as to the receipt and interchange of messages would be reached as soon as we could respond with the number "four," say, in reply to the signal "one, two, three." (Tesla 1901, 5)

When older, Tesla was still confident of getting the Guzman Prize thanks to a system he was developing for the interplanetary transmission of energy:

> I have devoted much of my time during the year to the perfecting of a new small and compact apparatus by which energy in considerable amounts can now be flashed through interstellar space to any distance without the slightest dispersion. I am expecting to put before the Institute of France an accurate description of the devices with data and calculations and claim the Pierre Guzman Prize of 100,000 francs for means of communication with other worlds, feeling perfectly sure that it will be awarded to me. The money, of course, is a trifling consideration, but for the great historical honor of being the first to achieve this miracle I would be almost willing to give my life.
>
> I am just as sure that prize will be awarded to me as if I already had it in my pocket. They have got to do it. It means it will be possible to convey several thousand units of horsepower to other planets, regardless of the distance. This discovery of mine will be remembered when everything else I have done is covered with dust. (Tesla 1937)

Unfortunately, Tesla had no opportunity to put his idea into practice and didn't get the prize. Very few details of his project were known before he died in 1943. The only thing we know for sure is that he wished to use transmission of energy (of about several thousand horsepower) to establish contact between the planets to convey information such as pictures, mathematical, and geometrical data:

> There is a solid foundation for a systematic attempt to establish communication with one of our heavenly neighbors, as Mars, which through some inventions of mine is reduced to comparatively simple problem of electrical engineering. (Tesla 1921)

Chief among his influences was the "revelatory work" of Percival Lowell about the so-called Martian canals (Tesla 1907). Convinced by the possibility of life on other planets, Tesla predicted that interplanetary communications could "become the dominating idea of the century that has just begun" (Tesla 1901, 4).

In fact, Tesla was not far off: the new century to which he alluded (the twentieth) was going to see the birth of SETI. The thing Tesla did not achieve was using radio waves to communicate over very great distances. Even on Earth, he believed that communication by extremely powerful electrical discharges was possible. Concerning the cosmos, the common thought was that the space between planets and stars was filled with a medium (ether) that made this type of communication feasible. Tesla was definitely a pioneer in telecommunications, even if he planned to transmit electrical power to convey information instead of using radio waves (Raulin Cerceau 2010).

Guglielmo Marconi, Tesla's competitor in the field of radio invention, also detected strange extraterrestrial phenomena. He was convinced he had intercepted wireless messages from Mars in 1919 when he was on his yacht *Electra*: he caught wireless signals "with wavelengths far in excess of those used by the highest powered radio stations in the world" (Marconi in Anonymous 1921).

A scientific controversy followed this announcement, since it was generally believed that these mysterious signals were caused by atmospheric disturbances (Anonymous 1922). This conclusion did not satisfy Marconi because the signals in question, he explained, "were intercepted regularly, regardless of other interference" (Marconi in Anonymous 1921). Moreover, the signals seemed to repeat the letter "V" of Morse code (Anonymous 1921; 1922).[8]

In the early 1920s, radio telecommunication was developing on Earth and made it possible to "listen" to other planets, especially our fascinating neighbor, Mars. As early as 1909, the American astronomer David Peck Todd (1855–1939) intended to test for the presence of inhabitants on Mars, as mentioned in the correspondence of *Scientific American*. He assumed that if Mars had inhabitants, they might attempt to communicate with the Earth when it is closest, and that they might employ Hertzian waves for this purpose (Anonymous 1909a). Are these suggestions pioneering views of SETI?

On August 21, 1924, Mars entered an opposition very close to Earth. In the United States, a "National Radio Silence Day" was promoted for all radios to remain quiet for five minutes every hour. At the United States Naval Observatory, a radio receiver was lifted three kilometers above the ground in a dirigible tuned to a wavelength between eight and nine kilometers, using a radio-camera developed by Amherst College and Charles Francis Jenkins. David Todd led the program with help from Admiral Edward W. Eberle (Chief of Naval Operations), with William F. Friedman (chief cryptographer of the US Army), assigned to translate any potential Martian messages. During this first "listening experiment," though, no signals were intercepted from Mars.

Modern radio astronomy was born in 1930 when Karl Jansky, working for the Bell Labs, accidentally discovered strong radio impulses coming from the center of our galaxy (Jansky 1933). American radio engineer Grote Reber became interested in Jansky's observation and decided to construct a 10 meter parabolic antenna in his backyard in order to focus the incoming radio waves, and started to investigate the sky in these wavelengths. During the 1940s, several universities started to build parabolic radio telescopes, and in a few decades radio astronomy became one of the most important areas in astronomy. In 1945, the first signal sent to the Moon was bounced off and received back on Earth. In 1959, the radar location of Venus was determined (Shklovskiy 1963). Thanks to Cocconi and Morrison's famous 1959 paper, Drake's Ozma experiment (Drake 1961) soon followed, and the foundations of modern SETI were laid. In Chapter 12, Seth Shostak explains how radio techniques have been applied to SETI searches.

In the USSR, Konstantin Tsiolkovsky was well known in the 1930s for his contributions to the theory of space flight. Yet he also wrote philosophical manuscripts about extraterrestrial life that were ignored and consigned to restricted archives for many years. The manuscripts became freely available to scholars about ten years ago. In these manuscripts, Tsiolkovsky imagined that the cosmos was filled with intelligent beings in various stages of evolution (Finney, Finney, and Lytkin 2000) but suggested no scheme to communicate with them.

Researchers in the USSR were long interested in detecting signals from extraterrestrial intelligence. As early as the 1960s, the Soviet Union named its program "the CETI problem" and hoped to establish communication with extraterrestrial civilizations ("Soviet CETI Report" 1977). In the Soviet Union, work comparable to Project Ozma was carried out at the Gorky Radiophysics Institute. Major contributions aimed at formulating and discussing the CETI problem were presented in 1964 at the Soviet National Conference on the Problem of Communication with Extraterrestrial Civilizations held at the Byurakan Astrophysical Observatory. A Soviet-American CETI conference was held at Byurakan in September 1971.[9] Even if the CETI problem was "searching for" instead of "sending" signals, the central question was as focused on the possibility of contact between cosmic civilizations as it was for nineteenth-century pioneers. The language used for establishing contact between "intelligent" systems was among the many topics that were discussed ("Soviet CETI Report" 1977).

Conclusion

During the nineteenth century, pioneering attempts to communicate with extraterrestrials focused on the possibility of interplanetary communication. One major difference with our modern SETI is that today we focus on interstellar distances. Nineteenth-century advocates believed that neighboring planets could be reached by visual signals, either through huge figures subject to change or powerful lights combined with parabolic mirrors. Obviously, visual signals were the only means of remote communication known at that time. Light signals aimed at other planets were thought of as a celestial telegraphy with optical flashes using a code. We presume that there is a link between the first steps in terrestrial electrical telegraphy and the first proposals in celestial telegraphy (see Raulin Cerceau 2006). However, no relationship seems to have been established between the first techniques proposed during the nineteenth century for interplanetary communication and those carried out since the 1960s to reach the stars. In 1924, David Todd undertook pioneering radio experiments that were similar to SETI, but they concerned interplanetary communication.

The light flashes that fascinated our nineteenth-century pioneers could perhaps be compared to optical pulses (OSETI), despite the fact that they present a major difference: the first case concerns messages *sent to* aliens while the second one is about messages *expected from* aliens. As early as the 1960s, optical contact with lasers and masers was discussed by Schwartz and Townes (1961) and Shklovskii and Sagan (1966, 399).

The first attempts also used early stage reflectors to deliver a universal "language" through the learning of a code: data are first delivered in numerical or geometrical form, and become more and more complex in order to progressively give pieces of information about human knowledge. This was a pedagogic method intended to lead to a dialogue between two communicating planets. The interplanetary messages were supposed to contain information of progressive difficulty in order to be more easily understandable. Series of numbers, logics and mathematics, and geometry were selected to represent basic (or even universal) language to communicate with aliens. The feature common to today's interstellar radio messages is the transmission of numbers and drawings.[10] Naturally, the messages proposed by the pioneers grew more complex following a very rudimentary beginning and were not able to constitute a complete language.

These forerunners raised a significant problem that is still valid today. Whatever the techniques employed, the intentional nature of the signal has to be obvious, because the most important goal is being detected as an "intelligent" civilization. Considering this last point, the early proposals are closer to METI (messaging for extraterrestrial intelligence)[11] than to (passive) SETI. METI deals with the creation of messages *to* aliens, instead of searching for messages *from* aliens (Zaitsev 2006), exactly as in the nineteenth-century proposals. Philosophical motivations concerned these pioneers, from Gauss to Tesla, who wished "to overcome the Great Silence in the Universe," as mentioned by Zaitsev (2006). The pioneering approach is also somewhat similar to the CETI problem because, in each case, a potential "contact" is considered between two civilizations. Two-way contacts between supposed close neighbors – such as residents of our solar system – were thought possible by pioneers.

Let us leave the last word to Tesla, for whom celestial communication was an imperative:

> The desire to know something of our neighbors in the immense depths of space does not spring from idle curiosity nor from thirst for knowledge, but from a deeper cause, and it is a feeling firmly rooted in the heart of every human being capable of thinking at all. (Tesla 1901, 4)

References

Anonymous. 1826. "The Moon and its Inhabitants." *New Philosophical Journal* 1: 389–90.

Anonymous. 1909a. "More about Signaling to Mars." *Scientific American* 100(May 15): 371.

Anonymous. 1909b. "Signalling to Mars, A Heliograph Proposal Quarter of a Mile of Mirrors." *Wairarapa Daily Times* 61(June 5, no. 9389): 6.

Anonymous. 1921. "Marconi Sure Mars Flashes Messages." The *New York Times,* September 2, p. 1.

Anonymous. 1922. "Progress in Wireless." *The Argus*, February 23, p.10.

Anonymous. 1975. "The Soviet CETI Program," *Soviet Astronomy-AJ* 18(5): 669–75.

Antoniadi, Eugène M. 1909. "Note on Mr. Lowell's Drawings of Mars." *Journal of the British Astronomical Association* 20: 42–43.

Antoniadi, Eugène M. 1930. *The Planet Mars.* English translation by Patrick Moore. Chatham: W & J Mackay, 1975. Translated from *La Planète Mars.* Paris: Hermann, 1930.

Basalla, George. 2006. *Civilized Life in the Universe: Scientists on Intelligent Extraterrestrials.* New York: Oxford University Press.

Cocconi, Guiseppe, and Philip Morrison. 1959. "Searching for Interstellar Communications." *Nature* 184: 844–46.

Cros, Charles. 1869. *Etude sur les moyens de communication avec les planètes.* Paris: Gauthier-Villars.

Crowe, Michael J. 1986. *The Extraterrestrial Life Debate 1750–1900: The Idea of a Plurality of Worlds from Kant to Lowell.* Cambridge University Press.

Drake, Frank. 1961. "Project Ozma." *Physics Today* 14: 140–46.

Drake, Frank, and Dava Sobel. 1994. *Is Anyone Out There?* London: Simon & Schuster.

Finney, B., L. Finney, and V. Lytkin. 2000. "Tsiolkovsky and Extraterrestrial Intelligence." *Acta Astronautica* 46(10–12): 745–49.

Flammarion, Camille. 1862. *La pluralité des mondes habités*, Paris: Mallet-Bachelier.

Flammarion, Camille. 1888. "Idée d'une communication entre les mondes." In *Rêves étoilés*, 167–84. Paris: Marpon et Flammarion.

Flammarion, Camille. 1891. "Idée d'une communication entre les mondes à propos d'un testament astronomique." *L'Astronomie* 10: 282–87.

Forrest, D. W. 1974. *Francis Galton: The Life and Work of a Victorian Genius*. London: Elek.

Galton, Francis. 1896. "Intelligible Signals between Neighbouring Stars." *Fortnightly Review* 60: 657–64.

Jansky, Karl. 1933. "Electrical Disturbances Apparently of Extraterrestrial Origin." *Proceedings of the IRE* 21: 1387–98.

Lehti, Raimo. 1998. "Edvard Engelbert Neovius, an Early Proponent of Interplanetary Communication." *Acta Astronautica* 42(10–12): 727–38.

Lemarchand, Guillermo. 1998. "Is There Intelligent Life Out There?" *Scientific American*, spec. ed., November 4: 96–104.

Lowell, Percival. 1895. *Mars*. Boston: Houghton Mifflin.

McKim, Richard. 1993. "The Life and Times of E. M. Antoniadi, 1870–1944: Part 2, The Meudon Years." *Journal of the British Astronomical Association* 103(5): 219–27.

Meunier, Stanislas. 1869. "Séance annuelle du 5 juillet 1869 de l'Académie des Sciences, Présidence de M. Bernard." *Revue Cosmos*, 3rd series, 5: 50–51.

Mercier, A. 1899. *Communications avec Mars*. Orléans: Imprimerie Orléanaise.

Nieman, H. W., and C. Wells Nieman. 1920. "What Shall We Say to Mars? A System for Opening Communication Despite the Absence of Any Common Basis of Language." *Scientific American* 122: 298, 312.

Pickering, William H. 1909. "Signalling to Mars." *Cambridge Chronicle*, May 1, p. 7.

Proctor, Richard. 1870. *Other Worlds than Ours: The Plurality of Worlds Studied under the Light of Recent Scientific Researches*. London: Longmans, Green.

Raulin Cerceau, Florence. 2006. *A l'Ecoute des Planètes*. Paris: Editions Ellipses.

Raulin Cerceau, Florence. 2010. "The Pioneers of Interplanetary Communication: From Gauss to Tesla." *Acta Astronautica* 67(11–12): 1391–98.

Raulin Cerceau, Florence, and Bénédicte Bilodeau. 2012. "A Comparison between the Ninteenth-Century Early Proposals and the Twentieth–Twenty-First Centuries Realized Projects Intended to Contact Other Planets." *Acta Astronautica* 78(Sept.–Oct.), Special Issue, *Searching for Life Signatures,* ed. John Elliot, 72–79.

Schwartz, R. N., and Charles H. Townes. 1961. "Interstellar and Interplanetary Communication by Optical Masers." *Nature* 190: 205–8.

Shklovskiy, Iosif S. 1963. "Is Communication Possible with Intelligent Beings on Other Planets?" In *Interstellar Communication,* ed. A. G. W. Cameron, 5–16. New York: W. A. Benjamin.

Shklovskii, I. S., and Carl Sagan. 1966. *Intelligent Life in the Universe*. New York: Dell Publishing.

"Soviet CETI Report." 1977. Available at http://history.nasa.gov/SP-419/s3.11.htm.

Tesla, Nikola. 1901. "Talking with the Planets." *Collier's Weekly* 26(Feb. 9): 4–5.

Tesla, Nikola. 1907. "Signalling to Mars: A Problem of Electrical Engineering." *Harvard Illustrated*, March, pp. 119–21.

Tesla, Nikola. 1921. "Interplanetary Communication." *Electrical World* 78, no. 13: 620.

Tesla, Nikola. 1937. "Sending of Messages to Planets Predicted by Dr. Tesla on Birthday." *Interview, New York Times*, July 11, p. 13.

Vakoch, Douglas A. 1998a. "Constructing Messages to Extraterrestrials: An Exosemiotic Perspective." *Acta Astronautica* 42(May–June): 697–704.

Vakoch, Douglas A. 1998b. "Signs of Life beyond Earth: A Semiotic Analysis of Interstellar Messages." *Leonardo* 31(4): 313–19.

Zaitsev, Alexander. 2006. "Messaging to Extra-Terrestrial Intelligence." Available at http://arxiv.org/pdf/physics/0610031v1.pdf.

Notes

1 Information about Gauss and Von Littrow's proposals is analyzed in Crowe 1986, chapter 4.

2 Schiaparelli called the lines he observed in 1877 "canali," which he apparently simply meant to describe the channel-like shape of these surface markings. However, this term was interpreted to literally mean artificial "canals."

3 Very little information is available about him.

4 Douglas A. Vakoch, Director of Interstellar Message Composition at the SETI Institute, discussed Galton's proposals in historical overviews about the early (at that time only interplanetary) message formats in Vakoch (1998a; 1998b).

5 Very little is known about him. He was a garage mechanic in Orléans-France, and no mention is made of his first name in his publication.

6 And, a posteriori, it has some similarities with the construction of the 1974 Arecibo Message.

7 In reality, he more probably intercepted natural radio signals caused by lightning strikes.

8 It seemed highly unlikely to Marconi that a natural source would use Morse code. In fact, Marconi himself had used the letter "V" of Morse code many years previously in one of his first wireless tests and was puzzled by these strange signals that seemed to send his own code back to him.

9 The CETI Soviet plans have been summarized in Anonymous (1975).

10 For example, Arecibo, 1974; Cosmic Call 1, 1999; Teen Age Message, 2001; Cosmic Call 2, 2003; and A Message from the Earth, 2008.

11 A term coined by Alexander Zaitsev (Zaitsev 2006).

12

Fraction of civilizations that develop a technology that releases detectable signs of their existence into space, f_c, 1961 to the present

Seth Shostak

Abstract

We consider estimates of the value of f_c – the fraction of intelligent species that can make themselves visible to other societies – by examining the recent capabilities of terrestrial technology. The value of this Drake Equation parameter is important for evaluating SETI experiments because the development of radio or other technology that would permit interstellar communication is hardly inevitable, even when intelligence is present. For example, note that *Homo sapiens* lacked advanced technology throughout most of its history. In estimating f_c, we are attempting to gauge whether intelligent species frequently become detectable.

Unfortunately, any evaluation of f_c must deal with the fact that communication depends on both the technology of the listener and the speaker, and therefore even deciding whether our own species should be counted among those that are visible depends on assumptions about the receiving capabilities of others. For consistency, we assume those capabilities are similar to our own.

We consider the strongest microwave emissions from Earth and find that – with the exception of the Arecibo radar – our own broadcasts into space would be too weak to be found by our current SETI experiments at a distance of 100 light-years. Similarly, neither our inadvertent optical emissions (street lighting) nor the type of large-scale artifacts we've built on Earth are detectable by terrestrial telescope technology at this distance. Therefore, we conclude that *Homo sapiens* has not yet attained a value of $f_c = 1$. We could not find our own society at the distances of even relatively nearby stars.

This situation might be short-lived, however, and in any case should not dissuade us from doing SETI experiments. This is because both our visibility

The Drake Equation, ed. Douglas A. Vakoch and Matthew Dowd. Published by Cambridge University Press.

and our sensitivity to signals from others will likely increase in the near future. In other words, although our communication abilities are still inadequate in the context of interstellar signaling, even a modest extrapolation of today's technology suggests that the terrestrial value of f_c will change from 0 to 1 within a few centuries. If we assume that our species will survive that long – and that most other societies can do at least as well – we conclude that the best estimate for the cosmic value of f_c is 1.

Introduction

In the series of terms in the Drake Equation, f_c is all but last. It marks the dividing line between factors that are determined by astronomy and biology, and those that are societal. Unless and until we succeed in finding intelligence elsewhere, the only data we can expect in determining a value for f_c must be found either in our own, terrestrial experience or by speculation.

The definition of f_c is the fraction of intelligent species that, at any time, develop the ability and desire to communicate with other civilizations. For a technologically competent species, once f_c becomes one, then the clock begins ticking on the last term of the Drake Equation, L (the lifetime in a communicating state). The variable f_c is a filter applied to those worlds where intelligent creatures evolve, and it separates those worlds into two categories:

(a) those that eventually make themselves visible to searches at interstellar distances, and
(b) those that never do.

If a species dies out before developing powerful radio, laser devices, or other means of becoming visible, or if its planet undergoes some natural or self-imposed catastrophe before this level of communication is reached, then this species lowers the average represented by f_c. Otherwise, it raises it.

From the perspective of our own accomplished culture, it is easy to assume that devices such as high-powered radio transmitters and lasers were inevitable once *Homo sapiens* made its appearance. Our history tempts us to believe that, sooner or later, any clever species will garner sufficient understanding of natural law to allow similar developments.

Nonetheless, we must acknowledge that humanity existed for 200,000 years without science. It wasn't necessary for survival. Indeed, given the singular – and worldwide – threats posed by modern technology, some might argue that such "advancements" are actually counter-productive to long-term existence.

However, such lugubrious thoughts are germane to the *last* term in the Drake Equation, L, not to f_c, and won't be considered here. Rather, we accept as highly probable that – with a long enough run – any intelligent species will stumble upon the practical benefits of understanding the workings of the universe in which it was born. Consequently, in this chapter we consider (a) how f_c should be defined, and (b) examine in some detail what its value might be for *Homo sapiens* during the time period from 1961 to the present. We then hazard a guess as to what its value might be for the cosmos beyond Earth.

Note that several intelligent species might arise (presumably serially) in the same parcel of cosmic real estate. Some might reach a communicative level while others might not. The variable f_c is dependent on species, not the worlds on which they arise. It is also relevant to point out that "worlds" are explicitly presumed to be synonymous with "planets" in Drake's original formulation, although it's now believed that large moons – even those orbiting planets that themselves are not habitable – might also spawn intelligence. This implies some obvious modifications to early terms in the Drake Equation.

Ambiguities in formulating f_c

One of the problems in estimating f_c – a parameter of clear relevance to any SETI experiment – is that its value depends on the technological sophistication of those doing the listening. For example, any extraterrestrial society having technology comparable to our own would be unable to detect 1961-era *Homo sapiens*. Many of the powerful transmitters we have today were not yet in operation (e.g., the Arecibo radar), and no *deliberate* transmissions to putative extraterrestrials had yet been made.

On the other hand, the situation is different for a society far beyond our level. If one of these cultures wielded radio antennas much larger than anything now found on Earth, or was able to deploy monitoring probes to other star systems, they would have been able to detect our 1961 technology.

Such considerations demonstrate that any value of f_c that we proffer must be tied to an assumed level of detection capability. Since the original purpose of the Drake Equation was to judge the prospects for our own SETI searches (as opposed to those conducted by extraterrestrials), we will reckon values of f_c based on terrestrial capabilities. We will match our broadcasts with our detection technologies during the past half-century to gauge whether we had reached a level at which f_c became one.

We will refer to this measure of our own communicative ability as terrestrial f_c. The "cosmic" value of f_c – that which applies to other intelligent species – can be inferred only by considering our own case. Of course it is the cosmic value that is relevant to the success, or otherwise, of our SETI searches, as it governs the communicative ability of others.

Note that our analysis has obvious limitations:

(a) Since we have stipulated that f_c can only be either one or zero at any time, this approach is very crude. When do we declare f_c equal to 1? Does that happen if a single, high-powered transmitter goes on the air? Or should we take a conservative tack, and define f_c for *Homo sapiens* as 0 until such point as we launch very high-powered signals to, say, 90 percent of the sky?

Suppose we do nothing to build stronger transmitters, but construct a very large receiving antenna that could find terrestrial leakage from many light-years distant? In other words, such an instrument would give us the capability to find a society with leakage at our level. Can we say that terrestrial f_c has reached 1 at that point?

These are ambiguities without clear resolution.

(b) By focusing on our own technology, we have the benefit of example. But it is a single example, and consequently gives no indication of its precision. What we deduce from our own situation may not give useful insight into the prospects for SETI, the whole point of the Drake Equation.

In what follows, we consider the development of both our signaling and receiving capabilities from 1961 to the present. This will give some indication of the changing value of terrestrial f_c. We then briefly speculate on what implications our investigation has on the value of cosmic f_c.

Terrestrial communicative capability

In the early 1960s, industrialized societies were literally rocketing ahead with developments that would significantly increase their potential visibility from afar. The space age, generally dated from the launch of Sputnik in 1957, had just begun. In little more than a decade, many satellites had been placed in orbit, and humans had reached the moon. Although this crowning achievement of human spaceflight would not, by itself, be detectable by putative extraterrestrials, it sparked the building of communication technologies such as the Deep Space Network and other telemetry transmitters that increased leakage into space.

At the same time, the production of the first commercial jet aircraft (the Boeing 707 appeared in 1958) initiated the rapid growth in aviation, with its concomitant and widespread deployment of radars. The Arecibo radio telescope, which even today sports the most intense radar in the world, was completed in 1963.

The interstate highway network, which is certainly one of the most visible earthly constructions, had been underway since roughly the launch of Sputnik, and street lighting was becoming more intense (low-pressure sodium vapor lamps were first introduced into the commercial market in 1932, while high-pressure lamps for street lighting came into vogue in the 1960s), producing a relatively narrow-band glow from large urban areas.

Would any of these inadvertent "signals" be easily detectable by terrestrial SETI technologies if they were put at light-years' remove? We consider the signal efficacy of representative samples below.

Signal generation and detection

Although the physics underlying radio was worked out 150 years ago, practical radio was developed only in the twentieth century. Almost immediately, it was realized that it offered an efficient and speedy means of communication through space. Early experimenters like Marconi and Tesla both thought they might have detected a civilization on Mars by using radio to tune in its signals (see Anonymous 1921; Corum and Corum 2003).

In retrospect, these claims were clearly faulty, not only because of the lack of intelligent life on the Red Planet, but also because these radio pioneers were using equipment operating at relatively low frequencies. The ionosphere severely affects radio below about 10 megahertz, making space communications at these frequencies problematic (Australian Space Academy 2014). Above about 150 gigahertz, water vapor blocks radio. Almost all SETI experiments – which have been so far confined to the surface of the Earth – operate in the ranges of 1–10 gigahertz, where these difficulties are less, and where natural noise from either high-speed electrons in the galaxy or quantum noise from individual radio photons is at a minimum. This is the so-called "microwave window."

Commercial radio and television

As noted, until the advent of FM and television roughly at the time of the Second World War, most commercial radio was at low frequency – below 1 megahertz – and strongly affected by the ionosphere. By 1961, however, that situation had changed dramatically. There were several hundred television transmitters in the United States, and FM radio, whose early adoption was delayed by the competitive maneuvering of those invested in AM stations, had begun a dramatic rise in popularity. Today, there are approximately 1,700 high-powered TV stations in the United States, and 6,600 moderate-to-high-powered FM transmitters.

Clearly, given their large number, their (mostly) 100 percent duty cycle, and their ability to transmit in many directions (the definition of broadcasting), these signals are excellent candidates for providing extraterrestrials with proof of our existence. But how intense are these transmissions?

Typical high-powered transmitters might run at 100,000 watts. Assuming a modest antenna gain of 10 (signals are aimed at the horizon) a television signal would have an average flux density over the 5 megahertz analog TV band of 2×10^{-37} watts/m^2-Hz at 100 light-years distance. A far more visible emission would be the narrow-band (~1 Hz) carriers within these signals, with perhaps one-tenth of the total transmitter power. These carriers will have a much higher flux density of 10^{-31} watts/m^2-Hz, and – when account is made of the difference in bandwidth – a signal-to-noise ratio that is two hundred times greater than for the full signal.

How do these fluxes compare to SETI detection sensitivities? By 1961, only one (microwave) SETI experiment had been conducted, the celebrated Project Ozma observations by Frank Drake himself (Drake 1961). A very rough estimate of the sensitivity of this experiment, using an 85-foot diameter antenna, is 10^{-21} watts/m^2-Hz over an approximately 100 hertz bandwidth.

However, since that time, SETI experiments have benefited from antennas having larger collecting areas, with low-noise amplifiers feeding multichannel receivers. The best searches – including Project Phoenix and SERENDIP (Werthimer et al. 2000; Tarter 1997) – have been able to detect narrow-band (~1 Hz) signals at around 10^{-24}–10^{-25} watts/m^2-Hz, typically at 1.4 gigahertz frequency. These efforts are too insensitive by six or seven orders of magnitude to find TV and FM signals at 100 light-years, even assuming they were tuned to the correct frequencies.

The new LOFAR (Low-Frequency Array) radio telescope being constructed in Europe operates at the lowest frequencies useful for observations from Earth's surface and has considerable collecting area. LOFAR's sensitivity at 120 megahertz, close to the band used by FM and broadcast TV, is estimated to be 2×10^{-27} watts/m^2-Hz over 3.6 megahertz of bandwidth after eight hours of integration (ASTRON 2014). Again, this is still inadequate for detecting our own commercial broadcasting at 100 light-years – or even at the distances of the nearest star systems.

While TV and FM radio broadcasting have been signaling our presence to the entire sky, twenty-four hours a day for approximately seventy years, these markers of our presence would be too weak by 6–10 orders of magnitude to be found by equipment similar to the best we currently use for our own SETI searches. They do not budge our terrestrial value of f_c at the present time above zero.

Radar

The rapid spread of radar following the Second World War has produced the most obvious communicative signal from Earth. We measure the visibility of radar signals to extraterrestrial receivers as the effective radiated power per unit bandwidth. This depends on (a) transmitter power, (b) duty cycle, (c) instantaneous frequency spread, and (d) the beam size of the transmitting setup.

While many early radars emitted continuous signals, most today are pulsed. Pulsing a radar signal allows for quiet intervals to receive the return echo, but it also permits higher pulse power for a given average transmitter power, thus increasing range.

Clearly, for a given transmitter power, the detectability of a radar signal – from the extraterrestrials' point of view – will increase as the duty cycle (the space between pulses) decreases. This assumes that those trying to detect these emissions can time-slice their receivers, rather than integrating incoming signals for long (many seconds) periods of time. If they don't do this (and many of our own SETI experiments do not), then the detectability is governed only by the average transmitter power, and not the duty cycle.

Examples of some powerful radar systems constructed since 1961 include:

(1) The NEXRAD network, currently consisting of 160 S-band weather radars used by the US National Weather Service, and blanketing the country (NOAA 2014). This system was developed beginning in the 1970s. They generally work at near 3 gigahertz, with pulse widths of a few microseconds and peak power of 750 kilowatts (average power of 300–1,300 watts) and a beam width of ~1 degree. At 100 light-years, this would produce fluxes of 10^{-27} watts/m^2 during the pulses, or 10^{-30} watts/m^2 average. (Note the difference between flux and flux density. The latter is the intensity per unity frequency interval.)

(2) Long-Range Air Route Surveillance Radar (ARSR), used to monitor aviation around the borders of the United States and dating from approximately 1990 (Weber 2000). There are approximately 100 of these, operating at about 1.3 gigahertz and ranging in peak power from a few tens of kilowatts to more than a megawatt. Pulse widths are from 1 to 150 microseconds, and the beam width is 1.4 degrees. Again, at 100 light-years, these would generate fluxes of 5×10^{-28} watts/m^2 during the pulses, or 10^{-29} watts/m^2 average.

(3) The Air Force Space Surveillance System (the so-called "Space Fence"), put in place in 1961 and consisting of three powerful radars in an east–west line crossing the United States (Chien 2012; Thomson 1996). It was used to monitor orbiting satellites. The most powerful of the three, in Lake Kickapoo, Texas, operated at 210 megahertz as a continuous-wave radar with power output of 770 kilowatts (the other two had transmitter powers of 40 kilowatts). With a two-mile-long dipole array, the north–south beam width was approximately 0.02 degrees, although the east–west component of the fan beam was nearly horizon to horizon. The maximum flux at 100 light-years would be 2×10^{-28} watts/m^2. Due to sequestration, this system was shut down in the fall of 2013. A replacement system is expected to be deployed by 2017.

(4) HAARP. Operated by the University of Alaska in Fairbanks, HAARP (High Frequency Active Auroral Research Program) illuminates – either in CW or pulse mode – a small section of the ionosphere at frequencies between 2.7 and 10 gigahertz ("High Frequency Active Auroral Research Program" 2014). Maximum power is 3.6 megawatts, and antenna gain is 31 decibels, for an effective isotropic radiated power (EIRP) of 5.1 gigawatts. When activated, this device would produce a flux at 100 light-years of 10^{-28} watts/m^2.
(5) Arecibo. With its 305 meter (1,000 foot) reflector, this is the largest single-dish antenna in the world. It is used in a radar mode for ionosphere, planetary, comet, and asteroid studies (Nolan 2014). The transmitter is 1 megawatt, operating at 2.4 gigahertz, with a beam of ~ 0.03 degrees. Consequently, the flux at 100 light-years is 1 x 10^{-24} watts/m^2, making this the most powerful signal from Earth today. Note, however, that it covers only about 2×10^{-8} of the sky at any time.

The intensities given for the radar transmitters above are fluxes. To convert these to flux densities, one has to consider their instantaneous bandwidth. Pulsed radars are intrinsically broadened to a bandwidth ~ $1/\tau$, where τ is the pulse width. SETI receiver sensitivity decreases for any given transmitter power as the bandwidth increases, and by a factor of $\tau^{1/2}$. Consequently, the detection threshold for the best SETI searches, if looking for radars with a microsecond pulse, will be ~10^{-21}–10^{-22} watts/m^2 over a second or longer (the minimum integration time of virtually all SETI experiments).

The bottom line for the detectability of terrestrial radar can be summarized as follows:

(a) In 1961, we had not conducted any SETI experiments that were sensitive enough to pick up radar transmissions from societies 100 light-years away at our own technological level (or even at the ~12 light-year distance of the star systems targeted by Project Ozma).
(b) By 2014, the best SETI equipment could achieve sensitivities adequate to find Arecibo (albeit under the assumption that our receiving equipment is oriented to intercept the beam of such a radar), but is inadequate to find other emitters on the above list by at least six or seven orders of magnitude.

Given that our easily detectable radar signals are highly constrained in space and time, we suggest that these too have not increased the terrestrial value of f_c beyond zero as of the present.

Note that we have chosen 100 light-years as the nominal distance for determining these sensitivities, and have done so on the basis of estimates for the number of detectable galactic societies (Dick 1998, 217). If there are as many as 1 million such societies transmitting now, then the average separation of these worlds will be about 100 light-years. If the number of transmitting societies is lower – say 10,000 – then the average separation is 1,000 light-years, and sensitivity figures should be adjusted accordingly.

Optical signals

Electromagnetic signaling for terrestrial messaging was first done at visible wavelengths, using mirrors and semaphores. In the twentieth century, optical communication was superseded by the rapid development of radio. However, with the eventual rise of lasers and high-speed photon detectors, optical signaling has become important for both communications and, in a limited way, for SETI efforts.

A simple thought experiment illustrates the attractiveness of optical SETI. If we were to beam a high-powered laser towards the sky – comparable in intensity to that used by the National Ignition Facility in California (National Ignition Facility 2014) – and focus it with a mirror of several meters diameter, its nanosecond burst of light ($>10^6$ joules) would momentarily outshine the Sun at the laser's wavelength, even at 100 light-years remove.

To search for such deliberate flashes from others, optical SETI practitioners employ small conventional telescopes fitted with sensitive detectors that slice time into nanosecond intervals. Such short bursts would betray an artificial source – and could easily be distinguished from the gentle background glow of stars and nebulae. An experiment at Harvard University, using a 1.8 meter optical telescope, has been relentlessly scanning much of the sky visible from Massachusetts, searching for ultra-brief laser pings (The Planetary Society 2014)

There have not yet been any deliberate attempts to send powerful laser flashes to putative recipients on other worlds. Consequently, we have made no intentional efforts that would help us estimate the value of f_c due to optical signaling capability.

There is, however, a certain amount of optical leakage. A city the size of New York has ~3×10^5 streetlamps, each putting out approximately 5×10^3 lumens or 50 watts in a narrow part of the visible spectrum. If 10 percent of the light is sent skyward, then the photon flux at 100 light-years is ~5×10^{-12} photons/m^2-sec. However, this weak signal is also an inadequate one from Earth. It would be thoroughly swamped by light from the Sun, which even in the narrow-band emission regime of high-pressure sodium vapor lamps still amounts to more than 10^5 photons/m^2-sec at this distance.

However, for advanced technologies, the use of gravitational lensing coupled with occulting techniques to block starlight might allow detection of this type of leakage. Note that it would not be necessary to image the city itself; the street lamps' unusual spectral distribution would give away the technological origins of this light.

Optically Visible Artifacts

It is said that the most visible evidence of *Homo sapiens*, when seen from orbit, consists of large constructions such as the Great Wall of China, airport runways, or major highways. What would it take to see similar artifacts of intelligence on a world 100 light-years away? For a highway having a 100 meter breadth, the required aperture size to resolve this structure at optical wavelengths is roughly 5×10^4 kilometers, or more than three times the diameter of Earth, far beyond our capabilities.

Another possibility is that we might detect large orbiting structures that periodically transit their home star and produce dips in the star's brightness. The Kepler space telescope can detect such dips at a level of 10^{-4} the brightness of the star (NASA Ames Research Center 2014). For the favorable case of a small red dwarf with 0.2 the diameter of the Sun, this sets a minimum size for a detectable object of several thousand kilometers, considerably larger than any space structures that we have either built or contemplated.

None of the above suggests that we have reached a level at which f_c equals one.

Implications of $f_c = 0$

We have noted that even our best SETI experiments would be unable to detect terrestrial signals, both radio and optical, at 100 light-years, with the exception of the most powerful of our radars. But the latter might never be noticed, given their miniscule sky coverage.

Consequently, at the present time (and certainly in 1961), we conclude that the terrestrial value of f_c is 0. We could not find our own society at the distances of even relatively nearby stars.

This isn't to say, however, that this inability should be a disincentive for ongoing SETI searches, nor does it mean that others couldn't find us. This is because of (a) the relentless improvement in our SETI detection capability, (b) the very real possibility that other, more advanced societies are producing far stronger signals than we are, and (c) the increasing visibility of our species as a consequence of technical progress.

With the development of broadcasting and radar, the leakage from Earth has increased substantially over the last seventy years. If past is prologue, this growth in signal strength will continue. However, some SETI practitioners have pointed out that the trend to more efficient dissemination of information (direct satellite broadcast, for example) as well as the growth of spread-spectrum transmissions is actually reducing the detectable leakage from Earth.

For certain communication activities, this is true. But one could also plausibly argue that the necessity for high-powered signals will never disappear. Imagine

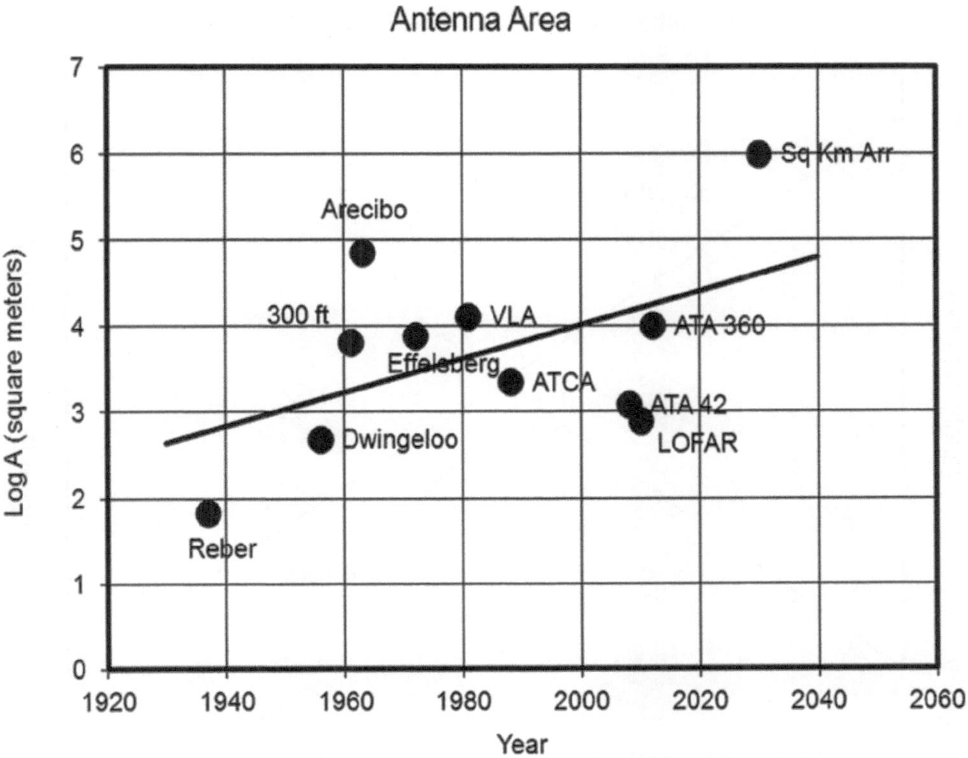

Figure 12.1 Collecting areas of existing and planned radio telescopes. Note that the Allen Telescope Array is plotted twice, both in its current configuration (42 antennas) and in the design goal configuration of 350 antennas. The overall trend amounts to two orders of magnitude growth in collecting area per century. *Credits*: Matthew F. Dowd

an effort to map the Oort Cloud with radar, motivated by the threat posed to our existence by long-period comets. Such a project is hardly infeasible, and might involve transmitters far more powerful than our current radar installations. This simple example suggests that we should be cautious in interpreting the short-term diminution in emission detectability occasioned by digital and spread-spectrum technologies as indicative of a long-term trend. It could just as well be merely a bump in the road of relentlessly increasing leakage.

On the receiving side, it seems certain that our abilities to detect signals from other worlds will only improve. Figure 12.1 plots the collecting areas of existing and planned radio astronomy antennas over the last eighty years. While there is considerable scatter, the clear growth in sensitivity is evident, and amounts to an increase of ~2 orders of magnitude per century.

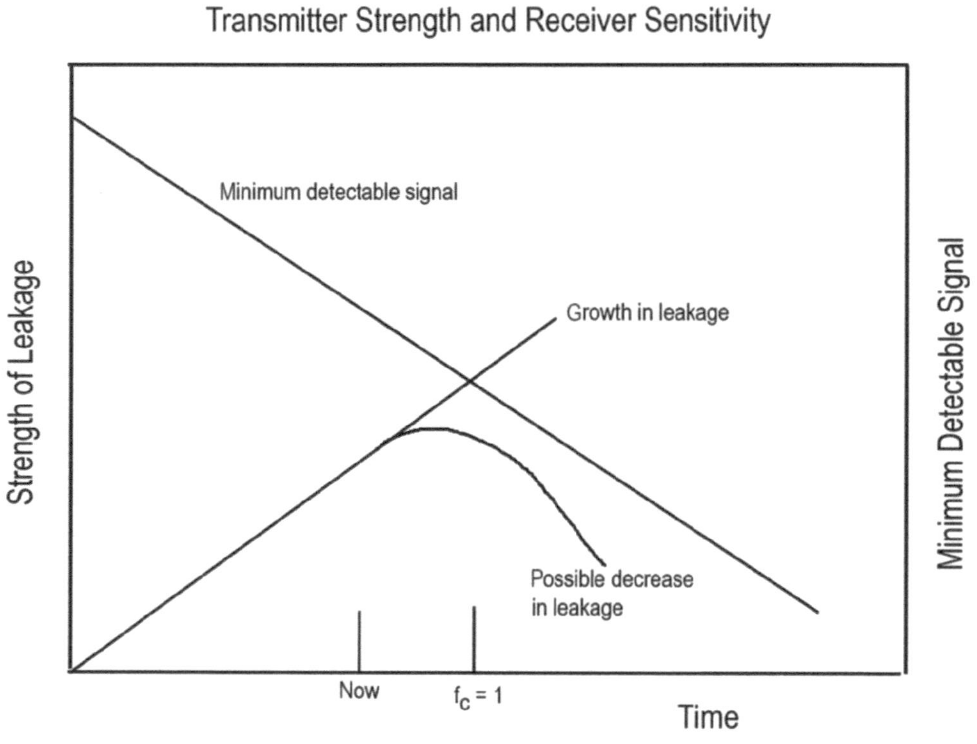

Figure 12.2 Schematic representation of the two technologies that determine the terrestrial value of f_c. The increasing curves show the strength of leakage emissions from Earth. The straight line assumes that these will continue to increase, while the bent line indicates what might happen if – for reasons of energy economy – we discontinue all high-powered, narrow-band transmissions in favor of direct broadcasting, fiber networks, and spread-spectrum radio. The descending line traces the continued improvement in our technologies for detecting transmissions from others (see Figure 12.1). At the point when these two curves cross, the terrestrial value of f_c switches from 0 to 1.
Credits: Matthew F. Dowd

These two lines of development are schematically depicted in Figure 12.2, which shows improvements in SETI sensitivity as well as two scenarios for the future strength of emissions from Earth. A change in the terrestrial value of f_c from zero to one occurs if and when the two lines cross. As described in this chapter, we are still in the regime to the left side of that crossover.

However, and despite the fact that *Homo sapiens* has not yet qualified as a "civilized" species as defined by the Drake Equation, the rapid pace of technology suggests that, if our cultural history is typical of other species, at least some of them will have progressed beyond our present circumstances and achieved the ability to produce signals that will be detectable with our present or future receiving capabilities. This is, after all, the motivation for SETI searches: These experiments are not designed to find our doppelgängers amongst the stars – as noted, we do not yet have the ability to do this – but to discover societies that are several centuries or more beyond our abilities to transmit.

Conclusion

The term f_c is maddeningly difficult to estimate because (a) the data are limited to the experience of our own species, so no "average value" can be computed, and (b) the abilities to transmit and receive that determine the value of f_c change rapidly with time, once communicative technologies such as radar and the laser are developed.

At present, we would be unable to detect nearly all of the unintentional leakage from our own planet even with the best of our SETI experiments at a distance of 100 light-years. An additional 3–10 orders of magnitude improvement in our receiving equipment is required to change that situation. So the value of f_c we would deduce strictly on the basis of our present capabilities is zero.

However, of greater immediate relevance to SETI experiments is the trend of our technology. The strength of our transmissions into space has increased in the last century. Yes, this trend might abate as neither leakage nor deliberate transmissions (which could be significantly stronger) are guaranteed to grow. But historically, our planet has become noisier, not quieter. If the same is true for other societies, our SETI experiments could find proof of extraterrestrial intelligence at any time, including next week or next year.

Even if that's not the case, it is also clear that our own receiving capabilities are changing rapidly, and within one to two centuries (assuming no civilization-ending catastrophes), we will have the requisite sensitivity to detect emissions comparable to those leaking off Earth today from star systems 100 light-years or more distant. Even if extraterrestrials make no deliberate transmissions nor increase the noise they produce beyond what *Homo sapiens* manages a scant half-millennium after the beginnings of modern science, we could find proof of extraterrestrial intelligence.

In other words, although our own communication abilities are still inadequate in the context of interstellar signaling, even a modest extrapolation of our own likely developments suggests that the terrestrial value of f_c will change from zero to one within a few centuries. Unless extraterrestrial societies are routinely destroyed short of their own scientific renaissance, it seems that the best estimate for the cosmic value of f_c derives from this modest assessment of our own technological trajectory. And that implies that the best estimate for f_c is 1.

References

Anonymous. 1921. "Marconi Sure Mars Flashes Messages." The *New York Times*, September 2, p. 1.

ASTRON. 2014. "Sensitivity of the LOFAR Array." Available at www.astron.nl/radio-observatory/astronomers/lofar-imaging-capabilities-sensitivity/sensitivity-lofar-array/sensiti/.

Australian Space Academy. 2014. "The Ionosphere and Radio Astronomy." Available at www.spaceacademy.net.au/env/spwx/raiono.htm.

Chien, Philip. 2012. "How to Listen to RADAR Signals from Space." Available at www.neatinformation.com/science/space%20radar.html.

Corum, Kenneth L., and James F. Corum. 2003. "Nikola Tesla and the Planetary Radio Signals." Available at www.teslasociety.com/mars.pdf.

Dick, S. J. 1998. *Life on Other Worlds*. Cambridge University Press.

Drake, F. D. 1961. "Project Ozma", *Physics Today* 14(4): 40–46.

The Planetary Society. 2014. "Optical SETI." Available at www.planetary.org/explore/projects/seti/optical-seti.html.

"High Frequency Active Auroral Research Program." 2014. Available at www.princeton.edu/~achaney/tmve/wiki100k/docs/High_Frequency_Active_Auroral_Research_Program.html.

NASA Ames Research Center. 2014. "Kepler: A Search for Habitable Planets." Available at http://kepler.nasa.gov/.

National Ignition Facility. 2014. "How NIF Works." Available at https://lasers.llnl.gov/about/how-nif-works.

NOAA. 2014. "NEXRAD Technical Information." Available at www.roc.noaa.gov/WSR88D/Engineering/NEXRADTechInfo.aspx.

Nolan, Mike. 2014. "Arecibo Planetary Radar." Available at www.naic.edu/~nolan/radar/.

Tarter, J. C. 1997. "Results from Project Phoenix: Looking Up from Down Under." In *Astronomical and Biochemical Origins and the Search for Life in the Universe,* eds. C. B. Cosmovici, S. Bowyer, and D. Werthimer, 633. Bologna: Editrice Compositari.

Thomson, Allen. 1996. "US Naval Space Command Space Surveillance System." Available at www.fas.org/spp/military/program/track/spasur_at.htm.

Weber, M. E. 2000. "FAA Surveillance Radar Data as a Complement to the WSR-88D Network." Available at www.ll.mit.edu/mission/aviation/publications/publication-files/ms-papers/Weber_2000_ARAM_MS-14191_WW-10147.pdf.

Werthimer, D., S. Bowyer, J. Cobb, M. Ledofsky, and M. Lampton. 2000. "The SERENDIP IV Arecibo Sky Survey." *Bioastronomy 99: A New Era in Bioastronomy*, ed. G. L. Lemarchand and K. Meech, 479–83. ASP conference Series 213.

13

Length of time such civilizations release detectable signals into space, L, pre-1961

David Dunér

Abstract

This chapter is an overview of the prehistory of L, how people across the globe from our earliest sources to 1961 have tried to understand the beginning and end of history, and the rise and fall of civilizations. Factor L is put into a longer historical context of human conceptions about history, time, and civilization. In focus is the question of how it was possible to formulate L in its modern version, as embodied in the Drake Equation. This was not possible, I argue, until the end of the nineteenth century. L required a number of philosophical, scientific, and technical discoveries and inventions before it became possible to discuss the longevity of extraterrestrial technical civilizations. Of special significance was the "discovery of time," the emergence of a set of ideas for understanding human temporality: first, linear time, time that has a beginning and an end, and in which nothing is forever; second, long time lines, in which there was a time before humans and human civilization, and that the history of our civilization is only a fraction of the history of universe; and third, that time has a direction, that humans are historical beings – that is, knowledge, culture, and society are not something preexisting but something created by humans, evolving, that rests on the experiences and actions of previous generations in a cumulative process leading to the development of knowledge, behavior, and life conditions, or what is sometimes called the "idea of progress."

The first section concerns the beginning of time: notions of the dawn and age of the world, thoughts about the history of the Earth and humankind, when humans entered the history of the universe, and the emergence of the notion that human civilization has existed for only a fraction of the total age of the universe. The second section concerns the direction of time: where we are heading, ideas about

The Drake Equation, ed. Douglas A. Vakoch and Matthew Dowd. Published by Cambridge University Press.

how societies emerge, the rise of civilizations, and the notion of advancement, the thought that civilization is not a given but something created by humans. The third section puts forward notions of the end of time: doom, cataclysms, the meaning of history, and how and why civilizations and empires – or the whole world – fall. Finally, I conclude that L is a measure of the civilizing or socialization process, and the variables that underlie it: biocultural coevolution and the interaction between the evolution of cognition and socialization.

> Once upon a time, in some out of the way corner of that universe which is dispersed into numberless twinkling solar systems, there was a star upon which clever beasts invented knowing. That was the most arrogant and mendacious minute of "world history," but nevertheless, it was only a minute. After nature had drawn a few breaths, the star cooled and congealed, and the clever beasts had to die.
>
> Friedrich Nietzsche, *Über Wahrheit und Lüge im außermoralischen Sinne [On Truth and Lies in an Extra-Moral Sense]* (1873)

Introduction

L and human existence

The factor L – the longevity of civilizations – is one of the great existential questions. In a more technical sense, L could be formulated as the number of Earth years that a cognitively flexible extraterrestrial life form can manage to maintain a social organization that enables it to voluntary transmit electromagnetic radiation that we can detect. As with all forms of communication, interstellar communication is not just about the transmitter's ability to emit signals, but it also concerns the recipient's ability to perceive and interpret these signals as containing a meaning, or at least to be a sign of an artificially produced signal containing a meaning that is supposed to be received and interpreted (Dunér 2011). In that sense, Drake's equation and the factor L are not just about possible life forms out there, but also about ourselves and our ability to perceive and interpret our surroundings. In other words, it is a human, existential problem (Figure 13.1).

The various factors in the Drake Equation, which are meant to capture the essential conditions for any detectable intelligent life form, move gradually from the astronomical, geological, chemical, and biological conditions for life as we know it to emerge. With the factor f_i, the fraction of living planets that also evolve intelligent life, we arrive at a property that we assign our own species, the modern subspecies *Homo sapiens sapiens*. The factor f_i is thus a problem for cognitive science: what mechanisms and environmental conditions are needed in order for living organisms to evolve intelligence, or what I would call cognitive flexibility: in other words, the mental ability to adapt to changes in the physical and social

Figure 13.1 Death and civilization in the universe. A churchyard with crosses, and above gleaming stars.
Credit: Bischof (1791, table I of the Swedish edition, 1796)

environment (Dunér 2014). As such, intelligence is a result of a biocultural coevolution. Intelligence, or cognitive flexibility, is not only a capacity for rational thinking – for example, to draw logical conclusions, to be able to use mathematics, and so forth – but also a mental capacity to imagine things and events that are not presently in time and space, that have occurred, will occur, or never will or can occur. A particularly characteristic feature of intelligent life is the ability to become familiar with other individual minds, to understand, interpret, feel, and anticipate what others are feeling and thinking. This is precisely what is at the core of the research in the humanities: how meaning emerges and is transmitted and interpreted.

The last two factors in the Drake Equation, f_c and L, are mainly problems within the humanities and social sciences. They imply the emergence of culture, namely, knowledge and experience that can be passed on from generation to generation in a way that is not genetically encoded but is mediated through learning and memory by using language or other communicative and artificial, external artefacts. We are thus dealing with cultural evolution. The factor f_c is about the extent to which intelligent life succeeds in gaining the technical and scientific capability to transmit artificial electromagnetic radiation. To assign a value to this factor, we have to obtain an idea of how advanced science and technology can occur and what factors make them possible. In other words, we approach this question within the history of science and technology. In addition to cognitive flexibility, another factor also forms the basis of advanced scientific and technological capability: the ability to cooperate, which has as a prerequisite a sufficient social and communicative complexity (Dunér 2014). A technologically advanced civilization must have developed a social and communicative organization that facilitates cooperation, regulates and prevents conflicts that could lead to the collapse, and is able to adapt itself or its environment to prevailing physical conditions – that is, it must have been able to create a sustainable society in equilibrium with the physical and biological environment around it so that it can survive for a time period long enough to significantly increase the probability of detection. Thus we arrive at the final factor in the Drake's equation, L, the longevity of technological civilizations, which is the focus of this chapter.

Of all the factors in the Drake Equation, L is often described as the most difficult to estimate (Denning 2013; Dominik and Zarnecki 2011; Penny 2011; Shostak 2009). We know only one global advanced technological civilization – ours – and we have not yet seen the end of it. Is L 200, 2,000, 200,000, or 2,000,000 years? We do not know. The difficulty has to do not only with the fact that we lack empirical data about other technological civilizations but also that we are dealing with very complex questions about how self-conscious beings organize their social structure. They can manipulate and change their way of life, and are not subject only to given, more or less determining physical, chemical, geological, and biological laws of nature. Clues to the factors behind the longevity of civilizations might be found in our own history. The attempt to estimate L is in its essence a historical problem that confronts us with sociocultural issues concerning how advanced social systems are organized, what social and cultural factors make them possible, and what factors are involved in their eventual breakdown.

L in human history

Factor L captures, as we have noted, the great existential questions: Where do we come from? Who are we? Where are we heading? What will happen in the future? When will the last days come? Or are we already there? It also addresses a number of fundamental philosophical, conceptual, and ontological-metaphysical questions: What is time? What is history? What characterizes culture, civilization, and human action? L raises historical questions, seemingly "eternal" macrohistorical, global-historical, or universal-historical ones that humanity has tried to answer for thousands of years. When did it all start? What is the origin of human beings? How old is the Earth, life, the human race, our culture, and civilization? How and why did modern society appear? How do complex societies and empires develop, and how do advanced science and technology arise? Does history have an end, and if so, when and why will it come? L is also about social issues. How are communities organized, how are they maintained and developed, and how do we prevent and handle crises in a society? Why do some civilizations fall? To a large extent, L is also about definitions of what we mean by "time," "history," "culture," and "civilization" (see also Chick, Chapter 14).

Mankind has pondered all of these questions as far back in time as we can find written sources or other cultural, material artifacts. In this chapter, I will give an overview of the prehistory of L, how people across the globe from our earliest sources to 1961 have tried to understand the beginning and end of history, and the rise and fall of civilizations. I will thus put factor L into a longer historical context of human conceptions about history, time, and civilization, about the longevity of civilizations and the prerequisites for their emergence, and the threats to their existence. In focus is the question of how it was possible to formulate L in its modern version, as embodied in the Drake Equation. I shall argue that this was not possible until the end of the nineteenth century, after thousands of years of human thinking. L required a number of philosophical, scientific, and technical discoveries and inventions before it became possible to discuss the longevity of extraterrestrial technical civilizations. Particularly, I will here highlight what I call the "discovery of time" (see also Toulmin and Goodfield 1965), the emergence of a set of ideas for understanding human temporality: first, linear time, time that has a beginning and an end, and in which nothing is forever; second, long time lines, in which there was a time before humans, before human civilization, and that the history of our civilization is only a fraction of the history of universe; and third, that time has a direction, that humans are historical beings – that is, knowledge, culture and society are not something preexisting but something created by humans, evolving, that rests on experiences and actions of previous generations in a cumulative process leading to the development of knowledge, behavior, and life conditions, or what is sometimes called the "idea of progress."

Next, I will, in an attempt to make an historical-philosophical synthesis, highlight a few of the huge number of ideas formulated through the millennia of human history that can be said to form the prehistory of L. It is an attempt to create a *mappa mundi* of human understanding of the beginning and end of time. The first section concerns the beginning of time, notions of the dawn and age of the world, thoughts about the history of the Earth and the humankind, when humans entered the history of the universe, and the emergence of the notion that human civilization has existed for only a fraction of the total age of the universe. The second section concerns the direction of time, where we are heading, ideas about how societies emerge, the rise of civilizations, and the notion of advancement, the thought that civilization is not a given but something created by humans. And finally, in the third section, I discuss notions of the end of time, doom, cataclysms, the meaning of history, and how and why civilizations and empires – or the whole world – fall.

The age of the world

Time and creation myths

The most tangible experience of time's limits is the knowledge of our own mortality. We are born; we age; and then we die. In cultures without writing there was, and still is, the experience that ancestors have preceded them and that others will follow when they leave this life (Clark 1992). There is a past and a future. Early remains of Neanderthals show that the dead were buried with certain ceremonies that suggest that they had an idea about the course of time. Even clearer examples of a notion of a time after death are the graves and stone chamber tombs containing grave gifts from the Neolithic era that can be found in northwest Europe, and other remnants of burial customs that can be found in almost all ancient cultures from all parts of the world. Early cultures also established genealogies, which preserved knowledge of ancestors and put the present in a temporal dimension. An early example of this is the chronicle of Egyptian rulers down to the end of the Hyksos period. It could be argued that some oral cultures without a writing system lived and live in a single moment, in a recurring forever. For example, the Amondawa tribe in the Amazon forest still speaks a language that lacks a way to express past and future (Sinha 2011). The lack of an abstract idea of time does not mean that they are unable to imagine the past and the future, but that they have no need in their daily lives to describe their situation in a temporal dimension. Members of the Amondawa tribe who learn Portuguese have no problem with speaking in past or future tense.

The conception of time as a dimension in which we can look back, see the present as a result of what has come before, and plan for the future, is a defining feature of our understanding of reality. When we look around, we notice change, shifting seasons, rhythmic changes of rivers, germination and fruitage of vegetation, a time to sow, a time to reap, the arrival and departure of birds, the conception and birth of animals. In ancient Egypt, people observed that the Nile overflowed with great regularity, and that this was foreseeable. The Greek poet Hesiod tells around 700 BC, in *Works and Days*, of the wise farmer who planned his work according to the changes of the seasons, the vegetation, and the starry sky. Humanity was included in the ecological cycles around it. Prehistoric people also noticed the repeated movements of the sky above them, the course of the Sun, the Moon, and the planets. Changes in the sky, and its significance for the measurement and perception of the passage of time, are found in many cultures, such as Mayan calculations of the solar year or the Stonehenge monument in Southern England, which was oriented toward sunrise during the midsummer solstice. Time was cyclical and recurring. This eternal return was a resistance to historical time by archaic societies (Eliade 1949).

Time was not yet an abstract dimension that flows at a steady rate, which could be measured in specific units, independently of human life. Time was rather something that depended on local conditions. An altered sense of time, a more abstract understanding of it, enters with the creation of various tools for time keeping. Measurement techniques became part of an abstraction of time, that is, the ability to imagine it as something independent of visible phenomena, which can be described mathematically. The most common method in ancient cultures, such as Egyptian and the Mediterranean, was to read the shadow cast by a rod, called a gnomon, on a board with marked hours (Steele 2007). Time describes change, and time could be measured. The calendar became a tool to quantify this universal human experience of change (Duncan 1998; Richards 1998). The Egyptian calendar, which some believe began in 4241 BC, can therefore be regarded as the first exact year in human history. This time-keeping technique, however, got out of sync with the solar year, which forced the Romans to correct the Egyptian year by adding a leap day every four years. The Julian calendar, which was introduced in 46 BC during the reign of Julius Caesar, became the system for the Christian world for a long time, but it too fell out of step. The Gregorian calendar, which was introduced in 1582 by Pope Gregory XIII, adjusted the length of the year by omitting the leap day in century years that were not divisible by 400. It took some time before this was adopted in Protestant and Orthodox countries: in Great Britain this happened in 1752, in Protestant Germany in 1775, and in Russia in 1918. Chronology is a convention, something we agree on, but at the same time something abstract and independent of human life. To the physicist Sir Isaac Newton, time was absolute, something that flows on its own fixed rate independently of human measurement, as he states in Scholium I in *Philosophiae naturalis principia mathematica* (*Mathematical Principles of Natural Philosophy*, 1687): "Absolute, true, and mathematical time, of itself, and from its own nature, flows equably without relations to anything external" (Newton 1687, 6). The German philosopher Immanuel Kant saw time instead as an priori category for understanding our sensory experiences. With the theoretical physicist Albert Einstein, time again became something relative, which depended on the observer. For others, such as the twentieth-century philosophers Henri Bergson and Martin Heidegger, time became something intimately connected with human existence and subjective experience.

In all cultures, there are stories about how and why things are as they are. Creation myths (Evans 2014; Leemings 1994) tell stories about why the current world looks like it does. Creation myths are narratives, stories that describe causality, connecting events in time in a cause-and-effect relationship that also explains why and how something happened. Our earliest creation myths often describe how something is created out of shapelessness – out of chaos, the primeval sea, emptiness, darkness, an abyss – a development from the disordered to the ordered, from chaos to cosmos. In the Babylonian creation epic *Enuma Elish*, which could date as far back as to the eighteenth century BC, Marduk, the patron deity of Babylon, kills the chaos monster Tiamat, the personification of chaotic power of the primeval sea, and shapes the world out of her body. In the *Theogony* from 700 BC, Hesiod portrays creation from the primeval chaos to the birth of Zeus. The Prometheus myth, treated by Hesiod and Aeschylus, describes how our first ancestors were created from clay, a myth that became a symbol of the creative mind. In Plato's *Timaeus*, written circa 360 BC, the demiurge creates, like a craftsman or builder, order out of disorder. Creation myths also pinpoint the beginning of time. More explicit than many other creation myths, the Mosaic creation in the Pentateuch (composed in the late seventh or sixth century BC) marks a starting point, when the history of Earth and humans begins through a divine act of will (Figure 13.2). Before that, there was nothing – no time, no matter, just darkness.

The starting point for a chronology is always an event in the past combined with knowledge of regular cycles, such as solar or lunar years. The calendar is not just a result of technological and scientific achievements, but has also been used to exercise political power and social control (Stern 2012). Jewish chronology begins on New Year's Day 3761 BC, when the world was created. The Greeks started from recurrent Olympiads and the list of winners from July 8, 776 BC. Instead of the Roman starting point, the founding of Rome in 753 BC, the early Christian church adopted the birth of Christ, *Anno Domini*, as the beginning of the calendar (Nothaft 2012). In the early sixth century AD, the Scythian monk Dionysius Exiguus estimated the birth year of Christ, probably by using tables prepared by

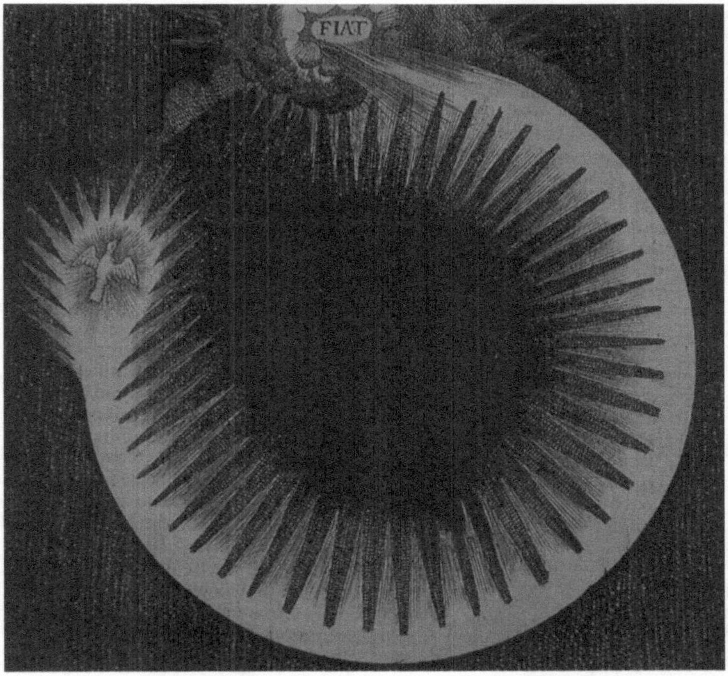

Figure 13.2 Fiat lux, let there be light, said God.
Credit: Fludd (1617, 49)

the Roman historian Eusebius of Caesarea two centuries earlier. Saint Bede the Venerable later used the new chronology in his *Historia ecclesiastica gentis Anglorum* (*The Ecclesiastical History of the English People*) from 731. The Islamic world adopted instead the prophet Mohammad's escape to Medina in 622 as year one and the beginning of the Hijri calendar.

The invention of linear time has commonly been associated with Christianity, even though Jewish thinking could have priority (Momigliano 1966; Patrides 1964). Time was in Christian context reinterpreted, not as something eternally recurring in cycles of birth and death, but as a salvation story between established coordinates: the creation, Christ's birth and death, Christ's second coming, and the final judgment. Time became limited, confined between the creation and the last judgment. Before the creation, there was nothing, only an eternal now. As the theologian and philosopher Augustine of Hippo wrote in his *Confessions* (397–98), only the present is real, the past no longer exists, and the future is not yet a reality (Flasch 2004). The past is present in memory, the present in observation, and the future in expectation. God is eternal, timeless, in the sense of "one now." In *De Civitate Dei* (*City of God*, 413–26) Augustine describes the history of mankind as determined by God. The conception of a cyclical time was contrary to the uniqueness of the appearance of Jesus in the world. With the Christian conception of time, historical events could be understood and explained by the providence of God – in a linear history. Human history was about a battle between the kingdom of God and the devil, between Jerusalem and Babylon. The Dominican friar and priest Thomas Aquinas, known for his synthesis of Aristotelian philosophy and Christian theology, aimed at overcoming the contradiction between Aristotle's idea of the eternity of the world – that nothing can come from nothing – and the standard Christian interpretation of time as having a beginning, that God created everything from nothing, as the doctrine of creation proclaims (Wissink 1990). In *De aeternitate mundi* (*On the Eternity of the World*, 1270) Aquinas discusses the eternity of the world, that it has a beginning and is created out of nothing, and demonstrated that the notion of an eternal created universe is not contrary to reason.

The church fathers put great effort in trying to calculate the time of the creation and also to predict the time of the last judgment. By combining astronomical calculations and genealogies in the Bible, they came up with remarkably precise dates for the beginning of time. Matthew 1:17 states that fourteen generations passed from Abraham to David, fourteen generations from David to the captivity, and fourteen generations from the captivity to Christ, which could be combined with Genesis 5 and 11, which says that it passed ten generations to the flood and ten generations from the flood to Abraham (Figure 13.3). When Adam, who reached the age of 930, was 130 years old, he had his son Seth, who reached the age of 912, who in turn had a son, Enosh, when he was 105, and Enosh was 90 years old when he became the father of Kenan, and then lived until he was 905 years old, and so on.

Christian chronologies combine the idea of a creation with a counting of generations, observations of the flourishing of nature during spring and autumn, and celestial phenomena. There are parallels between the Bible and human history. In the twelfth century, the Italian monk Joachim of Fiore divided human history into three ages: the age of the Father (Old Testament), the Son (New Testament), and the Holy Spirit, an age that begins at Christ's ascension into heaven. The three kingdoms were also seen as analogous to annual seasons, the kingdoms of winter, spring, and summer. In other words, the solar year mirrored the world year. The reformer Martin Luther, who assumed that the creation took place 3,960 years before Christ, expressed the idea that the world will last for 6,000 years – 2,000 years of emptiness, 2,000 years of law, and 2,000 years of the Messiah. The idea of the timespan of the world divided into six ages, and the seventh, the eternal Sabbath, analogous with the days of creation and man's six ages, had its origin in the writings of Saint Augustine. And it had support in the Bible: "A thousand years in your sight are like a day that has just gone by," it says in Psalm 90:4. The Lord will accomplish everything in 6,000 years, for one day is with him as 1,000 years, it says in the Epistle of Barnabas 15:4. James Ussher, Archbishop of

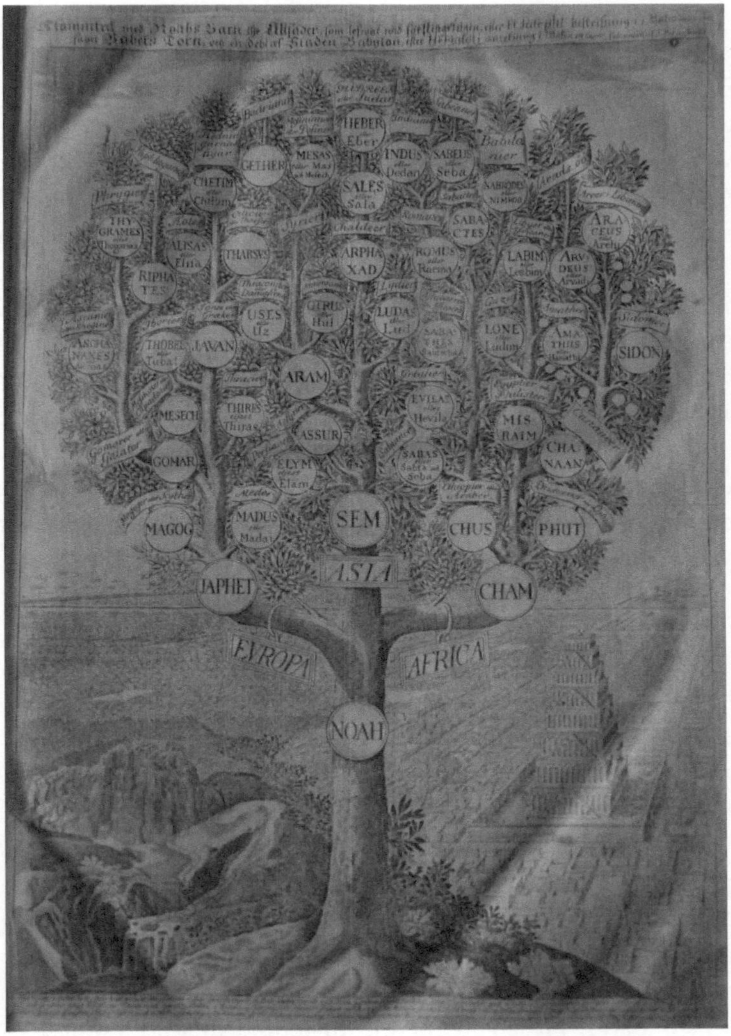

Figure 13.3 The family tree of Noah.
Credit: Peringskiöld (1713)

Armagh, calculated in *Annales Veteris Testamenti, a prima mundi origine deducti* (*Annals of the Old Testament, Deduced from the First Origins of the World*, 1650) that the Creation began at nightfall preceding Sunday, October 23, 4004 BC. According to the German biblical scholar Johann Albrecht Bengel, the world was created in 3943 BC, at the beginning of autumn, more specifically on October 10. Adam was created, accordingly, on the sixth day, October 15.

However, there were other circumstances mentioned in the Bible that contradicted these chronologies. According to Genesis 4:17, Adam's son Cain took a wife and built a city. But then a question arises: Where did she and all the inhabitants come from? The French theologian Isaac La Peyrère concluded, in his book *Prae-Adamitae* (1655), that there must have been people before Adam. Thus, there must have been two creations, first of the common people, then of Adam, who became the progenitor of the Jewish people. Historians and philosophers of the Enlightenment brought a growing effort to explain historical events, not in terms of providence but by reason. A rationalist text-critical movement, beginning in the seventeenth century by philosophers such as Thomas Hobbes and Baruch Spinoza, came increasingly to see biblical stories not as dictated words by God, but as a human creation that passed from generation to generation and has been defaced by time. Enlightenment philosophers like Rousseau, Montesquieu, and Voltaire tried to find the historical "truth" through a critical examination of historical sources. German history professor Leopold von Ranke presented the most clear-cut examples of this new critical thinking that won many followers in the nineteenth century and onwards. If you just study the sources critically, he believed, you could get at what really happened.

The prehistory of time

Still, the history of the Earth was the history of man. The strongest arguments against the Bible chronology, however, came not from theological criticism but from geological excavations (Albritton 1980; Gould 1987; Haber 1959; Richet 2007; Rudwick 2005; 2008). Fossilized seashells and other animals were found in bedrock, which were sometimes interpreted as sports of nature, *lusus naturae*, but usually explained as remnants of the flood. In 1665, the Dane Nicolaus Steno found shark teeth in the Tuscan mountains, suggesting that where it is now high mountains, it had once been a sea (Cutler 2003). Other findings seemed to have no counterpart in the living species (Figure 13.4). The Linnaean classification of plants and animals on Earth, presented for the first time in Carl Linnaeus's *Systema naturae* (1735), was a way of seeing similarities and affinities between species, which led to attempts in the next century to explain the diversity of species on the basis of a temporal understanding of change, which finally exploded the narrow time boundaries of Bible chronology. Geological time spans increased considerably during the late eighteenth century and the first half of the nineteenth, as geological layers, fossils, and extinct animals revealed the history of the Earth and past epochs.

Eight years after Bengel published *Ordo temporum* (*The Order of Time*, 1741), the French naturalist Georges Louis Leclerc, Comte de Buffon, presented his *Théorie de la terre et vues générales sur la génération et sur l'homme* (*A Theory of the Earth, a General History of Man*, 1749), one of the first estimates of Earth's geological history. He also divided the history of Earth into six epochs. The last,

Figure 13.4 Ichtyosaurus and Plesiosaurus, extinct life forms from the time before the deluge.
Credit: Figuier (1863; image from Swedish edition, 1868, 148)

the sixth, began when the Old and New World were separated 10,000 years ago. However, Buffon took his departure point from premises very different from Bible chronology. According to him, Earth consisted of solar matter that had been scattered after a collision with a comet. This hot, glowing mass took at least 74,832 years to cool to its current temperature, and another 93,000 years remain before life will perish from cold. A similar thought experiment had been carried out a few decades earlier, in 1722, by the Swedish inventor Christopher Polhem. He used the time it takes a tree root to burn, compared it with the size of the Sun, and concluded that the Sun would burn for at least a billion billion billion years (Dunér 2010). In *Allgemeine Naturgeschichte und Theorie des Himmels* (*Universal Natural History and Theory of Heaven*, 1755), Kant tells about a gradual order out of chaos that may have taken millions of years. The creation is not something instant. Millions of centuries will pass and many new worlds, one after another, will be formed and reach perfection.

The new cosmogony, geogony, and palaeontology of the late eighteenth and early nineteenth centuries opened up a completely different time scale, which became important for the conception of the history of life. The French naturalist Georges Cuvier counted four phases in vertebrate history but still believed in the idea of species as unchanging, which forced him to explain extinct species as a result of successive disasters. His compatriot, the naturalist Jean-Baptiste de Lamarck regarded species as the result of processes in which certain characteristics organisms acquired by the influence of the environment were transmitted to new generations. In Scotland in 1785, James Hutton, and later Charles Lyell, championed the idea that the history of Earth cannot be found in human stories, but instead in geological strata. Even the early Romantic philosopher Johann Gottfried von Herder, in *Ideen zu einer Philosophie der Geschichte der Menschheit* (*Outlines of a Philosophy of the History of Man*, 1784), joined this thinking about time-consuming processes before man. Human beings were put into a cosmic context in which they were preceded by a number of inorganic and biological processes. To Herder, history is organic and teleological, or directed toward certain purposes. Man is basically a historical creature, and each people or nation is collectively characterized by a unique spirit, *Volksgeist.*

In the philosophy of history, there is a tension between materialist explanations – in which natural and economic forces shape the course of history – and idealistic ones, especially prominent in Hegel and Herder, in which history is guided by cultural or spiritual forces. The course of history is also commonly interpreted as having a purpose, as a quest toward a definitive annihilation or redemption, or an eternal progress or receding and decay. The idea of human history as an organic development of society and law was introduced by Romanticism, in contrast to the rapid progression of the earlier contract theories. Prehistory could be divided in stages naturally following each other. In 1836, the Danish archaeologist Christian Jürgensen Thomsen divided early human history into three periods: Stone Age, Bronze Age, and Iron Age. A supporter of Darwin's theory of evolution, John Lubbock, 1 st Baron Avebury, divided in his *Pre-Historic Times* (1865) the Stone Age into the Palaeolithic and the Neolithic. Later on, more accurate tools for archaeologists and geologists were invented in order to place humans in time. The discovery of radioactivity in 1895 opened up a new method for determining time. Radioactive decay, Ernest Rutherford explained in 1905, could be used to measure geological time. In 1949, after the discovery of the radioactive carbon isotope ^{14}C, Willard Frank Libby introduced a method to determine the age of organic materials.

One important finding in paleoanthropology was made in a limestone cave outside Düsseldorf in 1856, in the Neander Valley. Workers found the skull of a human-like creature that was different from all contemporary people, with a crude, low forehead and powerful jaws. How could one explain these extinct humanoid life forms? In *The Origin of Species* (1859), Charles Darwin finally explained the mechanism of speciation brought about through the struggle for existence and natural selection. Combining his observations from a journey onboard the *Beagle* with careful study of animal and plant breeding and a reading of Thomas Malthus's

Essay on the Principles of Population (1798), Darwin formulated a theory of evolution. It held that multiplication of individuals increases faster than nutrient supply, which in turn means that only the most successful in the competition for food and partners survive and multiply. In *Evidence as to Man's Place in Nature* (1863), the English biologist Thomas Henry Huxley used embryological and morphological comparisons to show how close humankind is to other animals. He also searched for the missing link between modern humans and ape-like creatures in prehistory. Following Darwin, the British philosopher Herbert Spencer formulated a natural law for all forms of evolution that went from looser to tighter connection, from similar to the more disparate, from disordered to more ordered, and eventually reaching equilibrium. Humans as social beings will eventually achieve a state of happiness and prosperity. So-called social Darwinism applied the theory of evolution to the development of society and politics. In its most perverted forms it could be used to advocate the right of the strong to dominate the weak and the colonial suppression of other cultures.

The theory of evolution, which showed how humans have evolved from primates during a long process, created an entirely new conception of humankind's place in time. By describing humanity as a species, human history was finally bound together with the history of nature. It was a crucial discovery, which later made it possible to bind together f_i, f_c, and L with the other factors in the Drake Equation. The discovery of long timelines, that there was a time before humanity and that humanity was an integral part of the biological evolution, allowed humans to think about L.

The arrow of time

The problem of change

Human culture transfers memories, values, and ideas to future generations. The experience of living in time gave rise to attempts to explain the present in the context of the past, and to try to predict the future by using the present and past. A characteristic feature of human beings is their ability to envision what is not present in time and space. This notion of time – that it has a beginning, a present, and a future – implicated the ability to use knowledge of the past for the future, to anticipate events, to secure better future harvests in agriculture, or to prepare for coming periods of cold, drought, or famine. Humans see into the future and have always done so. The art of divination was important for the Babylonians in the second millennium BC. The Akkadians made predictions with the help of animal viscera, especially the liver. Portents, comets, and eclipses carried a message. For the Babylonians, astrology became an important tool for reading the present and predicting the future. In ancient Greece, one could ask the oracle at Delphi. The Roman augurs could determine the gods' attitudes toward certain actions by studying the flight of birds. These examples illustrate the human quest to interpret signs in the past and present, and our obsession with the future.

The problem of change has followed humankind throughout history. For Heraclitus of Ephesus around 500 BC, the world was a "becoming" rather than a present, or, as Plato puts Heraclitus's doctrine in *Cratylus*: "No man ever steps in the same river twice" (Plato 1926, 402a). All is subject to change, everything flows, *panta rhei.* For another pre-Socratic philosopher, Parmenides of Elea, beingness always existed. If a being exists, it must have been preceded by a nonbeing, but nonbeing does not exist, so the world has always existed and will never perish – it is eternal. No change can happen, Parmenides claimed, because if anything changes, it will be something that it is not, thus change requires both being and nonbeing, which is unthinkable. Likewise, Aristotle's world was limited in space but eternal, and was never created and will never perish.

For the Greeks, man faced the past with his back to the future, and time was flowing like a river through the present. The past was before us. The classical ideal was the myth of the golden age, which survived until the Renaissance. In the past, there was a golden time, with wise philosophers and a magnificent culture. Then came decline and deterioration. Ovid speaks of the four ages of man, *quattuor aetates*, in his *Metamorphoses* from the beginning of the first century AD: from the eternal spring of the golden age, over the silver and copper age, to the present iron age. The Renaissance revived classical ancient wisdom, the pinnacle of human achievement. Philosophers such as Aristotle, Plato, Plotinus, and other Greek and Roman thinkers had an unmatched wisdom that yet could be retrieved in preserved texts. In the Christian tradition there is also the idea of the original, paradisiacal state before sin entered the world. Then followed war, deceit, and discord. Even nature deteriorated; the soil is losing its fertility and will eventually turn into a sterile desert.

A critical historiography could be found in the works of Herodotus and Thucydides, who initiated a way of writing history that not merely described and retold what happened, but also tried to give an explanation and understanding of how and why something occurs or has occurred. Herodotus focused on conflicts between cultures, the wars between Hellas and Asia, Greeks and barbarians. A few centuries later, in the second century BC, Polybius sketched a sort of universal history when he tried to explain Rome's ascent to world domination, in which the history of isolated cultures and regions converge into a common history. A later attempt to present the history of humankind as a whole is, for example, Hartmann Schedel's *Weltchronik* (*World History*), the Nuremberg chronicle from 1493, where the history of the world is divided in seven ages, from the creation to the last judgment. Universal history embraces all times and nations.

The idea of progress

Eventually, a belief grew that our own time stood on the shoulders of giants. People born later in history could gaze longer than previous generations. During the Enlightenment, the idea of progress became widely spread. Modern, empirical, experimental science and technological advancement led to a more positive, optimistic mood. In the seventeenth century, some philosophers argued that contemporary life had reached far beyond the life of our ancestors. Knowledge was increasingly seen as cumulative, as Francis Bacon had proclaimed. Each generation added to earlier acquired experiences. For the Italian philosopher and historian Giambattista Vico, history is about what people have created and is a kind of self-understanding. History is more like a spiral than a circle, and returns constantly to the same phase but never repeats itself. Voltaire's *Essai sur les mœurs et l'esprit des nations* (*An Essay on the Manners and Spirit of Nations*, 1740) depicts the progress and setbacks of civilization and enlightenment. Even for Kant, in *Idee zu einer allgemeinen Geschichte in weltbürgerlicher Absicht* (*Idea for a Universal History with a Cosmopolitan Purpose*, 1784), the history of humanity is full of conflict and antagonism, by which man increasingly usurps reason and moral awareness. Enlightenment optimism gets full-fledged expression in Marquis de Condorcet's grand tribute to human progress in a book published just a few years after the French Revolution, *Esquisse d'un tableau historique des progres de l'esprit humain* (*Sketch for a Historical Picture of the Progress of the Human Spirit*, 1794). Humanity, he asserted, had step by step overcome the vagaries of natural forces, freeing itself from superstition, powerlessness, and fear, which wizards and priests had imposed. Thanks to science, the way was now open for unlimited progress and spiritual freedom. Romantic philosophers such as Friedrich Schelling also did not doubt the idea of continual progress. The French positivist Auguste Comte nourished a belief in the advancement of science and its ability to create a better world. Darwin's theory of evolution itself relied on a progression from "lower" to "higher" life forms. In the utopian genre, admirers of science, socialists, and liberals envisioned the future as a kingdom of joy, the perfect society.

The Mosaic creation myth became increasingly inadequate to explain not only the biological and geological processes but also the emergence of civilization, society, government, and commerce (Rossi 1984). Natural law – *lex naturalis*, the legal theory that there were certain inherent natural rights – gave rise to the idea of the social contract, which could explain the emergence of the state. Philosophers of law, such as Hugo Grotius and Samuel von Pufendorf, developed this idea in the seventeenth century. People, who had been born free and equal, sacrificed their natural rights and ceded power to a ruler in a sort of agreement. The treaty expressed the conventional character of social order, something that is established among people. Thomas Hobbes explained in *Leviathan* (1651) that in the state of nature, man was a selfish creature who thought only of his own advantage, which led to a war of all against all, *bellum omnium contra omnes*. In order to establish a peaceful existence, people were forced to unite and transfer power to a ruler. The Enlightenment philosopher Jean-Jacques Rousseau also explained the emergence of society with contract theory, but arrived at a more critical stance towards human civilization. His *Contrat social* (*The Social Contract*, 1762) opens with the epic words: "Man is born free; and everywhere he is in chains" (Rousseau 1762, 14).

Political and social thinkers in the nineteenth century saw a sort of internal logic to history. In conservative ideology, society could be seen, after Plato, as an organism, which meant that sudden changes, revolutions, were a threat, and that social order rested on the belief that every member of society had his or her given, predetermined place. This encouraged a static view of society. According to the dialectical logic of Friedrich Hegel, thesis was set against antithesis and converged into synthesis and was identical to the history of the world (Dudley 2009). History is driven by conflict and contradiction to eventually resolve into a harmonious final stage. This logic of history is manifested by the world spirit and can be found in religion, art, and politics. History is in some sense a reflection of the individual's development. The same stages that can be found in a person's life – childhood, youth, adulthood, and old age – are also found in humanity as a whole, from the barbaric state of the childhood to adulthood's autocracy, and to old age when discipline and freedom come together in the monarchy. History, according to Hegel, is a world event in which the parts cannot be understood without the whole.

Society followed a regular process of change for Hegel, as for Karl Marx, who together with Friedrich Engels developed a materialist version of Hegel's world history. There is a definite direction in history, they believed, and there is no possibility of a return to earlier stages of development. Socialist utopians saw a future dream society, and some, such as Henri de Saint-Simon, assigned to the state power over the lives of its citizens, while others proclaimed the collective. The Russian anarchist Pyotr Kropotkin proceeded from the Darwinian theory of evolution but focused on cooperation, not competition between individuals, as the factor that promoted survival. The most influential thinker for twentieth-century conceptions of the future, Karl Marx, envisioned a process in which the working class stood against the bourgeoisie, which led to revolution in which the working class took over the means of production. The civilizing process and long-term history – *la longue durée* – have continued to be the focus of much historical research during the twentieth century, with sociologists and historians such as Max Weber, Norbert Elias, and Fernand Braudel seeking to explain the emergence and development of modern society.

In summary, L assumes an idea of development and progress, that time has a direction, that the past is the cause of what is happening now and what will happen in the future. Thoughts about the future nearly always include beliefs and statements regarding the meaning of history and that there is a certain teleology embedded in time.

The end of time

The apocalypse

The question of when life on Earth will be wiped out – when the end of time will come and why – has to do with our concept of time and our view of the meaning of history. Visions and revelations about the end of world history, the apocalypse, occur in many cultures (Bull 1995; Hall 2009). The idea is found in Iranian religion and later in Judaism. Eschatology, the study of the ultimate destiny of humanity and the world, is a central part of many religions. Zoroastrianism anticipates a new creation of the world after its destruction, while Judaism expects a messianic kingdom of peace, Christianity a union with God in the coming kingdom, and Islam the arrival of the Mahdi, who precedes judgment, which will establish the righteous Islamic empire and unite the Muslims.

The idea of a punishment caused by human arrogance, *hubris*, and divine retribution, *nemesis*, are fundamental parts of Greek ethos. Greek mythology tells the story of Cassandra who warns but is ignored. Ancient literature expresses conceptions of the downfall of civilizations, such as the myth of Atlantis, which Plato describes in *Timaeus*, as an island beyond the "Pillars of Hercules" that in one day sank in the sea. In *The Republic*, Plato expresses a cyclical theory in which the world will survive 72,000 solar years before it ends in chaos. Stoics imagined a succession of worlds that perish when the larger cycles were completed. Mayan civilization in Central America developed an advanced timekeeping where a famous calendar prognosticates the end of the world when the cycle was completed. After 5,130 years – that is, December 21, 2012 AD – the present world would, according to a misguided popular belief, perish and a new one arise. For cyclical notions of time, the completion of the cycle was not seldom an especially critical moment. But the return to the beginning also gave the possibility of rebirth.

The end of paradisiacal life and the downfall of civilization is famously described in the Old Testament: Eve's eating of the forbidden fruit, the expulsion from the garden of Eden, and Noah and the flood (Figure 13.5). The end of time as a gigantic flood is also found in Akkadian and Babylonian texts, in the Vedas, and in Greek mythology. Another biblical example of the destruction of civilization is the tower of Babel (Genesis 11:3–9), in which human presumption and sin against God leads to the confusion of languages and the deprivation of the unity of humanity. The cause of the downfall was to be found in man, in sin, in the

Figure 13.5 Noah and the Flood. Etching by German artist Johann Christoph Sartorius from 1682–84.
Reprinted in *Läse-bok för barn*, Anonymous (1814, between pp. 20 and 21)

violation of taboo. It was an internal threat in which humans are the cause of their own demise. In Christian linear time, where the world is finite but God infinite, the history of humankind becomes a part of the history of salvation. Time was measured. The time will come when Christ will return and God will judge the living and the dead for their sins. The book of Revelation depicts in vivid images the end of world history and the arrival of the new Jerusalem. A huge beast is rising up out of the abyss with a whore on its back with which the kings of the Earth have engaged in adultery. In the end of times, the battle of Armageddon will take place and Antichrist will be defeated. A medieval tradition also speaks of the fifteen signs that will precede the last judgment, *Quindecim signa ante judicium*, which may have its origin in the apocryphal apocalypse according to Thomas: water will rise over the mountains, then sink and disappear, and then return to its original position; all sea animals will accumulate on the surface, and the water will burn from east to west; plants and trees will be covered with dew and blood; all buildings will be destroyed; the stones will fight each other; great earthquakes, mountains, and valleys will be leveled to plains; people will come out of their burrows but no longer understand each other; the stars will fall from the sky; dead men's bones will come out of the tombs; all people will die; and finally, the Earth will burn. And on the fifteenth day occurs the last judgment.

In line with calculations of the beginning of time, many attempts were made to predict when the last hour will come. Chiliasm, the expectations of a kingdom of God that will last for a thousand years and end history, has roots in Jewish apocalyptic thinking. The millennium is mentioned in Revelation (20:1–6) where an angel descends from the sky and defeats the devil who is thrown into the abyss and sits in jail for a thousand years. The prophecy of apocalypse about the millennium was interpreted by Augustine as referring to the spiritual community of the Christian church. Other interpretations saw the millennial kingdom of peace that would arise at the end of time, and others saw the millennium as an earthly kingdom ruled by true Christians (Campion 1994; Landes 2000; Landes 2011). Chiliasm or millennialism has several varieties: premillennialism expects the second coming before the millennium, postmillennialism assumes that the second coming occurs just before the last judgment, and amillennialism interprets the coming millennium in a metaphorical sense. The year 1,000, when the last judgment was expected to happen, was thus awaited nervously (Landes 2003; cf. Thompson 1996).

The end of time, the last judgment, assumes a linear concept of time, that history has a beginning and an end. In these apocalyptic renderings, there is often an underlying dualistic perception of the powers of the world, a struggle between good and evil, between light and darkness for control of the world (Figure 13.6). The current age of darkness is portrayed as in the grip of evil, but there is also a hope for the future victory of good, the savior's return and a new paradise, a dream of a brighter future in freedom. The destiny of the world, society, and individual humans were intertwined. The downfall of the world was also about our own bodily demise. Time was often represented as death with an hourglass – time was meted out, both for man and the world. Remember that you will die, *memento mori*, echoed in human minds. During the late Middle Ages and the Reformation, chiliastic ideas were prevalent in the apocalyptic world in which humans lived (Bynum and Freedman 2000). One of the most famous seers, Nostradamus (Michel de Notre-Dame), in the sixteenth century made obscure prophecies about the future. At the same time, in Münster, Germany, Anabaptists imagined an upcoming age of joy of social revolutionary character.

The rise and fall of civilizations

Since antiquity, the rise and fall of civilizations and conflicts between civilizations had been a recurrent topic in historical writing (Bowden 2009; Münkler 2007).

Figure 13.6 The final downfall of Earth in fire and sulphure.
Credit: Scheuchzer (1731)

The Arabic historian Ibn Khaldun wrote an introduction to the history of the world, *Muqaddimah* (*Introduction*, 1377), where he tries to explain the fall of civilizations by the inherent human sense of social cohesion, *asabiyyah*. When this sense of community is eroded, civilization will perish. Downfall as a consequence of lost virtue, a lack of morality and ethics, is a recurring theme in historiography. A number of works were written during the Enlightenment that discussed the fall of civilizations in a secular context. Historians tried to understand the processes underlying the rise and fall of civilizations, often in an organic analogy, where civilizations grow, flourish, and die. Of particular interest was the Roman Empire's rise and fall, the mightiest empire in the history of the world that eventually perished. Historians and philosophers, such as Montesquieu and Adam Ferguson, discussed the issue. Ferguson was thinking of society as a path from savagery and barbarism to civilization. But growing trade could also lead to decline of virtue. Civilization could thus move toward a collapse like the one that hit Rome. The most famous depiction of the Roman Empire, Edward Gibbon's *The History of the Decline and Fall of the Roman Empire* (1776–88) speaks of the fall as a natural, inevitable result of declining virtue (Roberts 2014). Romans, who had a proud tradition of being heroic warriors, had become effeminate and weak, and thus fell victim to barbarian invasions.

Seeing human history in terms of organic life cycles is a common analogy. In the introduction to *Römische Geschichte* (*History of Rome*, 1854–56), the German historian of antiquity Theodor Mommsen describes how every civilization has its own path, like the human biological life cycle, from birth to maturity and old age, from creativity to saturation. Oswald Spengler in his famous work *Der Untergang des Abendlandes* (*The Decline of the West*, 1918–22) criticizes the traditional idea of world history as a coherent sequence divided into periods: antiquity, the Middle Ages, and modern times. To him, cultures are like organisms. The history of the world consists of eight cultures in cyclic course – they occur, have a youth and heyday, and then move toward decline and eventually die. It is not a coherent history of civilization, instead Spengler counts eight civilizations, each of which has arisen, flourished, withered, and died: Egyptian, Babylonian, Arabic, Indian, Chinese, Mayan culture, ancient Greek and Roman culture, and the West. The present is not the pinnacle of civilization but is understood from a more pessimistic viewpoint of history as part of the cycle of rise and fall. The final stage in which, Spengler claimed, the West now finds itself is distinguished by physical weakness, decadent urban culture, and weariness. The historian Arnold J. Toynbee examined the rise and fall of twenty-six civilizations as successful or unsuccessful responses to physical or social challenges, and declared that civilizations die not by murder, but by suicide.

A common theme in many of these declarations of the downfall of civilization is moral decay, that the downfall is the result of human weaknesses, sins, ambitions, and greed. Humanity is the cause of its own demise. There is a collective guilt for the downfall. Cataclysms are not due to some unavoidable natural phenomenon beyond human control, but are God's punishment for human sin. In modern times, there are also theories of external threats, natural disasters where the forces of nature are stronger than the modern technological society, such as volcanic eruptions, earthquakes, asteroids, or the real or imagined external threats from alien civilizations – for example, invasions from Mars and conflicts with Martians

Figure 13.7 The future heat death. The last human family is surprised by the cold and buried under a perpetual layer of ice.
Credit: Flammarion (1880)

in H. G. Wells's *War of the Worlds* (1898) or Ray Bradbury's *The Martian Chronicles* (1950). An inevitable doom was actually found within physics itself – the heat death of the universe. Based on the second law of thermodynamics, William Thomson, 1st Baron Kelvin, predicted in the 1850s an increasing entropy until everything ends in a thermodynamic equilibrium (Figure 13.7). The thermodynamic arrow of time directs inevitably towards death.

Nevertheless, the idea of an internal threat has dominated the twentieth century, with downfall being the result of human actions. The most obvious examples are world wars, atomic wars, nuclear winter, and in recent decades ecological crisis, climate crisis, and declining biodiversity. In 1896, Swedish physicist and chemist Svante Arrhenius made calculations of how human emissions of carbon dioxide could affect Earth's temperature, now known as the greenhouse effect and one of the most dangerous threats to our civilization (Crawford 1996). By the time the Drake Equation was formulated, these human self-inflicted threats were a reality in the midst of the cold war, accentuated by the Cuban missile crisis (1962) and the chemical threats to our environment discussed in Rachel Carson's *Silent Spring* (1962). Doom and threats to our civilization exposed human deficiency and inability to cooperate in order to fend off global crises. Historians of the future will look back and ask themselves: What happened, and why did they not do anything (Oreskes and Conway 2014)? Factor L expresses the insight that civilizations come to an end unless members develop a social system that enhances the ability to cooperate for survival.

Conclusion: L and the civilizing process

I have tried to follow some lines of thought that made it possible in 1961 to formulate the factor L. A prerequisite for L is what I have called the "discovery of time" – that time has a beginning and is linear, that there was a time before humans, that human civilizations are the result of time and progress, that time also has an end, leading to an apocalypse, and that the end of world history is intimately associated with how humans live and act and how civilizations succeed in managing physical and social threats. Characteristic of human understanding of history is that we incessantly assign it a meaning and a purpose, and that there is a teleological meaning and direction to the world. Historical events cannot be a sequence of random coincidences.

Thought about L was not possible without thousands of years of human thinking. More specifically, modern L was not conceivable until the end of the nineteenth century. Only when human history had been tied to the history of the Earth and the history of life, due to Darwin's *Origin of Species* in 1859, did it become possible to place the history of civilization in a cosmic evolutionary context, and to view L not as an isolated factor but as an integrated part of the history of the universe. In other words, a concept of time underlying the notion of L was now in place. The question then is why L was not formulated already in the late nineteenth century. An explanation must probably rest on both scientific-technological and cultural-social circumstances. First is obviously the scientific description of the electromagnetic radiation – which was also crucial for the entire idea of the Drake Equation and interstellar communication – that was clarified in theory in 1864 by James Clerk Maxwell and experimentally in 1887 by Heinrich Hertz. This paved the way for the radio transmission developed by Guglielmo Marconi in 1895 when he sent the first radio signals, and Karl Guthe Jansky's discovery of radio waves from the Milky Way in 1931. Understanding of the speed of light, c, as it was formulated 1905 in Einstein's special theory of relativity, is another scientific prerequisite. Still, L was not formulated until 1961, probably, apart from the immediate inspiration from the famous article by Giuseppe Cocconi and Philip Morrison in 1959, due to new cultural, political, and social effects of the post-war ambience such as the nuclear threat, ecological devastation, cold war, and space race. For the first time in history, humans became aware of their awe-inspiring self-sufficient capability of destroying their entire civilization without assistance of any god or force of nature.

Estimating L is about universal history, to see a civilization "from the Moon," a historical-philosophical search for the factors affecting L's length. One such factor, I suggest, is the history and evolution of globalization and cooperation. The factor L concerns the history of increased contacts between peoples and cultures, enhanced exchange of goods, technology, ideas, and conceptions about the world and what it is to live. The whole history of humanity is constantly broadening circles of experience and encounters with other cultures toward an increasingly integrated world where people have become more and more dependent on others. The Paleolithic hunter knew his hunting grounds, could make most of his tools himself, and depended only on the knowledge and experience of a rather small group of people for his survival. The "postmodern" human being is living a life of dependence on people on the other side of the globe, of the work, knowledge, and experience of large numbers of people, in order to live the life that he or she takes for granted.

The question one might ask is to what extent human history can tell us something about L. It is likely that longevity and the rise and fall of specific cultures of the Earth will not give us the answer. They are too dependent on human and earthly conditions that are unique to the thinking species that inhabit Earth. Civilizations, cultures, and empires in the plural sense are something other than civilization as a singular. The factor L is thus rather about the civilizing or socialization process, and the variables that underlie it – that is, the biocultural coevolution and the interaction between the evolution of cognition and socialization. Advanced technology is in cognitive terms relatively easier to develop than to create a sustainable advanced social system for survival. Constructing an atomic bomb, for example, is less cognitively demanding than carrying out peace negotiations that have to handle self-concious living beings. The lifetime of an advanced technological civilization seems to be connected to how cognitively flexible creatures are able to handle the power of their technology, find equilibrium with their environmental resources, and organize their society in order to prevent it from breaking down. Human history is largely about how to successfully manage the discrepancy between technological and ethical development: L is a measure of this ratio.

References

Albritton, Claude C. 1980. *The Abyss of Time: Changing Conceptions of the Earth's Antiquity after the Sixteenth Century*. San Francisco: Freeman.

Anonymous. 1814. *Läse-bok för barn, till första utwecklingen af deras förstånd samt religiösa och moraliska begrepp*. Örebro: Nils Magnus Lindh.

Bischof, Carl August. 1791. *Angenehme Unterhaltungen über den gestirnten Himmel zur vorlæufigen Kenntnis der Welt-und Himmelskörper*. Nürnberg: Rasp; Swedish edition, *För wettgirige af bägge könen: Angenäma tidsfördrif rörande stjernhimmelen, til grundläggning för närmare kunskap om werlden och himmelskropparne [. . .]*. Stockholm: Kumblinska, 1796.

Bowden, Brett. 2009. *The Empire of Civilization: The Evolution of an Imperial Idea*. University of Chicago Press.

Bull, Malcolm (ed.) 1995. *Apocalypse Theory and the Ends of the World*. Oxford: Blackwell.

Bynum, Caroline Walker, and Paul Freedman (eds.) 2000. *Last Things: Death and the Apocalypse in the Middle Ages*. Philadelphia: University of Pennsylvania Press.

Campion, Nicholas. 1994. *The Great Year: Astrology, Millenarianism, and History in the Western Tradition*. London: Penguin.

Clark, Grahame. 1992. *Space, Time and Man: A Prehistorian's View*. Cambridge University Press.

Cocconi, Giuseppi, and Philip Morrison. 1959. "Searching for Interstellar Communications." *Nature* 184(4690): 844–46.

Crawford, Elisabeth. 1996. *Arrhenius: From Ionic Theory to the Greenhouse Effect*. Canton, MA: Watson.

Cutler, Alan. 2003. *The Seashell on the Mountaintop: A Story of Science, Sainthood and the Humble Genius Who Discovered a New History of the Earth*. New York: Dutton.

Denning, Kathryn. 2013. "'L' on Earth." In *Civilizations beyond Earth: Extraterrestrial Life and Society*, ed. Douglas A. Vakoch, and Albert A. Harrison, 74–83. New York: Berghahn Books.

Dominik, Martin, and John C. Zarnecki. 2011. "The Detection of Extra-Terrestrial Life and the Consequences for Science and Society." *Philosophical Transactions of the Royal Society A* 369: 499–507.

Dudley, Will (ed.) 2009. *Hegel and History*. Albany: State University of New York Press.

Duncan, David Ewing. 1998. *The Calendar: The 5000-Year Struggle to Align the Clock and the Heavens and What Happened During the Missing Ten Days*. London: Fourth Estate.

Dunér, David. 2010. "Daedalus of the North: Swedenborg's Mentor Christopher Polhem." *The New Philosophy* 113(3–4): 1077–98.

Dunér, David. 2011. "Cognitive Foundations of Interstellar Communication." In *Communication with Extraterrestrial Intelligence*, ed. Douglas A. Vakoch, 449–67. Albany: State University of New York Press.

Dunér, David. 2014. "Interstellar Intersubjectvity: The Significance of Shared Cognition for Communication, Empathy, and Altruism in Space." In *Extraterrestrial Altruism: Evolution and Ethics in the Cosmos*, ed. Douglas A. Vakoch, 141–67. Berlin and Heidelberg: Springer.

Eliade, Mircea. 1949. *Le Myth de l'éternel retour: Archétypes et répétition*. Paris: Gallimard.

Evans, Gillian Rosemary. 2014. *First Light: A History of Creation Myths from Gilgamesh to the God Particle*. New York: IB Tauris.

Figuier, Louis. 1863. *La terre avant le déluge*. Paris: Hachette; Swedish edition, *Jorden före syndafloden*. Stockholm: E. Giron, 1868.

Flammarion, Camille. 1880. *Astronomie populaire: Description générale du ciel*. Paris: Marpon et Flammarion.

Flasch, Kurt. 2004. *Was ist Zeit?: Augustinus von Hippo, das XI. Buch der Confessiones: Historisch-philosophische Studie: Text, Übersetzung, Kommentar*. 2nd ed. Frankfurt am Main: Klostermann.

Fludd, Robert. 1617. *Utriusque cosmi maioris scilicet et minoris metaphysica, physica atque technica historia in duo volumina secundum cosmi differentiam diuisa*. Oppenheim: Johan-Theodori de Bry.

Gould, Stephen Jay. 1987. *Time's Arrow, Time's Cycle: Myth and Metaphor in the Discovery of Geological Time*. Cambridge, MA: Harvard University Press.

Haber, Francis C. 1959. *The Age of the World: Moses to Darwin*. Baltimore, MD: John Hopkins University Press.

Hall, John R. 2009. *Apocalypse: From Antiquity to the Empire of Modernity*. Malden, MA: Polity Press.

Landes, Richard (ed.) 2000. *Encyclopedia of Millennialism and Millennial Movements*. New York: Routledge.

Landes, Richard (ed.) 2003. *The Apocalyptic Year 1000: Religious Expectation and Social Change, 950–1050*. Oxford University Press.

Landes, Richard. 2011. *Heaven on Earth: The Varieties of the Millennial Experience*. Oxford University Press.

Leemings, David Adams. 1994. *Encyclopedia of Creation Myths*. Santa Barbara, CA: ABC-Clio.

Momigliano, Arnaldo. 1966. "Time in Ancient Historiography." *History and Theory* 6:1–23.

Münkler, Herfried. 2007. *Empires: The Logic of World Domination from Ancient Rome to the United States*. Cambridge: Polity.

Newton, Isaac. 1687. *Philosophiae naturalis principia mathematica*. London: Joseph Streater; English edition, *Sir Isaac Newton's Mathematical Principles of Natural Philosophy and His System of the World*, vol. 1, *The Motion of Bodies*, ed. Florian Cajori. Berkeley: University of California Press, 1962.

Nietzsche, Friedrich. 1873. *Über Wahrheit und Lüge im außermoralischen Sinne*; new edition *Nietzsche's Werke: Schriften und Entwürfe 1872 bis 1876*, ed. Fritz Koegel. Leipzig: Naumann, 1896; English edition, "On Truth and Lie in an Extra-Moral Sense." In *The Portable Nietzsche*, ed. Walter Kaufmann. Harmondsworth: Penguin, 1994.

Nothaft, C. Philipp E. 2012. *Dating the Passion: The Life of Jesus and the Emergence of Scientific Chronology (200–1600)*. Leiden: Brill.

Oreskes, Naomi, and Erik M. Conway. 2014. *The Collapse of Western Civilization: A View from the Future*. New York: Columbia University Press.

Patrides, Constantinos Apostolos. 1964. *The Phoenix and the Ladder: The Rise and Decline of the Christian View of History*. Berkeley: University of California Press.

Penny, Alan. 2011. "The Lifetime of Scientific Civilizations and the Genetic Evolution of the Brain." In *Civilisations beyond Earth: Extraterrestrial Life and Society*, ed. Douglas A. Vakoch, and Albert A. Harrison, 60–73. New York: Berghahn Books.

Peringskiöld, Johan. 1713. *En book af menniskiones slächt, och Jesu Christi börd; eller Bibliskt slächt-register, från Adam, til Iesu Christi heliga moder jungfru Maria, och hennes trolåfwade man Ioseph*. Stockholm: Olof Enæus.

Plato. 1926. *Plato in Twelve Volumes,* vol. 4, *Cratylus; Parmenides; Greater Hippias; Lesser Hippias*. Cambridge, MA: Harvard University Press.

Richards, Edward Graham. 1998. *Mapping Time: The Calendar and its History*. Oxford University Press.

Richet, Pascal. 2007. *A Natural History of Time*. University of Chicago Press.

Roberts, Charlotte. 2014. *Edward Gibbon and the Shape of History*. Oxford University Press.

Rossi, Paolo. 1984. *The Dark Abyss of Time: The History of the Earth and the History of Nations from Hooke to Vico*. University of Chicago Press.

Rousseau, Jean-Jacques. 1762. *Du contract social: ou, Principes du droit politique*. Amsterdam: Marc Michel Rey; English edition, *The Social Contract*. New York: Cosimo.

Rudwick, Martin J. S. 2005. *Bursting the Limits of Time: The Reconstruction of Geohistory in the Age of Revolution*. University of Chicago Press.

Rudwick, Martin J. S. 2008. *Worlds before Adam: The Reconstruction of Geohistory in the Age of Reform*. University of Chicago Press.

Scheuchzer, Johann Jacob. 1731. *Kupfer-Bibel: In welcher die Physica sacra, oder geheiligte Natur-Wissenschafft derer in Heil. Schrifft vorkommenden natürlichen Sachen*. Augsburg: Johann Andreas Pfeffel.

Shostak, Seth. 2009. "The Value of "L" and the Cosmic Bottleneck." In *Cosmos and Culture: Cultural Evolution in a Cosmic Context*, ed. Steven J. Dick, and Mark L. Lupisella, 399–414. Washington, DC: NASA History Series.

Sinha, Chris. 2011. "A Life without Time." *New Scientist* 211: 2833.

Steele, John M. (ed.) 2007. *Calendars and Years: Astronomy and Time in the Ancient Near East*. Oxford: Oxbow Books.

Stern, Sacha. 2012. *Calendars of Antiquity: Empires, States, and Societies*. Oxford University Press.

Thompson, Damian. 1996. *The End of Time: Faith and Fear in the Shadow of the Millennium*. London: Sinclair-Stevenson.

Toulmin, Stephen, and June Goodfield. 1965. *The Discovery of Time*. London: Hutchinson.

Wissink, J. B. M. (ed.) 1990. *The Eternity of the World: In the Thought of Thomas Aquinas and His Contemporaries*. Leiden: Brill.

14

Length of time such civilizations release detectable signals into space, L, 1961 to the present

Garry Chick

Abstract

The final variable in Drake's equation for estimating the number of such civilizations in the Milky Way is L, the lifetime of communicating civilizations. Drake's initial estimate for L was 10,000 years, but others have suggested different values or even defined the variable differently. Unfortunately, for empirical information on how to estimate L, we have a sample of only one, ourselves. But even those data are incomplete, since our own technology that permits interstellar communication is less than 100 years old and there is no obvious end to our civilization in sight. So, we have not only no idea of what the value of L should be, we also have little idea of how to go about estimating it. Additionally, like each of the other variables in the equation, L is almost certainly the product of many other variables. Therefore, the first goal of this chapter is to examine efforts at estimating values for L since 1961, including how those estimates were developed. Second, I will consider the concept of civilization in order to determine if the historical study of earthly examples can inform us about L. Third, I will examine variables that might influence the value of L by analyzing lists of existential risks proposed by a sample of individuals and organizations. Finally, I summarize these lines of thought in order to determine if they can inform us regarding possible values of L.

In Egypt's sandy silence, all alone,
Stands a gigantic Leg, which far off throws
The only shadow that the Desert knows: —
"I am great OZYMANDIAS," saith the stone,
"The King of Kings; this mighty City shows
"The wonders of my hand."—The City's gone, —
Nought but the Leg remaining to disclose

The Drake Equation, ed. Douglas A. Vakoch and Matthew Dowd. Published by Cambridge University Press.

The site of this forgotten Babylon.
We wonder, —and some Hunter may express
Wonder like ours, when thro' the wilderness
Where London stood, holding the Wolf in chace,
He meets some fragment huge, and stops to guess
What powerful but unrecorded race
Once dwelt in that annihilated place.

Horace Smith, *Ozymandias* (1818)

Horace Smith's sonnet, *Ozymandias*, is less well known than Percy Bysshe Shelley's[1] of the same name but, like Shelley's, captures the theme that great men of the past and their empires were transitory. Unlike Shelley, however, Smith foretold that present civilizations – such as the British Empire of 1818 – will also fade into oblivion. Both Smith and Shelley seemingly understood that empires and civilizations are ephemeral, but how ephemeral are they? And can knowledge of the durations of earthly civilizations inform us regarding the lifetimes of possible extraterrestrial civilizations in the Milky Way? The final term, L, in astronomer Frank Drake's equation for determining the number of civilizations in the Milky Way capable of interstellar communication (where $N = R^* f_p n_e f_l f_i f_c L$) refers to the lifetime of such civilizations.

The difficulty in determining values for the right-hand variables in Drake's equation increases from left to right. As Dominik and Zarnecki (2011, 501) point out, the three "'astronomical factors' R^*, f_p, and n_e, are rather well understood compared to the 'biological factors' f_l and f_i, while the 'technological factor' f_c and, even more, the 'societal factor' L are the great unknowns." The purpose of this chapter is to examine what we might think about, and how we might think about it, in the hope of improving estimates of L.

In the years following Drake's presentation of his equation, numerous individuals have assigned a variety of values to L. Most are little more than guesses and not of particular interest here. Indeed, in a recent paper, Drake (2011, 637) himself indicated, "If you take plausible estimates and guesses for the factors in the Drake equation, as suggested by many SETI workers informally, and crudely estimate the number of detectable civilizations in our Galaxy, you arrive at a number of the order of 10,000. That is a big number. Ten thousand detectible civilizations, and by the way, that value depends on civilizations on average being detectable for 10,000 years or so, *which is little more than a guess*" (emphasis added).

Others, mostly astronomers and astrophysicists, have provided mathematical or statistical analyses or augmentations of the Drake Equation in an effort to increase its rigor and utility. These analyses have involved the development of mathematical methods to determine, for example, the possible distributions of variable values along with their error variances, the refinement of variables in the equation, or the addition of others. For the most part, however, these are useful only when estimates of the values of the variables are already available, so the estimates themselves remain conjectures.

Still others have discussed the trajectories and lifespans of civilizations, although not with the Drake Equation in mind (see Chapter 13 on pre-1961 considerations of the lifespans of civilizations). These include individuals with an interest, first, in the nature of "civilization" and, second, in what brings about its rise, lifetime, and ultimate demise. Unfortunately, in recent years, the "hard" scientists with their equations and the "soft" scientists with their histories appear to have had little communication with regard to how each might inform the other. While guesses are of little value, estimates for L derived from an examination of civilizations here on Earth or on a mathematical or statistical basis are of interest since they provide at least some empirical or logical foundation, however minimal, for estimating L. Therefore, one of my goals is to see what we might learn about L from the study of the only examples of civilizations we have, those that have existed, or currently exist, on Earth. This includes examining, first, how those estimates were developed since this information may be relevant to the evolution of extraterrestrial civilizations. A second goal is to examine how astronomers, astrophysicists, and others have analyzed the Drake Equations and, in some cases, extended it in ways that might make knowledge of the lifecycles of earthly civilizations useful in the determination of L. Finally, I will examine "existential risks" that currently face humanity as these may generalize to other possible civilizations in the Milky Way.

To pursue the first goal, we need to know several things. What is a "civilization" and how does it arise? What are some examples of civilizations and how long have they endured? Are there any common factors that precipitate the decline or collapse of civilizations?

It should be no surprise that definitions of civilization abound, and there is no clear consensus, especially across disciplines, on what the term means. The lack of a consensus definition of civilization compounds the problems of determining examples of civilizations, how long they last, or what contributes to their ultimate demise. With respect to L, we also have no information regarding civilizations other than those that have existed or currently exist on Earth. Moreover, the industrial-technological civilization that currently exists on Earth is unprecedented and, therefore, appears to be unique even among known civilizations. But, first, what is a "civilization?"

The concept of civilization

In 1871, Edward B. Tylor, a pioneer in the development of anthropology, defined culture as "that complex whole which includes knowledge, belief, art, law, morals, custom, and any other capabilities and habits acquired by man as a member of society" (Tylor 1871, 1). Notably, he began his definition with "Culture, or civilization" – that is, Tylor did not distinguish between culture and civilization, a common practice in the nineteenth century and one that is continued by some, but not others, to this day. Oswald Spengler (1880–1936) used a culture-based concept of civilization, claiming "Culture is the prime phenomenon of all past and future world history" (Spengler 1980, 104–5) and that all civilizations rise and fall in accord with a natural, but inevitable, cycle. Pitirim Sorokin (1889–1968) similarly regarded complex cultures or civilizations as massive assemblages of cultural phenomena (Sorokin 1950). Alfred L. Kroeber (1876–1960), one of the giants of American cultural anthropology during the first half of the twentieth century, chose not to distinguish cultures and civilizations: "Like many anthropologists, I use the word civilization almost synonymously with the word culture" (Kroeber 1957, 150). British historian Arnold Toynbee (1889–1975), in his celebrated twelve-volume *A Study of History* (1934–1961), described civilization as based in religion and identified twenty-six different civilizations, four of which he described as "abortive" and five others as "arrested." Four of the twenty-six remain in the twenty-first century, the Western, Islamic, Hindu, and the Far Eastern civilizations.

A problem with Toynbee's list of civilizations is that the demarcations of where one ends and others begin appear to be highly subjective, making the beginning and end points of the putative civilizations seem arbitrary. For example, he listed Sparta as separate (and as an "arrested civilization") from the rest of Hellenic civilization but categorized Roman civilization as Hellenic. As I pointed out in an earlier article (Chick 2014), it seems difficult to claim that the ancient Greeks and Romans were not influenced by earlier Babylonian culture. And, was Babylonian culture not influenced by even earlier Sumerian culture? Moreover, since much of the scientific knowledge of the ancient Greeks was maintained, augmented, and transmitted to Europe by the Arabs, how can their influence be discounted in claims for a Western civilization distinct from an Islamic civilization? While there is only contested evidence suggesting that New World civilizations, such as those of the Olmec, Aztec, Maya, or Inca, were influenced by Old World civilizations of Europe, the Middle East, or East Asia, it seems possible to trace our capability to both transmit radio signals from Earth and monitor ones that may originate elsewhere in the galaxy at least to the ideas and approaches to knowledge developed by the ancient Greeks and probably to earlier Sumerian civilization.

In his 1963 book, *The Rise of the West: A History of the Human Community,* historian William H. McNeill disputed Toynbee's and Spengler's ideas that civilizations developed independently and were subject to individual rise and fall. Instead, McNeill claimed that most of those that Toynbee regarded as distinct were connected through the trade of both materials and ideas. Political scientist David Wilkinson (1987) later claimed that an economic and military integration of Mesopotamia and Egypt around 1500 BC led to what he has termed the "central civilization." According to Wilkinson, the central civilization expanded over the ensuing millennia to include the Middle East and Europe and, via exploration and expansion, much of Africa, the Americas, South Asia, and East Asia. In his view, the central civilization is now global and about 3,500 years old. It is also the home of the technological capability of broadcasting, and listening, to the stars.

According to computer scientist and civilization theorist Andrew Targowski (2011, 83), early definitions of civilization described them as:

1. large societies extended in time and space;
2. culture-oriented;
3. religion-oriented;
4. having a rise, flowering, and decline.

In addition to culture, cities and urbanization are prominent in many recent definitions. In a definition prepared for a review by Targowski (2009), David Wilkinson (2009, 85) claimed, "A civilization is a city-state, cities-state, or tightly linked politico-military network of such states that are not a part of a larger such network." Similarly, and also composed for Targowski's 2009 review, Laina Farhat-Holzman, an historian, asserted, "A civilization must have a concentration of people in one or more urban areas. It must have (at a minimum) division of labor and specialization (people supported by the community to perform professional specialties), and it must have a surplus of food (wealth) to be used in support of such specializations (army, priesthood, centralized governance)" (Farhat-Holzman 2009, 87). Charles Redman (1978), an anthropologist, defined civilization as a complex network of relationships of components that operates at numerous levels originating from urbanization. Hence, cities serve as nexuses for the institutions and organizations that delineate a civilization. Mears (1996, 1) characterized civilizations as having "a high degree of societal complexity – a complexity sustained by hierarchically structured organizational mechanisms."

Targowski (2009) summarized civilizations as having the attributes of:

1. being large societies with labor specialization but sharing the same knowledge system;
2. occupying an extended space and time;
3. having a cultural system;
4. having an infrastructure that involves at least one of the following: urbanization, agriculture, industry, or information technology;
5. being cycle-driven, that is, experiencing a rise, growth, decline, and fall over time.

Based on these attributes, Targowski provided the following definition:

> Civilization is a large society living in an autonomous, blurry reification (invisible-visible) which is not a part of larger one and exists over an extended period of time. Labor is specialized and a civilization is differentiated from other civilizations by the development of its own advanced cultural system driven by communication, religion, wealth, power, and the sharing of the same knowledge system within complex urban, agricultural infrastructures, and other infrastructures such as industrial and information ones. It also progresses in a cycle of rising, growing, declining and falling. (2009, 94)

Targowski (2011, 75) then distinguished what he regarded to be "eight-well established religion-oriented civilizations and one business-oriented" civilization at the start of the twenty-first century. These include, "the Chinese (3,500 years old), Hindu (2,600), African (2,500), Eastern (2,325), Buddhist (1,400), Japanese (1,350), Western (1,200), Islamic (1,400), and Global (10+, business-oriented)."

Samuel P. Huntington, a political scientist who wrote the influential book, *Clash of Civilizations* (1996), offered the following definition: "A civilization is the highest cultural grouping of people and the broadest cultural identity people have short of that which distinguishes humans from other species. It is defined by common objective elements, such as language, history, religion, customs, institutions, and by the subjective self-identification of people" (Huntington 1996, 43). Huntington claimed, according to his definition, that nine civilizations exist in the post-1990 era. These are the Western, which includes Western Europe and North America, Latin American, African, Islamic, Sinic, Hindu, Orthodox (which includes Russia, Ukraine, Belarus, Bulgaria, and Moldavia), Buddhist, and Japanese.

Van Sloan, who identified himself as a "graduate of Princeton and Stanford Universities," in 1987 proposed a list of twenty-six "leading civilizations" through history. He divided these into three groups: Ancient, Middle Ages, and Modern. The twenty-six civilizations "represent mankind's highest overall achievement during each period. The time periods generally correspond to what is considered a civilization's golden age." He also listed the fifteen "top civilizations as well as fourteen that were important but did not achieve the highest rank." The nine societies Sloan categorized as Ancient averaged 519.8 years as the world's leading civilization. Those he categorized as the leaders of the Middle Ages ($n = 7$) lasted only 137 years, while those he categorized as Modern ($n = 10$) existed for only 71.4 years (although the American civilization, the most recent leader, still exists).

Even secondary educational organizations have entered the fray. According to the New York State Department of Education (Latasa 2008), for example, civilizations share eight features. These are:

1. cities;
2. complex religions;

3. social classes;
4. art and architecture;
5. organized central governments;
6. job specialization;
7. writing;
8. public works.

A problem evident in many of these definitions and sets of attributes of civilization is that the authors do not distinguish between culture, as an ideational system, and cultures, as phenomena. Tylor's (1871) definition of both culture and civilization, for example, is primarily ideational. However, conflation of the knowledge possessed by a group of people with the group itself only confuses, rather than clarifies, the nature of civilization.

Factors in the rise of civilizations

The rise of civilizations is generally associated with the end of the Neolithic and began in several areas of the world starting around 10,000 BC, approximately at the end of the last glacial period (Morris 2010). As archeologist Ian Morris (2010) indicates, while the poles remained cold and equatorial regions hot, the beginning of the current interglacial period opened up about a half-dozen areas around the world where warmer weather combined with local geography to produce plant and animal species that could be domesticated. Plant and animal domestication led to larger, although not necessarily more nutritious, food supplies that, in turn, led to increases in human populations. Professor of geography and physiology, Jared Diamond, in his award-winning book, *Guns, Germs, and Steel* (1997) provides a similar account of the rise and domination of Western societies over others. Hence, in accord with these views, the development of civilizations in several areas of the world were not due to any innate superiority of the people involved but a product of being in the right place at the right time. According to Diamond, societies shifted from hunting and gathering for subsistence to a relatively fixed agrarian mode when the local environment provided opportunities to do so. These opportunities included, first, the existence of grains and other vegetables that are high in protein and can be stored from one harvest to the next; second, a climate that is dry enough to permit such storage; and third, animals that can be domesticated and will reproduce in captivity. Domesticated animals large and strong enough to serve as beasts of burden were a bonus.

As populations rise, population density increases and permanent settlements develop. With permanent settlements, other foundations of civilization arise, including occupational specialization, full-time religious specialists, and bureaucrats, in addition to craftspeople, writing, urbanization, social stratification, full-scale agriculture, and political integration (Murdock and Provost 1973). I have explored the concept of cultural complexity and its history in detail elsewhere (Chick 1997) and recent studies, including experimental research (Caldwell and Millen 2010; Muthukrishna et al. 2013), support the idea that cultural complexity is linked to the size of social networks available to individuals. According to Muthukrishna and his colleagues (2013, 1), "technological sophistication may depend on sociality, on the size and interconnectedness of populations" (see also Kline and Boyd 2010).

Since each term in Drake's equation depends on the previous one having a positive value, L, "the length of time such civilizations release detectable signals into space," depends immediately on f_c, "the fraction of civilizations that develop a technology that releases detectable signs of their existence into space" (SETI Institute 2014). Therefore, the rise of civilizations is imperative for L to be meaningful (see Chapter 12 for a discussion of f_c), and the length of time between their rise and possible demise is critical for estimating L. In the last 10,000 years, many cultures, civilizations, and empires have risen, existed for varied periods of time, and fallen on Earth. Are there common factors that precipitated their demise and, if so, what are they?

Factors in the decline and fall of civilizations

Ibn Khaldun (1332–1406) argued that civilizations decline and fall because new powers tend to establish themselves on their peripheries and have greater social solidarity than the established societies. In his *Muqaddimah* (2005), Ibn Khaldun referred to community cohesion as *asabiyyah.* It is strongest in a people's nomadic period and decreases with the advance of civilization. Hence, their stronger *asabiyyah* allows peripheral societies to supplant the rulers of larger, presumably more powerful, empires (Tausch and Heshmati 2009). Rome, for example, was sacked by "barbarians" from the fringe of its empire. In more recent history, the British Empire has been supplanted by its former colony, the United States. Morris (2010) provides numerous additional examples from East Asia, the Middle East, and the West.

Edward Gibbon's *The History of the Decline and Fall of the Roman Empire*, published in six volumes between 1776 and 1789, is the most famous study of the most famous collapse of a civilization in history. Gibbon ascribed the fall of the Roman Empire largely to the decline in the civic character of Roman citizenry over time (Pocock 1976). That is, the previously strong and militaristic Romans had become weak and effeminate. Moreover, the rise of pacifistic Christianity in the empire furthered the turn away from militarism. Gibbon's analysis of the collapse of the Roman Empire was an early example of an enduring fascination with the decline and fall of complex civilizations among members of such civilizations. The fundamental reason for this interest, of course, is because if it happened to them, it could happen to us (Tainter 1988; Morris 2010).

Joseph Tainter, an archeologist, defined civilization in his well-known book, *The Collapse of Complex Societies* (1988, 41), as "the cultural system of a complex society." Moreover, "complex societies are problem-solving organizations in which more parts, different kinds of parts, more social differentiation, more inequality, and more kinds of centralization and control emerge as circumstances require" (1988, 37). Finally, "a society has collapsed when it displays a rapid, significant loss of an established level of sociopolitical complexity" (1988, 4). Tainter (1988, 42) claimed that the literature provides eleven primary themes as explanations for the collapse of civilizations. They are:

Figure 14.1 The Forum, dating to the eighth century BC, was in the middle of ancient Rome and served as a marketplace, center for the city's public and political life, and home to temples, statues, and other monuments. After the last Roman emperor was deposed in 476 AD, the metal clamps that held stone joints together were stripped, the stone and marble was reused for other building purposes, and the area became a pasture (Watkin 2009).
Credit: Garry Chick.

1. depletion or cessation of a vital resource or resources on which the society depends;
2. the establishment of a new resource base;
3. the occurrence of some insurmountable catastrophe;
4. insufficient response to circumstances;
5. other complex societies;
6. intruders;
7. class conflict, societal contradiction, elite mismanagement, or misbehavior;
8. social dysfunction;
9. mystical factors;
10. chance concatenation of events;
11. economic factors.

Tainter (1988, 74) discusses and critiques each of these themes in terms of specific examples from ancient Mesopotamia, Egypt, Mesoamerica, and the Roman Empire and suggests that some, such as "mystical factors," which include "decadence" or "senility," have no value as explanations as they rely on value judgments and intangibles. Others, such as "intruders," should be regarded, not as explanations themselves, but as circumstances to be explained. That is, why was the civilization unable to repel invaders and why would invaders "destroy a civilization worth invading in the first place" (1988, 89)? Tainter settles on an economic explanation as the ultimate causal factor in the collapse of civilizations. In particular, he was concerned with diminishing returns on investments in social complexity and provided statistics (pre-1988, of course) showing that the marginal returns on modern investment in energy, education, and technological innovation were in decline. He did allow that technological innovations may increase marginal productivity, at least in the short run, and possibly in the longer term, meaning that collapse is not absolutely inevitable.

William Ophuls, a former US Foreign Service officer and political scientist, appears to lack even Tainter's modest optimism. In his short book, *Immoderate Greatness: Why Civilizations Fail* (2012), he claims that civilizations are "hard-wired" for self-destruction due to two overarching factors. These are the biophysical limits of environments and human error. Biophysical limits include (1) ecological exhaustion, (2) exponential growth, (3) expedited entropy, and (4) excessive complexity, while human error involves (5) moral decay and (6) practical failure. The factors involved in biophysical limits refer, generally, to the idea of sustainability. Ophuls explains this succinctly:

> The city is an ecological parasite. It arrogates to itself matter and energy that do not naturally belong to it by sucking resources away from its hinterland. So, the central institution of civilization exists, and can only exist, by systematically exploiting its rural

and natural periphery. It is this exploitation that supports the higher level of social and economic complexity that characterizes civilization. (Ophuls 2012, 7)

According to Ophuls (2012, 7), damage to the environment is an important part of what brings down civilizations: "Thus every known civilization has caused environmental harm and ecological degradation to some degree." This is painfully familiar territory to the environmentally minded. Ophuls's second general category, human error, is more abstract and reminiscent of the work of Ibn Khaldun and Edward Gibbon. But Ophuls relies specifically on a short essay by Sir John Bagot Glubb (1978), also known as Glubb Pasha, a British soldier and scholar of the Middle East who examined the lifecycles of historical empires and claimed that they traverse six stages:

1. the Age of Pioneers;
2. the Age of Conquest;
3. the Age of Commerce;
4. the Age of Affluence;
5. the Age of Intellect;
6. the Age of Decadence. (Glubb 1978, 24)

According to Glubb,

Perhaps the most dangerous by-product of the Age of Intellect is the unconscious growth of the idea that the human brain can solve the problems of the world. Even on the low level of practical affairs this is patently untrue. Any small human activity, the local bowls club or the ladies' luncheon club, requires for its survival a measure of self-sacrifice and service on the part of the members. In a wider national sphere, the survival of the nation depends basically on the loyalty and self-sacrifice of the citizens. The impression that the situation can be saved by mental cleverness, without unselfishness or human self-dedication, can only lead to collapse. (Glubb 1978, 12)

Glubb claimed that the lifespan of empires tends to be about ten generations, or about 250 years. Moreover, the six-stage pattern appears to be independent of whether governments are democratic or despotic or of the civilization's technological capabilities. Ophuls characterizes Glubb's analysis largely as a demonstration of the human capacity for hubris. Moreover, the fate of civilizations is inevitable. We are not only doomed, ultimately, but well into our decline.

In a recent analysis, Samuel Arbesman (2011), of the Institute for Quantitative Social Science at Harvard University, used data on the lifetimes of forty-one empires that existed between approximately 3000 BC and 600 AD developed originally by Taagepera (1978; 1979). He found that their lifetimes fit an exponential distribution, $\lambda = e^{-\lambda t}$ where $\lambda = 0.46$, giving a mean lifetime of approximately 220 years. The question remains, however, whether any of these estimates are appropriate for our civilization, the only one in Earthly history capable of both communicating with the stars and blowing ourselves into oblivion.

Previous estimates of L and extensions to the Drake Equation

My second goal is to look at post-1961 estimates of L as well as alterations made to the equation itself. Initial estimates for L by Drake and his colleagues at the world's first meeting on the search for extraterrestrial intelligence, held in Green Bank, West Virginia, in November 1961, ranged from 1,000 to 10,000,000 years (Drake and Sobel 1992). These appear to have been based on little more than guesses. Other estimates for the number of communicating civilizations in the Milky Way based on the Drake Equation soon began to appear and ranged just as widely.

In a paper originally published in *Science* and later in a collection edited by A. G. W. Cameron, astronomer Sebastian von Hoerner derived equations to permit

Figure 14.2 Uxmal, a late classical period Mayan city located in the northwestern Yucatán Peninsula. The city is believed to have been the capital of a regional state between 850 and 950 AD with a population of as many as 25,000 at its height. Major construction ended around 1000 AD and the city was abandoned by the mid-fifteenth century (Mayan Ruins 2014).
Credit: Garry Chick.

the estimation of the most likely age of technical civilizations at the moment of contact but also noted that his analyses are straightforward up to the point where he made numerical values for the longevity of technical civilizations. These, along with the probabilities of the possible fates of such civilizations, for von Hoerner (1963, 275), "become a matter of personal opinion." He did, ultimately, arrive at an estimate for L of 6,500 years. However, he noted "It is purely a personal belief if we think that L will not be higher than, say, a million years and that it will probably be much less than that" (1963, 279). In the same book, A. G. W. Cameron (1963) offers a much more optimistic estimate of L of 10^6 years but, once again, one that is purely speculative.

Astronomers Iosef Shklovskii and Carl Sagan, in their classic *Intelligent Life in the Universe* (1966, 412), wrote, "The present technical civilization of the planet Earth can be traced from Mesopotamia to Southeastern Europe, to Western and Central Europe, and then to Eastern Europe and North America." This perspective differs from those of most of the "civilizationists" discussed above but is in line with the view of Wilkinson (1987) and the one that I took in an earlier essay (Chick 2014). They stated that, unfortunately, there is not a single terrestrial example that can tell us much about the lifetime of a "technical civilization possessing both the interest and the capability for interstellar communication" (Shklovskii and Sagan 1966, 412). But they also claimed, optimistically, that if a civilization perseveres for more one hundred years after reaching the communicative phase, "it will be unlikely to destroy itself afterwards" (1966, 413). Shklovskii and Sagan (1966, 314) provided two choices for L: "$< 10^2$ years and $>> 10^8$ years." Given their belief that there is a high number of technological civilizations, to start with, and not all will destroy themselves, they adopt L ~ 10^7 years as "an average for all technical civilizations" (1966, 413).

In his later book, *Cosmos* (1980), Sagan altered Drake's initial definition of L. Instead of Drake's "the length of time such civilizations release detectable signals into space," Sagan (1980, 299) defined L as "the fraction of a planetary lifetime graced by a technical civilization." Sagan discussed values between approximately a millionth of a percent and one percent, both of which give very low numbers of communicating civilizations in the Milky Way. His reformulation of L also makes comparing his values with those of others difficult since it made L a fraction rather than a whole number. However, Sagan speculated that the other values on the right side of the equation are high, which leaves N = L, a result that Drake had considered earlier (Drake and Sobel 1992).

Michael Shermer, editor of *Skeptic* magazine, disputed the claims of others that L is the most difficult of the variables in the equation to determine. For his own calculation, he compiled a database of

> the durations of 60 civilizations (years from inception to demise or the present), including Sumeria, Mesopotamia, Babylonia, the eight dynasties of Egypt, the six civilizations of Greece, the Roman Republic and Empire, and others in the ancient world, plus various

civilizations since the fall of Rome, such as the nine dynasties (and two republics) of China, four in Africa, three in India, two in Japan, six in Central and South America, and six modern states of Europe and America. (Shermer 2002, 33)

Shermer then calculated an overall average lifetime of these civilizations, resulting in a value of 420.6 years for L. The value for modern, technological civilizations (the twenty-eight since the fall of Rome) equaled 304.5 years. It should be pointed out, however, that the losses in cultural complexity following the demise of the civilizations (excepting those that remain in existence) in Shermer's sample were far from complete. That is, new civilizations that followed retained much, if not all, of the technology of earlier ones. Again, the problem of demarcating civilizations is apparent.

In a more mathematically sophisticated approach, astronomer Duncan Forgan (2008) used a Monte Carlo realization technique to estimate the distributions of variable values in the Drake Equation including error estimates. With respect to L, he posits, "Once a technologically capable civilization has formed, it must move through a 'fledgling phase': it is susceptible to some catastrophic event caused partially or fully by its own actions (e.g., war, plague, catastrophic climate change, botched macro-engineering projects)" (2008, 10). If a civilization survives this phase, it presumably moves to a stage of sufficient advancement where it can and will prevent such self-destructions. Parameters for Forgan are "the timescale for a civilization to move from 'fledgling' to 'advanced,' the probability that a fledgling civilization will destroy itself, and the lifetime of any signal or leakage from a civilization" (2008, 10).

Maccone (2008) derived a statistical version of the Drake Equation by applying the central limit theorem to equation parameters. The problem Maccone identified was that the equation did not account for error in the seven variables on its right side. He sought, therefore, to "transform the classical Drake equation into a statistical product of seven (or more) input random variables, each of which is assigned with its mean value and standard deviation, and each of which is supposed to be uniformly distributed since the uniform distribution has the largest uncertainty (entropy). Then, the number N of communicating extraterrestrial civilizations becomes a random variable also" (2008, 1). Maccone refers to his reformulation as the "Statistical Drake Equation."

Several authors, including Drake (e.g., Cirkovic 2004; Drake and Sobel 1992; Forgan 2008), point out that the Drake Equation, in its original formulation, fails to account for the possibility that the values of the parameters used to estimate N may change over time. Glade, Ballet, and Bastien (2012) propose stochastic processes that permit the modeling of evolutionary aspects of the equation parameters. However, while their application of Poisson stochastic process theory permits them to estimate the average time for new advanced civilizations to appear, it does so only if estimates for L are already available.

Prantzos (2013) proposed a framework for analyzing both the Drake Equation and the Fermi paradox jointly using a simplified version of the former and the strong version of the latter. Prantzos indicated that the product of all of the variables on the right side of the equation, with the exception of L, can be reduced to P, interpreted as the production rate of communicating civilizations in the Milky Way. While he presents estimates of the required lifetimes of civilizations that would permit them to communicate, he does not provide an actual estimate of L.

Astrophysicist J. Richard Gott developed an intriguing method for determining the lifespan of the human species, one that is clearly relevant to L. His technique is based on the idea that scientists generally accept hypotheses supported by data at high levels of confidence, usually at the 95 percent level. Based on the Copernican principle that there is nothing special about the Earth or about us, Gott assumed that we are living during the middle 95 percent of the time that humans, in our present form at least, will exist on the Earth. Modern humans appeared approximately 200,000 years ago. So, according to Gott in a 1999 interview for the *New Yorker* magazine,

> That's how long our past is. Two and a half percent is equal to one-fortieth, so the future is probably at least one-thirty-ninth as long as the past but not more than thirty-nine times the past. If we divide two hundred thousand years by thirty-nine, we get about fifty-one hundred years. If we multiply it by thirty-nine, we get 7.8 million years. So if our location in human history is not special, there's a ninety-five percent chance we're in the middle ninety-five percent of it. Therefore the human future is probably going to last longer than fifty-one hundred years but less than 7.8 million years. (Gott 1999)

Gott developed his argument based on a 1969 visit to the Berlin Wall and Stonehenge. At that time, the wall was eight years old and Stonehenge was estimated to be 3,868 years old. Gott assumed that he was a random observer, randomly located between the beginning and the end (meaning that either the items had been destroyed or there was no one left to observe them) of each. Since his argument (which he called "Delta *t*" [Gott 1993, 315]), asserts "that the length of time something has been observable in the past is a rough measure of its robustness not only against the calamities of the past, but also against whatever calamities may affect its observability in the future." All that is necessary is "that in the end *your* position as an observer turns out not to have been special." The fall of the Berlin Wall twenty-years later was within the 95 percent confidence interval predicted by Gott's technique as was the fact that Stonehenge was still observable in 1999. Gott (1999) also applied his technique to the determination of L. Since each of us was born into a civilization capable of transmitting signals across interstellar space, he indicated that we each represent a randomly chosen individual from the set of individuals in the set of transmitting civilizations. As he estimated the length of time that the capability has existed as 105 years, Gott calculated that there is a 95 percent probability that L is less than 12,100 years and the number of civilizations in the Milky Way with transmission capability is less than 121.

Gott's method has been challenged, however. Mathematicians and statisticians Anthony Ledford, Paul Marriott, and Martin Crowder (2001, 309) described Gott's technique, "If you observe that something has survived to age *X*, and provided that there is nothing special about the timing of your observation, then there is a 95% chance that its future longevity will be between *X*/39 and 39*X*."

Based on an examination of its underlying probability theory, Ledford, Marriott, and Crowder conclude that Gott's technique "either requires additional assumptions to be made, or is appropriate only for systems in which the chance of imminent failure decreases with time in a very particular way" (2001, 311). Similarly, Arbesman (2011), whose analysis of the lifespan of empires is described above, claimed that their longevity is independent of their current age, rendering use of the Copernican principle, and therefore Gott's (1993) method, inappropriate.

Existential risks

If we accept the Copernican principle, however, extraterrestrial civilizations should be subject to the same risks we face on Earth. Philosopher Nick Bostrom (2002, 7) refers to the most extreme of these as "existential risks," defined as those "where an adverse outcome would either annihilate Earth-originating intelligent life or permanently and drastically curtail its potential." Existential disasters would terminate human civilization by either eliminating humans or by so adversely affecting them that technological civilizations could not redevelop even in the future. Bostrom indicates that, with the exception of a major meteor or asteroid impact such as ended the age of dinosaurs, "there were probably no significant existential risks in human history until the mid-twentieth century" (2002, 10). Astronomer Martin Rees (2003, 25) writes:

> Throughout human history the worst disasters have been inflicted by environmental forces – floods, earthquakes, volcanoes and hurricane – and by pestilence. But the greatest catastrophes of the twentieth century were directly induced by human agency: one estimate suggests that in the two world wars and their aftermath, 187 million perished by war, massacre, persecution, or policy-induced famine.

Thus, existential risks come in two general varieties: nonanthropogenic, such as pandemics, volcanic eruptions, and asteroid strikes, and anthropogenic, including overpopulation, resource depletion, and atomic warfare. Existential risks place an upper limit on L.

The detonation of the first atomic bomb brought about the first human-made existential risk in that some thought that it might result in a runaway chain reaction.

While this turned out to be incorrect, a genuine existential risk appeared with the cold war expansion of nuclear arsenals in the United States and the Soviet Union. While the end of the cold war and subsequent reductions in nuclear arms are positive developments, more than enough nuclear warheads remain to keep their existential risk very real. Unfortunately, additional existential risks have emerged in recent years that may be even greater threats than nuclear war to the future of humankind and, in accord with the Copernican principle, to intelligent species that may exist in other parts of the galaxy. Doomsday prophesies have long been part of human history, with one of the most famous in recent years being the alleged Mayan prediction of either the end of the world or of human civilization with the completion of the thirteenth *baktun* (a cycle of about 400 years) on December 21, 2012. As we are aware, the world did not end, and in any case, the Mayans had never made any such prediction but only expected the rebirth of a new cycle of time (e.g., Webster 2007).

With the increase in anthropogenic risks from human technologies, such as the atomic bomb, speculation on both the number and kind of risks, as well as their likelihood, has become much more common since the turn of the twenty-first century. Again, if highly developed extraterrestrial civilizations face the same existential risks as we do, one way to approach my third goal for this paper is to examine and compare lists of such risks formulated by individuals, including Bostrom (2002), Rees (2003), and the astrophysicist Neil deGrasse Tyson (2014), and organizations, such as the Future of Humanity Institute, the University of Cambridge Centre for the Study of Existential Risk, and the Future of Life Institute. To do so, I have employed, with minor modification, a technique known as free listing.

Free listing is a data collection method commonly used in psychology, linguistics, and anthropology to elicit terms or concepts that populate cognitive, linguistic, and cultural domains (Borgatti 1998; Quinlan 2005). For example, one might ask a small sample of informants (typically twenty to thirty are sufficient) to list the names of all of the birds they can recall. Analyses involve the examination of the frequencies with which items are listed (e.g., in North America, informants will likely list "robin" more frequently than "kiwi," while informants from New Zealand might do the opposite) and the order in which they are listed (e.g., again, North American informants will likely list "robin" and "sparrow" before they list "kiwi" and "emu"). The salience of items can then be determined using Smith's S (J. Smith 1993). Smith's S is calculated as the frequency in which items are listed weighted inversely by the rank of the item in each informant's list.

I assembled the lists from articles or books by twenty-two authors or co-authors, the websites of eight institutions, and the Wikipedia article on Existential Risks as it appeared in 2014 when I wrote this chapter.[2] While some of these had risks in list form, in most cases, I had to develop lists using what I will term "embedded free lists," that is, lists that can be gleaned from texts. In some cases, items are located near each other, while in others they are dispersed in the texts, sometimes across the length of a book. Such lists require considerable data cleaning and interpretation because some individuals or institutions listed items at different levels of specificity or used different words to mean the same thing. So, some listed "terrorism" while others mentioned "bioterrorism," much as individuals listing birds might provide a more generic term, such as "eagle," or a specific species, such as "bald eagle." Hence, I combined things such as "water pollution" and "soil toxification" as "ecological degradation," and "solar disturbance," "nearby supernova," and "gamma ray burst" (which would likely come from a nearby supernova) as "energetic cosmic event." On the other hand, while I reduced "asteroid strike," "meteor strike," and "comet strike," to "asteroid strike," I did not combine them with "energetic cosmic event" as they seemed different both to me and to the authors of the lists. While this method prevents listing items mentioned only once or a couple of times, it requires my interpretations to be similar to those of the original authors.

After compiling the lists, I used Anthropac 4.92 (Borgatti 1992) for analysis. The results are presented in Table 14.1, with risks ordered in terms of how frequently they were listed. Several of the existential risks listed in the table require elaboration. While nuclear war is still seen as the major threat facing humankind, malign artificial intelligence was mentioned as frequently. Malevolent superintelligent computers have been a staple of science fiction (e.g., Arthur C. Clarke's HAL in *2001: A Space Odyssey* [1968]) but may be nearing reality. In a recent installment of the Science Channel program, *Through the Wormhole* (June 11, 2014), for example, host Morgan Freeman asked whether it is possible for the Internet to become self-aware and what might happen if it did.

Nanotechnology involves precision manufacture at molecular or atomic levels. While this will most likely be a boon to humankind, especially with respect to its potential use in medicine, it might also be used to make autonomous weapons or "gnatbot" types of surveillance systems used to control populations. The implications of malicious use of biotechnology, possibly resulting in a biopandemic, are similar. High-energy accidents involving particle accelerators include the possible creation of tiny black holes or the formation of stable "strangelets" that accrete matter, changing it to strange matter. The Earth could be destroyed in both scenarios.

I have kept "nuclear war" and "war" separate, even though the former is clearly a kind of the latter, primarily because they were discussed separately, sometimes by the same authors. However, conventional wars, while increasingly deadly in recent history, have never accounted for large percentage declines in the human population and are unlikely to do so. A nuclear exchange or the introduction of new forms of weapons, including chemical and biological agents, could change

Table 14.1 *Existential risks cited in lists developed by thirty-one individuals or institutions*

Rank	Existential Risk	Response Frequency	Response Percentage	Average Rank	Smith's S
1	Nuclear war	21	68	3.00	0.53
2	Malign artificial intelligence	21	68	4.74	0.36
3	Climate change	20	65	4.50	0.42
4	Pandemic	19	61	5.68	0.31
5	Asteroid strike	17	55	4.12	0.36
6	Nanotechnology	15	48	4.40	0.29
7	Supervolcano	10	32	4.80	0.20
8	Biotechnology	9	29	3.11	0.22
9	Ecological degradation	8	26	5.50	0.15
10	High energy accident	8	26	5.88	0.15
11	Energetic cosmic event	7	23	6.57	0.13
12	Worldwide famine	6	19	5.17	0.11
13	Hostile aliens	6	19	10.67	0.05
14	Resource depletion	5	16	10.00	0.05
15	Bioterrorism	4	13	3.25	0.08
16	Overpopulation	4	13	8.25	0.08
27	Unknown unknowns	4	13	6.25	0.06
18	War	4	13	10.50	0.05
19	Genetic transformation	4	13	11.00	0.05
20	Technological accident	4	13	7.75	0.03
21	Dystopia	4	13	11.00	0.02
22	Terrorism	3	10	3.00	0.07
23	Worldwide economic collapse		10	5.67	0.05
24	Cultural collapse	3	10	9.67	0.02
25	Cyberattack	2	6	2.50	0.05
26	Biopandemic	2	6	2.50	0.05
27	Chemical contamination	2	6	3.00	0.05
28	Bioweapons	2	6	3.00	0.05
29	Superearthquake	2	6	8.00	0.04
30	Ozone layer depletion	2	6	6.50	0.04
31	Loss of biodiversity	2	6	4.00	0.04
32	Energy failure	2	6	4.00	0.03
33	New ice age	2	6	9.50	0.03
34	Universe simulation halted	2	6	6.50	0.03

Table 14.1 (*cont.*)

Rank	Existential Risk	Response Frequency	Response Percentage	Average Rank	Smith's S
35	Migrations	2	6	6.00	0.02
36	Megatsunami	2	6	13.50	0.02
37	Earth's polarity reversal	2	6	15.00	0.02
38	Synthetic biology	1	3	1.00	0.03
39	Excessive complexity	1	3	4.00	0.02
40	Breakdown of natural cycles	1	3	7.00	0.02
41	Chemical weapons	1	3	4.00	0.01
42	Nuclear terrorism	1	3	8.00	0.01
43	Nemesis	1	3	17.00	0.01
44	Change in cosmic constants	1	3	18.00	0.01

that, however. I also kept bioweapons and chemical weapons separate, although they could reasonably be included under the more general category of "war." If the various forms listed are combined, war becomes the most commonly indicated existential risk.

"Technological accident" refers to things that could happen to extant technologies, including catastrophic breakdown of electrical grids or interconnected information systems, such as the Internet. This differs from "cyberattack" in that the latter is planned, either by terrorists or hostile governments. Resource depletion includes natural resources such as rare earths, increasingly important in the manufacture of electronics, as well as more common essentials, such as oil.

"Dystopia" refers to an Orwellian *1984*-type scenario wherein human progress is thwarted by governance failure, such as an indefinitely persisting totalitarian regime or, on the other hand, by long-term anarchy. Either of these could go hand-in-hand with worldwide economic collapse or culture collapse. Kroeber (1944) envisioned a version of the latter that he termed "cultural fatigue," that is, the view that the capacity for culture change is limited once a civilization is established. Possible examples of cultural fatigue include the rapid collapse of Aztec religion following the Spanish conquest and the decline of the Ohio Hopewell culture (Griffin 1952). Melko (1969) projected the end of Western civilization as a form of cultural fatigue that would follow from a widespread loss of interest in problem solving through technological innovation.

"Breakdown in natural cycles" designates the water, nitrogen, phosphorus, and other Earth cycles that are crucial to the maintenance of life. "Chemical contamination" of Earth could occur by design or by accident, while "energy failure" refers to a scenario where the cost of producing energy outstrips returns on investment. The primary theme of Tainter's 1988 study is that while disease, invasion, climate change, or degradation of the environment may have been proximate causes of the collapse of civilizations such as the ancient Maya, the Western Roman Empire, and Chaco culture of the American southwest, the ultimate cause was declining return on investments in cultural complexity. Tainter sees marginal returns on investments in education, energy production, and technology happening today. While "excessive complexity" was mentioned specifically only once as a present existential risk, several sources (e.g., Ophuls 2012) address it indirectly. Our world is increasingly complicated and interconnected, and accidental breakdowns or sabotage would have worldwide implications, particularly in systems with little or no redundancy.

Two risks exist only in the realm of theory. "Universe simulation stopped" deals with recent speculation that our universe exists only as a simulation on some form of computational device. If who- or whatever is running the simulation becomes bored with it, the result could be our demise. Finally, one author suggested possible risk from the hypothesized dwarf star or planet, Nemesis, sometimes called "Planet X." Allegedly, from its distant and eccentric orbit, Nemesis periodically disturbs objects in the Oort Cloud, sending comets toward the inner solar system, several of which collided with Earth in the past, causing mass extinctions. However, recent data from NASA's Wide-Field Infrared Survey Explorer (WISE) spacecraft produced no evidence of any such planet or companion star while locating approximately four thousand other, far more distant objects, including brown dwarfs and other stars (Redd 2014). Nemesis is evidently not our nemesis.

While many of the risks listed, including energetic cosmic events, such as a nearby supernova or the death of the sun, are either highly unlikely or far in the future, a recent study, funded by NASA's Goddard Space Flight Center, suggests that global civilization could collapse in the near future due to the unsustainable exploitation of resources and the unequal distribution of wealth (Motesharrei, Rivas, and Kalnay 2012). Similarly, members of the Cambridge Centre for the Study of Existential Risk, which includes cosmologist Stephen Hawking, astronomer Martin Rees, and evolutionary geneticist George Church, have expressed concern that human technology poses extinction-level risks to humanity. Fortunately, Motesharrei and his colleagues, among others, feel that avoiding such a collapse is possible should we come to understand the emergency and choose to overcome it. Such choices may not be planned, however. Kelly (2013) and others have pointed out that neither the population and resource crisis envisioned by Malthus (1798) nor the population explosion of the 1990s predicted by Ehrlich (1968) have happened.

The free listing exercise above should be viewed with some skepticism. First, from a methodological standpoint, the sample of sources is neither exhaustive nor randomly chosen. Hence, the results cannot be generalized to any population of speculations about existential risks. Second, while none of the lists are identical, several were likely influenced by the work of Bostrom, who has written extensively on existential risk, including the definition given earlier in this chapter from his seminal 2002 article. Finally, we should also consider one of the risks listed above, "unknown unknowns," a phrase made famous, or infamous, by then-US Secretary of Defense Donald Rumsfeld at a Department of Defense briefing in 2002. The nuclear bomb as an existential risk was an unknown unknown prior to Einstein's work on the equivalence of matter and energy and perhaps a known unknown until the detonation of the first atomic bomb in July 1945 in New Mexico. Similarly, although the Swedish physical chemist Svante Arrhenius proposed in 1896 that burning fossil fuels could result in a greenhouse effect (Maslin 2008), global warming, along with its causes and implications, has become a matter of public concern only recently and is still disputed by many, including some climate scientists. What this means is that the ensemble of known existential risks changes over time and will likely continue to do so in the future. Unfortunately, we have no better projections of which risk, or risks in combination, might ultimately end our civilization, or their probabilities of doing so, than we have estimates of L.

Factors in the estimation of L

In his 1941 science fiction classic *Nightfall*, Isaac Asimov described events on Lagash, a planet lit not by one, but six, stars. Residents of Saro City had never experienced the darkness of night nor seen the universe of stars beyond those that bathed them in constant daylight. The problem was that once every 2,049 years, an eclipse occurs that sends Lagash into darkness for more than half a day. And when it does, residents go mad, they burn their cities in desperation for the light from the fires, and their civilization collapses. On Lagash, the length of a civilization was therefore necessarily less than 2,049 years. However, determining a value for L depends not on a single, cyclic event as on Lagash but on multiple variables as described earlier in this chapter and in Chapter 13.

Where does all of this leave us? Ascertaining the lifespans of earthly civilizations depend on how "civilization" is defined. Statistical and mathematical approaches to L may help in some ways, such as shrinking the standard deviations around estimates, but, so far, are of little help in making the estimates themselves. Finally, if the existential risks that apply to Earth at present can be assumed to apply elsewhere in the galaxy, war, especially of the nuclear variety, malevolent artificial intelligences, climate change, pandemic, and several others may tell us why we have yet to hear from anyone. But we haven't been listening or sending for very long.

Radio waves emanating from the Milky Way were discovered by Karl Jansky in 1931, and Grote Reber constructed the first parabolic dish radio telescope in 1937 (National Radio Astronomy Observatory Archives 2014). Reber first detected extraterrestrial radio signals in 1938 (Spradley 1988). Marconi transmitted radio signals over a distance of about 1.5 miles in 1895 and claimed to have sent the first transatlantic radio transmission in 1901. However, these were likely too weak to penetrate the ionosphere and be detectable beyond Earth. Some think that the first electromagnetic signals created by humans, at a high enough frequency to penetrate the ionosphere and conceivably detectable beyond the Earth, could be the German television telecast of the 1936 Olympic Games opening ceremony in Berlin, featuring Adolf Hitler. This possibility is usually discounted because the transmission was low in power and nondirectional (Pomeroy 2013). The first intentional broadcast into space took place on November 16, 1974, at the Arecibo radio telescope in Puerto Rico and was aimed at the globular star cluster M13, some 25,000 light years distant (Cornell News 1999). Frank Drake and Carl Sagan, both participants in the Green Bank meeting in 1961, contributed to the design of the three-minute-long message. These figures place a lower boundary on the lifetime of civilizations capable of releasing detectable signals into space at something less than one hundred years, depending on which of our early transmissions may have actually escaped Earth. The upper bound remains the problem.

But how can it be estimated? While we have evidence of past civilizations on Earth, different experts (some self-appointed) give very different numbers, ranging between one and approximately sixty, of how many civilizations have existed on Earth. Indeed, the number of present civilizations is disputed. Estimates of the average lifespan of earthly civilizations range widely, as well, between as few as twelve years (Sloan's 1987 claim for an Austrian civilization) to approximately 5,500 years if we regard our own civilization as having descended from early Mesopotamian cities, such as Uruk (Morris 2010). While there is some agreement in terms of how a civilization should be defined (see, e.g., Targowski 2009), putting a definition into practice has been difficult. While lifespans in the 200- to 500-year range seem to be common, the estimated lifetimes of civilizations correlate inversely with the number of civilizations proposed by those reckoning their durations: the more civilizations demarcated, the shorter their lifespans.

Archaeologist and anthropologist Kathryn Denning (2013, 26) asked, "Do Earth-based examples really matter?" Indeed, as she pointed out, it may be that the whole concept of civilization is irrelevant. Astronomer Alan Penny (2013) suggested a novel alternative. He considered the possibility that the "conceptual thinking" required to develop and support a scientific civilization capable of emitting electromagnetic signals into space may involve genetic changes that could occur relatively abruptly and vanish just as quickly.

Four of the sources for the free listing exercise indicated genetic change, either through Darwinian processes or genetic engineering, as an existential risk. However, even punctuated equilibrium models of evolution (Ayala 2005; Gould and Eldridge 1977) suggest that punctuations involving species change require some 50,000 to 100,000 years, rapid in a geological sense but hardly so in human terms. Genetic engineering, on the other hand, is now commonplace, so the possibility of a biopandemic designed by sophisticated terror groups or deranged individuals seems increasingly real. Bioengineering ourselves may not be far off, as well.

Examinations of the collapses of previous civilizations suggest a set of factors common to most. Some of these were discussed above, but there are others. For one, some researchers have found a relationship between one important existential risk, climate change (specifically, a rise in temperatures) and human violence, including both interpersonal violence, such as murder, assault, rape, and domestic violence, and intergroup violence, including riot, warfare, and institutional breakdown to the point of the collapse of civilizations (e.g., Scheffran et al. 2012; Zhang et al. 2007). In a cross-cultural comparative study of twenty-seven modern societies, Hsiang, Burke, and Miguel (2012) found that all exhibited a positive relationship between warmer temperatures and increased violence. A change of one standard deviation in the direction of warmer conditions increased the likelihood of interpersonal violence by 4 percent and intergroup violence by 14 percent. On the other hand, others have found no such relationship (e.g., Tol and Wagner 2010).

Finally, the Earth-based information that we do have regarding the lifespans of civilizations fails to address our own, which remains the only one in the history of our planet with the technology to communicate with extraterrestrials. It seems likely that a society in possession of that capability, as well as the capacity to exterminate itself using weapons of mass destruction, is qualitatively different from those that lacked such technologies.

Mathematical and statistical efforts to enhance the Drake Equation suffer from the fact that they still require estimates of L, and such estimates remain little more than guesses. Collaboration between students of earthly civilizations and those who have enhanced the Drake Equation mathematically might lead to improved estimates of L. As shown in the free listing exercise above, there is considerable agreement among the sources sampled with respect to the most salient existential risks humanity currently faces. Understanding these risks – and predicting their likelihood – would be of tremendous help in developing estimates of L. At the very least, collaborations among historians, social scientists, astronomers, astrophysicists, mathematicians, and philosophers should raise the level of thought regarding the evolution of civilizations, the perils they face, and the methods required to understand and predict their fates. As Denning (2013) noted, realizing such collaboration would be a big project but that has not stopped SETI in the past.

References

Arbesman, Samuel. 2011. "The Life-Spans of Empires." *Historical Methods: A Journal of Quantitative and Interdisciplinary History* 44(3): 127–29.

Asimov, Isaac. 1941. "Nightfall." *Astounding Science Fiction* 28(1): 9–34.

Ayala, Francisco J. 2005. "The Structure of Evolutionary Theory: On Stephen Jay Gould's Monumental Masterpiece." *Theology and Science* 3: 97–117.

Borgatti, Stephen P. 1992. *ANTHROPAC 4.0*. Columbia, SC: Analytic Technologies.

Borgatti, Stephen P. 1998. "Elicitation Techniques for Cultural Domain Analysis." In *The Ethnographer's Toolkit, vol. 3*, ed. Jean J. Schensul and Margaret D. LeCompte, 1–26. Walnut Creek, CA: Altamira Press.

Bostrom, Nick. 2002. "Existential Risks: Analyzing Human Extinction Scenarios and Related Hazards. *Journal of Evolution and Technology* 9: 1–111.

Caldwell, Christine A., and Ailsa E. Millen. 2010. "Human Cumulative Culture in the Laboratory: Effects of (Micro) Population Size." *Learning and Behavior* 38: 310–18.

Cameron, A. G. W. 1963. "Future Research on Interstellar Communication." In *Interstellar Communication*, ed. A. G. W. Cameron, 309–15. New York: W. A. Benjamin.

Chick, Garry. 1997. "Cultural Complexity: The Concept and Its Measurement." *Cross-Cultural Research* 31: 275–307.

Chick, Garry. 2014. "Biocultural Prerequisites for the Development of Interstellar Communication." In *Archaeology, Anthropology, and Interstellar Communication*, ed. Douglas Vakoch, 203–26. Washington, DC: NASA History Series Publications.

Cirkovic, Milan M. 2004. "The Temporal Aspect of the Drake equation and SETI." *Astrobiology* 4: 225–31.

Clarke, Arthur C. 1968. *2001: A Space Odyssey*. New York: New American Library.

Cornell News. 1999. "It's the 25th Anniversary of Earth's First (and Only) Attempt to Phone E. T." *Cornell Chronicle*, Nov. 12. Available at www.news.cornell.edu/stories/1999/11/25th-anniversary-first-attempt-phone-et-0.

deGrasse Tyson, Neil. 2014. "The Immortals" (television series episode). In *Cosmos: A Spacetime Odyssey*. Cosmos Studios, Fuzzy Door Productions, and the National Geographic Channel.

Denning, Kathryn. 2013. "'L' on Earth." In *Civilizations Beyond Earth: Extraterrestrial Life and Society*, ed. Douglas A. Vakoch and Albert A. Harrison, 74–83. New York: Berghahn Books.

Diamond, Jared. 1997. *Guns, Germs, and Steel: The Fates of Human Societies*. New York: Norton.

Dominik, Martin, and John C. Zarnecki. 2011. "The Detection of Extra-Terrestrial Life and the Consequences for Science and Society." *Philosophical Transactions of the Royal Society A* 369: 499–507.

Drake, Frank. 2011. "The Search for Extra-Terrestrial Intelligence." *Philosophical Transactions of the Royal Society A* 369: 633–43.

Drake, Frank, and Dava Sobel. 1992. *Is Anyone Out There?* New York: Delacorte Press.

Ehrlich, Paul R. 1968. *The Population Bomb*. New York: Ballantine Books.

Ehrlich, Paul R., and Anne H. Ehrlich. 2013. "Can a Collapse of Global Civilization Be Avoided?" *Proceedings of the Royal Society B* 280(175). Available at http://rspb.royalsocietypublishing.org/content/280/1754/20122845.full.pdf+html.

Farhat-Holzman, Laina. 2009. "Civilization." In Targowski (2009, 87).

Forgan, Duncan H. 2008. "A Numerical Testbed for Hypotheses of Extraterrestrial Life and Intelligence." Available at http;//arxiv.org/pdf/0810.2222.pdf.

Freeman, Morgan, et al. 2014. "How to Collapse a Superpower" (television series episode). In *Through the Wormhole: With Morgan Freeman*. Silver Spring, MD: Discovery Communications.

Gibbon, Edward. 1776–88. *The History of the Decline and Fall of the Roman Empire*. 6 vols. London: W. Strahan and T. Cadell.

Glade, Nicolas, Pascal Ballet, and Olivier Bastien. 2012. "A Stochastic Process Approach of the Drake Equation Parameters." *International Journal of Astrobiology* 11: 103–8.

Glubb, Sir John Bagot. 1978. *The Fate of Empires and Search for Survival*. Edinburgh: William Blackwood & Sons.

Gott, J. Richard, III. 1993. "Implications of the Copernican Principle for Our Future Prospects." *Nature* 363,(6427): 315–19.

Gott, J. Richard, III. 1999. "How to Predict Everything." Interview by T. Ferris in *The New Yorker* 75(18; July 12): 35.

Gould, Stephen Jay, and Niles Eldredge. 1977. "Punctuated Equilibria: The Tempo and Mode of Evolution Reconsidered." *Paleobiology* 3: 115–51.

Griffin, James B. 1952. *Archeology of Eastern United States*. University of Chicago Press.

Hsiang, Solomon M., Marshall Burke, and Edward Miguel. 2013. "Quantifying the Influence of Climate on Human Conflict." *Science* 341: 1212.

Huntington, Samuel P. 1996. *The Clash of Civilizations and the Remaking of World Order*. New York: Simon & Schuster.

Ibn Khaldun. 2005. *The Muqaddimah: An Introduction to History*. Translated and introduced by Franz Rosenthal; abridged and edited by N. J. Dawood with a new introduction by B. B. Lawrence. Princeton University Press.

Kelly, Michael J. 2013. "Why a Collapse of Global Civilization Will Be Avoided: A Comment on Ehrlich and Ehrlich." *Proceedings of the Royal Society B* 280. Available at http://rspb.royalsocietypublishing.org/content/280/1767/20131193/.

Kline, Michelle A., and Robert Boyd. 2010. "Population Size Predicts Technological Complexity in Oceania." *Proceedings of the Royal Society B* 277: 2559–64.

Kroeber, Alfred L. 1944. *Configurations of Culture Growth*. Berkeley: University of California Press.

Kroeber, Alfred L. 1957. *Style and Civilizations*. Ithaca, NY: Cornell University Press.

Latasa, Marc. 2008. "Eight Features of Civilization." New York State Education Department, Global History and Geography Online Resource Guide. Available at www.p12.nysed.gov/ciai/socst/ghgonline/projects/features.html.

Ledford, Anthony, Paul Marriott, and Martin Crowder. 2001. "Lifetime Prediction from Only Present Age: Fact or Fiction?" *Physics Letters A* 280: 309–11.

Maccone, Claudio. 2008. "The Statistical Drake Equation." Presentation at the 59th International Astronautical Congress (IAC), Glasgow, UK, Sept. 29–Oct. 3.

Malthus, Thomas R. 1798. *An Essay on the Principle of Population*. London: J. Johnson.

Maslin, Mark. 2008. *Global Warming: A Very Short Introduction*. Oxford University Press.

Mayan Ruins. 2014. "Uxmal." Available at http://www.mayan-ruins.org/uxmal/.

McNeill, William H. 1963. *The Rise of the West: A History of the Human Community*. University of Chicago Press.

Mears, John. 1996. "Methodological Considerations for the Comparative Study of Civilizations." *Comparative Civilizations Review* 34: 1–10.
Melko, Matthew. 1969. *The Nature of Civilizations*. Boston: Porter Sargent.
Motesharrei, Sata, Jorge Rivas, and Eugenia Kalnay. 2012. "A Minimal Model for Human and Nature Interaction." Available at www.aosc.umd.edu/~ekalnay/pubs/lyapbredamsfinal_files/handy-paper-for-submission-2.pdf.
Morris, Ian. 2010. *Why the West Rules—For Now*. New York: Farrar, Straus and Giroux.
Murdock, George P., and Catherine Provost. 1973. "Measurement of Cultural Complexity." *Ethnology* 12: 379–92.
Muthukrishna, Michael, Ben W. Shulman, Vlad Vasilescu, and Joseph Henrich. 2014. "Sociality Influences Cultural Complexity." *Proceedings of the Royal Society B: Biological Sciences* 281. Available at http://rspb.royalsocietypublishing.org/content/281/1774/20132511.short.
National Radio Astronomy Observatory Archives. 2014. "NRAO Timeline." Available at www.nrao.edu/archives/Timeline/timeline.shtml.
Ophuls, William. 2012. *Immoderate Greatness: Why Civilizations Fail*. North Charleston, SC: CreateSpace Independent Publishing Platform.
Penny, Alan. 2013. "The Lifetimes of Scientific Civilizations and the Genetic Evolution of the Brain." In *Civilizations beyond Earth: Extraterrestrial Life and Society*, ed. Douglas A. Vakoch and Albert A. Harrison, 60–73. New York: Berghahn Books.
Pocock, J. G. A. 1976. "Between Machiavelli and Hume: Gibbon as Civic Humanist and Philosophical Historian." *Daedalus* 105(3): 153–69.
Pomeroy, Ross. 2013. "Will Hitler be the First Person that Aliens See?" *Real Clear Science*, September 20. Available at www.realclearscience.com/blog/2013/09/will-hitler-be-the-first-person-that-aliens-see.html.
Prantzos, Nikos. 2013. "A Joint Analysis of the Drake Equation and the Fermi Paradox." *International Journal of Astrobiology* 12: 246–53.
Quinlan, Marsha. 2005. "Considerations for Collecting Freelists in the Field: Examples from Ethnobotany. *Field Methods* 17: 1–16.
Redd, Nola Taylor. 2014. "Where's Planet X? NASA Space Telescope Discovers Thousands of New Stars, But No 'Nemesis.'" Space.com, March 7. Available at www.space.com/24989-planet-x-search-nasa-telescope.html.
Redman, Charles L. 1978. *The Rise of Civilization*. San Francisco: W. H. Freeman.
Rees, Sir Martin. 2003. *Our Final Hour*. New York: Basic Books.
Sagan, Carl. 1980. *Cosmos*. New York: Random House.
Scheffran, Jürgen, Michael Brzoska, Jasmin Kominek, P. Michael Link, and Janpeter Schilling. 2012. "Climate Change and Violent Conflict." *Science* 336: 869–71.
SETI Institute. 2014. "The Drake Equation." Available at http://www.seti.org/drakeequation/.
Shermer, Michael. 2002. "Why ET Hasn't Called." *Scientific American* 287(2): 33.
Shklovskii, Iosef S., and Carl Sagan. 1966. *Intelligent Life in the Universe*. San Francisco: Holden-Day
Sloan, Van. 1987. "The Old Age of the United States." Available at www.sq.4mg.com/ageUS.htm#z/.
Smith, Horace. 1818. "Ozymandias." *The Examiner 527*, Feb. 1, p. 73. London: John Hunt. www.potw.org/archive/potw192.html.
Smith, Jerome J. 1993. "Using ANTHROPAC 3.5 and a Spreadsheet to Compute a Freelist Salience Index." *Cultural Anthropology Methods Newsletter* 5(3): 1–3.
Sorokin, Pitirim. 1950. *Social Philosophies of an Age of Crisis*. London: Adam & Charles Black.

Spengler, Oswald. 1980. *The Decline of the West*. Translated by Charles F. Atkinson. New York: Knopf. Orig. pub. 1932.

Spradley, Joseph L. 1988. "The First True Radio Telescope. *Sky & Telescope* 26(1): 28–30.

Taagepera, Rein. 1978. "Size and Duration of Empires: Growth-Decline Curves, 3000 to 600 B.C." *Social Science Research* 7: 180–96.

Taagepera, Rein. 1979. "Size and Duration of Empires: Growth-Decline Curves, 600 B.C. to 600 A.D." *Social Science History* 3: 115–38.

Tainter, Joseph. 1988. *The Collapse of Complex Societies*. Cambridge University Press.

Targowski, Andrew. 2009. "Towards a Composite Definition and Classification of Civilization." *Comparative Civilizations Review* 60: 79–98.

Targowski, Andrew. 2011. "The Civilization Approach to Education in the 21st Century." *Comparative Civilizations Review* 65: 73–104.

Tausch, Arno, and Almas Heshmati. 2009. "*Asabiyya:* Re-Interpreting Value Change in Globalized Societies." Institute for the Study of Labor Discussion Paper No. 4459. Bonn. Available at http://ideas.repec.org/p/iza/izadps/dp4459.html.

Tol, Richard S. J., and Sebastian Wagner. 2010. "Climate Change and Violent Conflict in Europe Over the Last Millennium." *Climatic Change* 99(1–2): 65–79.

Toynbee, Arnold J. 1934–61. *A Study of History*. 12 vols. Oxford University Press.

Tylor, Edward B. *Primitive Culture*. London: Murray, 1871.

von Hoerner, Sebastian. 1963. "The Search for Signals from Other Civilizations." In *Interstellar Communication*, ed. A. G. W. Cameron, 272–86. New York: W. A. Benjamin.

Watkin, David. 2009. *The Roman Forum*. Cambridge, MA: Harvard University Press.

Webster, David. 2007. "The Uses and Abuses of the Ancient Maya." Presentation at the Emergence of the Modern World Conference, Otzenhausen, Germany, September 25. Available at http://anth.la.psu.edu/documents/Webster_GermanyMaya.pdf.

Wilkinson, David. 1987. "Central Civilization." *Comparative Civilizations Review* 17: 31–59.

Wilkinson, David. 2009. "Civilization." In Targowski (2009, 85).

Zhang, David D., Peter Brecke, Harry F. Lee, Yuan-Qing He, and Jane Zhang. 2007. "Global Climate Change, War, and Population Decline in Recent Human History." *Proceedings of the National Academy of Sciences* 104: 19214–19.

Notes

1 Ozymandias was a name sometimes used to refer to the Egyptian pharaoh Ramesses II (ca. 1303 BC–1213 BC). Shelley's version of *Ozymandias* appeared in the January 11, 1818, issue of *The Examiner,* while Smith's appeared in the same publication two weeks later in the February 1, 1818, issue (H. Smith 1818).

2 To save space, I have not included the list of sources or the data. I will supply these upon request, however.

Afterword

Paul C. W. Davies

Abstract

In the five decades since Frank Drake formulated his eponymous equation, our understanding of astrophysics and planetary science has advanced enormously. The first three terms of the equation refer to factors that are now known with reasonable precision, due in no small part to the discovery of enough extrasolar planets for meaningful statistics to be developed. Unfortunately, this progress has not been matched by a similar leap in understanding of the remaining factors – the biological ones. In particular, the probability of life emerging on an Earth-like planet, f_l, remains completely unknown. In the 1960s and 1970s, most scientists assumed that the origin of life was a freak event, an accident of chemistry of such low probability that it would be unlikely to have occurred twice within the observable universe. Today, however, many distinguished scientists express a belief that life will be almost inevitable on a rocky planet with liquid water – a "cosmic imperative," to use the evocative term of Christian de Duve. But this sentiment is based on little more than fashion. One may assign a probability to a process only after the mechanism that brings about that process is known. As we have little knowledge of how a nonliving mix of chemicals is transformed into a living thing, we can say almost nothing about its likelihood. Indeed, it is easy to imagine plausible constraints on the chemical pathway to life that would make its successful passage infinitesimally small. In the case of the fifth term in the Drake Equation – the probability that intelligence will evolve if life gets going – at least we have a well-understood mechanism (Darwinian evolution) on which to base a probability estimate (though it still remains deeply problematic). The same is true of the remaining terms. Thus, the uncertainty in the number of communicating civilizations in the galaxy, N, is overwhelmingly dominated by f_l. The uncertainty

The Drake Equation, ed. Douglas A. Vakoch and Matthew Dowd. Published by Cambridge University Press.

would be immediately removed, however, if a second, independent sample of life was discovered. The best current hope for establishing a value of f_l close to one would be the discovery of a "shadow biosphere" on Earth or a breakthrough in understanding the rules governing the organization of information in complex chemical networks.

Habitable versus inhabited

A perennial problem in astrobiology is the conflation of a necessary with a sufficient condition. NASA's "follow the water" mantra was widely misinterpreted to imply that if an extraterrestrial body had liquid water, then life there was likely. The reasoning, often stated explicitly, is that wherever there is liquid water on Earth there is life. (Actually, it's not true: water is liquid under pressure on the ocean floor even up to temperatures of 350 degrees Celsius on account of volcanic vent outflow, but there is no evidence of life above 125 degrees Celsius.) Therefore – so the flawed reasoning goes – where there is liquid water *beyond* Earth, there should be life there too. While it may be true that terrestrial organisms, if transported to the said extraterrestrial body, might survive there, it is a completely different matter to assert that life should have arisen *de novo* there. Nobody supposes that life on Earth has arisen *de novo* in all the places it is found; rather, it arose in one place and spread to all available niches that possessed liquid water, for the simple reason that Earth's biosphere is a contiguous system. But in the absence of a panspermia mechanism, the same cannot be said of extraterrestrial bodies; they are biologically quarantined.

This blindingly obvious reasoning, often sidestepped in discussions, amounts to the statement that water is a necessary condition for life (as we know it), but it is far from sufficient. And what is true of water is true for the broader but more diffuse concept of "Earth-like." In the important hunt for Earth-like extrasolar planets, astronomers are busy cataloging habitable real estate across the galaxy. The qualification "Earth-like" is admittedly vague: Beyond liquid water, it may include the requirement of a magnetic field, a stabilizing moon, a hot interior to drive plate tectonics, and so on. Nevertheless, it is clear that our galaxy alone contains millions if not billions of worlds that are Earth-like in some respect, and thus potential abodes for life. This prospect is very exciting for astrobiologists. However, the qualification "Earth-like" may well be a necessary condition for life to arise, but it is far from sufficient. "Earth-like" refers to a setting, not a process. To get life on an Earth-like planet, all the necessary physical and chemical steps have to happen, and as we don't know what those steps are, we are in the dark as to how many habitable planets do, in fact, host some form of life.

All too often, commentators slide sneakily from talking about "habitable" to discussing "inhabited" with scarcely a beat in between. Many examples may be found from the flurry of excitement that surrounded a 2013 guestimate of 8.8 billion habitable planets in the Milky Way, based on an analysis of data from NASA's Kepler satellite. NBC News carried this statement on its website: "As for what it says about the odds that there is life somewhere out there, it means 'just in our Milky Way galaxy alone, that's 8.8 billion throws of the biological dice,' said study co-author Geoff Marcy, a longtime planet hunter from the University of California at Berkeley" (Borenstein 2013). Marcy's comment succinctly identifies the key issue with his eloquent phrase "throws of the biological dice." In the context of the article, given most people's familiarity with games of dice, the statement makes it appear that, in the vast inventory of suitable planets, you don't need to go too far down the list to throw that combination of numbers that signals, "Life!" But suppose that the pathway from nonlife to life requires odds comparable to throwing a six, twenty times in a row? Odds like that are certainly not unreasonable when it comes to a specific sequence of difficult chemical reactions. Then, Marcy's 8.8 billion "throws of the biological dice" stand a chance of only about one in 3,000 of yielding the dream run needed to have just one other planet in the Milky Way with life.

Let me now dispose of another prevalent fallacy. Over recent decades, we have discovered that life on Earth can survive, or even flourish, under a much wider range of conditions than believed hitherto. For example, microbes can be found making a living at temperatures well below freezing, or up to more than 120 degrees Celsius. Others can tolerate very low or very high pH, or extremely saline conditions. Some microorganisms can be subjected to massive doses of radiation and remain viable. This has considerably widened the range of extraterrestrial bodies on which terrestrial life could survive, and thus increases the total number of habitable planets in the universe by, say, an order of magnitude. The resilience of terrestrial life to harsh conditions is often cited as a reason why we can expect life beyond Earth, even in such a hostile environment as Mars or Europa. This reasoning, while correct as far as it goes, does not seriously alter the argument in the preceding paragraph. If the odds of life forming from nonlife are infinitesimal, then ten times infinitesimal won't make a difference to the basic point of contention. In the "dice-throwing" example, it matters little whether there is one chance in three thousand or one chance in three hundred for life existing elsewhere in the Milky Way. In addition, it should be pointed out that the conditions necessary for life as we know it to survive may be quite different from the conditions needed for it to arise in the first place. It is likely that the cradle of life was very constrained physically and chemically, but that once life got going it was able to adapt and spread to a much broader range of habitats.

The probability of life emerging on an Earth-like planet is formalized by the fourth term in Drake's equation, f_l. Drake himself favors a number close to unity. That is, given an Earth-like planet, it is very likely that life will arise there. In Chapter 7, for example, Stéphane Tirard discusses the wide disagreements over the years concerning the value of f_l. In his book *Chance and Necessity*, Jacques Monod famously argued that it is infinitesimal and concluded, "Man at last knows that he is alone in the unfeeling immensity of the universe, out of which he emerged only by chance" (Monod 1972, 167). Conversely, Christian de Duve, writing two decades later, deemed life to be "a cosmic imperative," almost bound to occur wherever conditions were suitable (de Duve 1995). Unfortunately, these disagreements were based almost entirely on philosophical judgments rather than scientific evidence, for the simple reason that science remains largely silent on the specifics of the pathway from nonlife to life. Although expressed as a fraction, the term f_l, is a probability. One may estimate the odds of a process occurring only if the process is known. One cannot estimate the odds of an unknown process.

The "up-it-pops" fallacy

One argument often used for why the origin of life might be a rather probable affair (i.e., f_l is not very much less than one) is that life established itself on Earth fairly rapidly once conditions became favorable. During the epoch known as the Late Heavy Bombardment period 4.0–3.8 billion years ago, Earth's surface was mostly (but not always) hostile to life because of a succession of massive sterilizing impacts. Yet there is good evidence for microbial life by 3.6 billion years ago. Carl Sagan concluded from this that "the origin of life must be a highly probable affair; as soon as conditions permit, up it pops!" (Sagan 1995, 1).

While it is certainly the case that the rapid appearance of life on Earth is consistent with its genesis being probable, it is equally consistent with it being exceeding improbable, as was pointed out by Brandon Carter over three decades ago (Carter 1983, 347). The essence of Carter's argument is that any given Earth-like planet will have a finite "habitability window" during which life might emerge and evolve to the level of intelligence. On Earth, this window extends for about 4 billion years – from about 3.8 billion years ago to about 800 million years hence, when the Sun will be so hot that the planet will be an uninhabitable furnace. Suppose, reasoned Carter, life's origin is so improbable that the expectation time for it to occur is many orders of magnitude longer than this habitability window. And further suppose that, in addition to the (improbable) transition from nonlife to life, several other very hard steps are needed before intelligence is attained (for example, eukaryogenesis, sex, multi-cellularity, evolution of a central nervous system). If in all, there are n hard steps, each of which has an expectation time much longer than the habitability window, and each of which is necessary before the next step may be taken, then a simple statistical argument leads to a relationship between n and the duration of the window. Specifically, the window can be divided into n + 1 roughly equal intervals (roughly, because this is a statistical argument), with the final interval being the duration between now (the epoch of intelligent life on Earth) and the fiery armageddon 800 million years hence. Given that the intervals are roughly equal, and given the 800 million year figure, Carter is able to conclude that there are about five extremely improbable steps involved in attaining intelligent life on Earth. But the main force of his argument as far as this essay is concerned is that the first step is bracketed by a similar interval of 800 million years. That is, if the emergence of life was an exceedingly improbable process (but of course one that had to happen for humans to be here and ponder it) then probability theory predicts it should have happened fairly rapidly – within about 800 million years. Another way of expressing it is that, unless life had got going quickly, we would not be here to discuss it three billion years later.

A curious corollary of Carter's argument is that the duration between the improbable steps is roughly the same irrespective of just *how* improbable they are, so long as they are all so improbable that the expectation time for them to occur is far longer than the habitability window. Let me take a concrete example. Suppose the expectation time for life to form is a billion trillion years. What this means is that, in a fictitious world in which conditions on Earth remain the same for a billion trillion years, life would very likely arise there at some stage in that (long) window. Alternatively, it means that among a random sample of one trillion Earth-like planets, life would likely arise on one of them every billion years or so. Now, suppose that the expectation time for intelligence to evolve, once life is established, is much less – say, a mere billion billion years. That is, the probability for intelligence to evolve, while extremely small, is nevertheless a thousand times greater than the probability for life to emerge from nonlife. Then you might imagine that the first step – the emergence of life – would on average take a thousand times longer than the evolution of intelligence, since the latter is a thousand times less improbable. But in the subset of planets on which observers arise to witness the outcome of these improbable events, it is not so: both these steps are predicted by Carter's argument to take about the same time to happen, that is, about 800 million years.

A brief digression: if the first step took 800 million years, it would imply that Earth was habitable at least as long ago as 4.3 billion years (as we can trace life back 3.5 billion years). But many astrobiologists doubt that conditions on Earth were conducive for life prior to 3.8 billion years ago, on account of the bombardment. However, their doubts are not definitive. Life may have started on Mars as long as 4.3 billion years ago and come to Earth later in impact ejecta. Or it may have started on Earth that early and gained a refuge in solar orbit to ride out the sterilizing effects of the bombardment. Our understanding of early life is in fact quite consistent with Carter's 800 million year figure.

The reason why Carter's argument seems so counterintuitive is because we are used to thinking about probabilities of future events given present uncertainties. But in the case of intelligent life, events are *post*selected. That is, scientists attempt to use the fact of the existence of an intelligent species on Earth now (humans) to *retrodict* about past probabilities. And our intuition collapses. We need to imagine that our present existence entailed extremely improbable past steps, which nevertheless occurred, against huge odds, and without which we would not be here to reflect on those antecedent events necessary for our existence.

To summarize Carter's powerful line of reasoning, the rapid appearance of life on Earth is equally consistent with two scenarios. One of these is that the critical steps leading up to intelligence are all fairly likely, so that if Earth is typical, there should be very many other planets with intelligent life (the SETI optimist's scenario). The other is that there are several critical steps, each of which is so unlikely that the expectation time for them to occur is very much longer than the age of the Earth; in which case life will be exceedingly rare in the universe (the SETI pessimist's scenario). Given a sample of only one, we cannot discriminate between them.

How, then, may we make progress?

Improved understanding of life's origin

Perhaps we can guess a plausible value of f_l by studying the chemistry that underlies life? Attempts to re-create the chemical pathway from nonlife to life have been pursued since the pioneering work of J. B. S. Haldane (1929) and Oparin (1924) in the 1920s, and were boosted by the famous experiment of Stanley Miller in 1952 (Miller 1953). The basic idea of this work is that by attempting to take the first footsteps along the road to life we might gain insights into how easy or hard that process might be. After many decades of prebiotic synthesis work, the conclusion is that some essential organics (e.g., amino acids) form readily, others (e.g., some nucleotides) are not easy to make, and cellular structures form rather naturally. The general belief is that, in addition to the molecular building blocks, some sort of stable autocatalytic chemical cycles are critical to drive primitive metabolism. While such cycles may be observed in living cells and contrived in the lab, their spontaneous emergence in nature in a *pre*biotic setting seems very difficult to imagine.

These mixed results can be interpreted either way. The pessimistic view is that life is so complex that the results of prebiotic synthesis go only a tiny way down the long pathway to life and tell us little about potentially extremely hard chemical obstacles at later stages. On the other hand, the optimist may point toward the immense periods of time available for prebiotic chemistry to work its magic.

There is, however, a more serious issue lurking here. Life is clearly more than complex chemistry. Chemistry deals with concepts such as molecular shapes and binding strengths, reaction rates and thermodynamics. By contrast, when biologists describe life, they use terms like signals, coded instructions, digital error correction and editing, and translation – all concepts from the realm of information processing. While chemistry provides the substrate for life, the informational aspects (which are instantiated in the chemistry) require an origin story of their own. In a nutshell, prebiotic chemical experiments help us understand how the hardware of life *might* have come to exist, but so far have cast little light on the origin of the software aspect. Indeed, it is hard to see how an explanation for information processing and organization can ever emerge from an account cast in terms of molecules and forces, for these descriptions of life belong to completely different conceptual categories. So while we may conjure up ideas about prebiotic settings on Earth – or elsewhere in the universe – that could serve as incubators of biologically important molecules, we remain largely in the dark about how this molecular stuff can spontaneously organize into coded information management systems (Walker and Davies 2012). Although we can recognize and potentially quantify patterns of information flow in chemical networks, and study how in living systems information feedback and control constrain chemical pathways, the basic laws of information processing and their coupling to chemistry remain a subject in its infancy. Because life requires both specialized hardware *and* specialized software to come out of as-yet little-understood physical processes, we are very far from being able to quantify the likelihood of getting both in a plausible molecular soup. While prebiotic experiments and theories are an important line of investigation, they are a long way from illuminating the key question before us, which is whether $f_l << 1$ or not.

A second sample of life

If we were to be presented with a second sample of life, and we could be sure that it arose independently of known life, then the case would instantly be made that f_l is not infinitesimally small. To encounter evidence for *two* independent origins of life, while not totally conclusive, would nevertheless be strong evidence in favor of the "cosmic imperative" – that the transition from nonlife to life is not highly improbable, and therefore the universe should be teeming with life.

Various scenarios have been suggested for the discovery of a second sample of life.

SETI succeeds!

In the event that mankind obtains incontrovertible evidence of the existence of alien technology, then we could conclude that life must have arisen in at least one other location in the universe (Davies 2010a). (This conclusion assumes, of course, that the pathway to technology involves biology and intelligence. Logically, there is no reason why this has to be the case, but it is the default assumption.) Note that the conclusion would follow even in the absence of an actual message or signal from an alien civilization – the "gold standard" of SETI. It would be sufficient to discover signs of technology, such as nuclear waste dumped in the solar system, a Dyson sphere, or any other large-scale modification of an astronomical system. The most likely scenario, however, is that any putative evidence would be subject to many alternative natural explanations and would probably be totally convincing taken in isolation.

Synthetic biology

The burgeoning field of synthetic biology, in which new forms of life are engineered in the laboratory, is held by some to suggest that life is literally easy to make, and that it may manifest itself in a wide range of molecular forms. Although synthetic biology currently falls far short of constructing living organisms from scratch (as opposed to rewiring or reprogramming existing organisms), one may imagine that in the future this will be possible. Would we then conclude that the transition from nonlife to life is not especially difficult and therefore likely to be widespread in nature?

The answer is no. Creating life in the laboratory will demand a great deal of sophisticated scientific equipment, a host of purified substances, and a particular sequence of chemical and physical steps, each of which is likely to take place under tightly controlled conditions; indeed, under *different* conditions for each step. But above all, creating life in the laboratory entails the attentions of an intelligent designer – the scientist – who embarks on the venture with a particular end product in mind and a well thought-out sequence of steps to attain it. This element of teleology is absent in nature, where the necessary physical and chemical steps have to take place spontaneously, in the rough and ready environment of a planetary surface (presumably), without the attentions of a trained scientist, and without the end state of "life" being preprogrammed into the process. So it may turn out to be relatively easy (if expensive) for scientists to make life, but that does not mean it is also easy for nature to do so.

To give a clear and highly simplistic illustration, suppose the scientist discovers that life can be made in ten easy stages. Stage 1 (chemical reaction 1, say) requires a temperature of 10±2 degrees Celsius for one hour, stage 2 a temperature of 20±5 degrees Celsius for three hours, stage 3 a temperature of 5±2 degrees Celsius for two hours, and so on. Now ask on how many planets this sequence of temperatures will concatenate in the necessary order for the necessary durations. Although each individual step may not be very improbable, the product of all the steps taken together may reduce the overall probability to a very small number. Take into account that not just temperature, but pressure, pH, density, purity, and many other variables may also require fairly stringent bounds at each step, and that it may take not just ten steps but a hundred or a thousand, and it is easy to believe that the whole sequence may have been completed only once in the observable universe. So even if the individual chemical reactions are straightforward, their accumulated product may be exceedingly rare. Synthetic biology will undoubtedly cast light on the nature of the problem, but it simply does not follow that if humans can create life then it must be a probable affair.

Life on Mars

The best hope for finding life in the solar system seems, by common consent, to be on the planet Mars. Although life on the surface of Mars looks problematic (but perhaps not hopeless), in the subsurface (at a depth of 200 meters or more) where water is likely to be liquid because of the planet's internal heat, there may exist briny reservoirs that play host to microorganisms resembling those that inhabit the deep subsurface on Earth. A future mission to Mars may succeed in culturing these putative organisms and transporting them back to Earth for analysis. The significance of the discovery would hinge, however, on whether it could be established that we really were encountering on Mars a genuine second sample of life emerging from a second genesis independently of the one that produced terrestrial life. The problem is that Mars and Earth are not biologically quarantined from each other. Over the history of the solar system, these two planets have traded a prodigious amount of material. The cause is the continual bombardment of the planets by comets and asteroids, some of which strike with enough force to splatter rocks around the solar system. Cocooned inside a rock, a microbe would be shielded from the harsh environment of interplanetary space, and might remain viable for thousands or even millions of years – in any case, for long enough to make the journey from Mars to Earth or vice versa. The existence of many known Mars meteorites demonstrates that rocky ejecta can arrive on another planet relatively unscathed, and the same could be assumed about any hitchhiking organisms. Given this traffic of material over billions of years, it seems very likely that if life were to have arisen on Mars, it would very soon be transported to Earth, to seed our planet. The same could be said of terrestrial life seeding Mars. The resemblance between the two planets before about 3.5 billion years ago, when Mars was warm and wet and had a thick atmosphere, implies that such planetary cross-contamination was all but inevitable (Davies 1996; 2010a; 2010b). So finding life (past or present) on Mars would not of itself demonstrate a second genesis.

Extrasolar planets

The complication described above vanishes in the case that we find evidence for life outside the solar system. The probability of a rock blasted off Earth (or Mars) ever hitting an Earth-like planet outside the solar system is extremely low, and in any case the duration of the journey across interstellar space is likely to be far longer than the survival times of the ensconced organisms against irreversible radiation damage. However, establishing the presence of life on an extrasolar planet is challenging. Instruments capable of detecting biologically associated atmospheric gases are being planned, but it may be decades before we have that capability.

The best immediate hope for establishing that life is a "cosmic imperative" lies much closer to home.

A shadow biosphere on Earth

If, in accordance with the current fashion of thought, life does form readily in Earth-like conditions, then we might expect it to have started many times on Earth itself. After all, Earth is the most Earth-like planet we know. It therefore makes sense to seek evidence for a second form of life on our home planet. All known life on Earth is interrelated, with a common genetic code, and a common biochemical scheme involving the same suite of nucleotides and amino acids, the manufacture of proteins by ribosomes, and several other specific universal features. This commonality enables biologists to organize all known species, from microbes to humans, on a single tree of life and extrapolate back to a common origin. The discovery of just a single microorganism so biochemically distinct from known life (i.e., so alien) that it could not belong to this familiar tree would be powerful evidence for an independent genesis. It would be immaterial whether the precursors of known life – call it Life A – and of the second form of life – Life B – both arose on Earth, or both on another body (e.g., Mars) and came to Earth, or whether A and B originated on different planets and became mixed together. All that matters is that life started more than once.

The question before us, then, is whether there exists, or has existed in the past, a sort of "shadow biosphere" on Earth, consisting of alien (in the sense of having a radically different biochemistry) organisms cohabiting with familiar terrestrial organisms (Davies and Lineweaver 2009; Benner et. al 2009). Could Life B exist all around us, intermingled with Life A? As it is extremely unlikely that biologists would have overlooked alien multicelled life on Earth, the denizens of the putative shadow biosphere would presumably be microbes only. In fact, almost all known species are microbial, and at the present time scientists have only scratched the surface of this microbial realm. The vast majority of microorganisms have yet to be classified, let alone cultured and sequenced. Thus, there is plenty of room at the bottom for microbes biochemically weird enough to qualify for Life B.

How might one go about searching for Life B? Easiest would be if A and B were sufficiently different that they preferred distinct biological niches. In that case, Life B would stand out because it would be found in habitats beyond the reach of all currently known life. For example, archaea living around the scalding effluent of deep ocean volcanic vents seem to be restricted to regions in which the temperature is below about 130 degrees Celsius. If microbes were detected living in regions in the temperature range, say, 160–180 degrees, then they would invite closer scrutiny. To generalize this concept, known life occupies a connected region of parameter space, spanned by temperature, pressure, pH, salinity, radiation, heavy metal concentration, and size. If microorganisms – preferably an ecosystem – were identified occupying a disconnected region of this parameter space, then it would be strong evidence in favor of Life B. In addition to deep-sea volcanic vents, Life B could be sought in high UV environments such as mountaintops or the upper atmosphere, in heavily polluted rivers and lakes, and in deep subsurface rocks where the pore spaces are too small to accommodate microorganisms with ribosomes (Benner et. al. 2009).

It would be much harder if Life A and Life B are intermingled in similar habitats, especially if B were present at very low density. The problem is then to establish the separate identities of A and B. You can't normally tell by looking what makes a microbe tick, biochemically speaking. You have to analyze its innards. Without a way to filter out A, B might continue to lurk inconspicuously under our very noses. A possible filter would be either physical conditions or a chemical substance customized to kill, or at least metabolically arrest, all Life A. Anything that continued to metabolize would deserve closer scrutiny. Historically, the bacterium *Deinococcus radiodurans* was discovered this way, being the only organism left alive after the administration of massive doses of radiation. Although that particular organism definitely belongs to the domain of Life A, the discovery does establish a filter protocol that might serve as a strategy to seek Life B. A pilot study along these lines was conducted by Hoover and his colleagues (Pikuta et. al. 2006) using a nutrient medium of reversed chirality organics. Life A uses predominantly left-handed amino acids and right-handed sugars, so it generally finds right-handed amino acids and left-handed sugars unpalatable. By contrast, if Life B were "mirror life," utilizing organics of the opposite chirality, they could be expected to thrive in the reversed-chirality medium. In the pilot, organisms were discovered that were able to metabolize the medium successfully, but they turned out to belong to the domain of Life A. Evidently the chirality story is more subtle than simply accepting one form and rejecting another. But the principle of the experiment as a discriminator and filter remains a good one.

Should any of the foregoing stratagems identify a likely candidate, it may prove hard to establish that it really does belong to Life B, that is, that it really is a descendant of an independent second genesis. The issue would hinge on just how biochemically different it is from known life. In the example cited above – mirror life – the case would be fairly convincing, as it is difficult to envisage an achiral common ancestor lineage bifurcating into opposite chiralities after life itself got going. By contrast, an organism that used the same set of nucleotides and amino acids, but a different genetic code, would not be convincing, because one could easily imagine a simpler precursor code common to all organisms that evolved into separate coding assignments. One would thus be dealing with two widely divergent branches on a single tree of life, with a single origin, and not two separate trees. In between these two examples is a sliding scale of differentness, and it is likely that a great deal of careful analysis would need to be carried out to establish the claim that a second sample of life had been found.

It is possible that life did begin many times on Earth, for example, starting anew in each quiescent period between successive sterilizing impacts between about 4.0 and 3.8 billion years ago (Maher and Stephenson 1988; Sleep et. al. 1989). Because those same impacts would propel prodigious quantities of material into solar orbit, some of which would return to Earth after millions of years, a natural mechanism for reseeding our planet (and others) exists. Perhaps Life A began in one of these quiescent windows, and was then expunged from the Earth's surface by the heat pulse of a giant impact, but survived in rocky ejecta to return to Earth much later, by which time Life B would have got going. Life A and B would then be cohabitants. This cycle of renewed genesis followed by reseeding of preceding microbial forms may have occurred many times in succession. The default assumption seems to be that only one genesis (the last?) left descendants that survived to the present epoch, but we have no reason to assume that. The argument that the Darwinian struggle for survival would produce only one "winner" is not convincing. Archaea and bacteria, which are genetically very distinct, have peacefully coexisted in similar habitats on Earth (e.g., in the human gut); nobody suggests that one form of microbe should have eliminated the other after billions of years of Darwinian struggle. And in the example that Life A and B occupy different niches, direct competition would not occur anyway. All this adds up to a strong case for seeking evidence of a shadow biosphere on Earth, which has the advantage over, say, looking for life on Mars, in being a lot simpler and cheaper.

Conclusion

The Drake Equation contains a string of terms spanning astronomy, physics, chemistry, evolutionary biology, and sociology. Each comes with error bars. Some of those error bars have narrowed dramatically in recent years, most obviously those pertaining to the number of Earth-like planets in the galaxy. The uncertainty in N, the number of communicating civilizations in the galaxy, remains dominated, however, by the error bars on f_l. With just a single sample of life at our disposal, we can conclude almost nothing about whether the formation of life from nonlife on an Earth-like planet is exceedingly improbable, implying that life is very rare in the universe, or almost inevitable, making life a "cosmic imperative."

How, then, may this problem be solved? In the absence of SETI succeeding, the best hope for progress on the experimental front is to find a second sample of life, a sample so biochemically distinct that it could not share a common ancestor with known life, and would therefore represent a second tree of life stemming from an independent genesis. Although it is natural to seek such a second sample on other astronomical bodies – Mars being the favorite – it is not inconceivable that we may discover it right here on Earth as a sort of shadow biosphere.

On the theory front, advances in understanding the flow and management of information in chemical networks, and the laws that govern these processes, may enable us to use computational models to estimate the probability of the "software" aspect of life emerging spontaneously. If there are basic laws of informational organization that apply generally across a wide class of chemical systems, it would be strong evidence in favor of a value of f_l not very much less than one. In that case, SETI would be given an enormous boost.

References

Benner, Steven, Paul C. W. Davies, Carol E. Cleland, Charles H. Lineweaver, Christopher P. McKay, and Felisa Wolfe-Simon. 2009. "Signatures of a Shadow Biosphere." *Astrobiology* 9: 241–49.

Borenstein, Seth. 2013. "8.8 Billion Habitable Earth-Size Planets Exist in Milky Way Alone." Available at www.nbcnews.com/science/space/8-8-billion-habitable-earth-size-planets-exist-milky-way-f8C11529186/.

Carter, Brandon. 1983. "The Anthropic Principle and Its Implications for Biological Evolution." *Philosophical Transactions of the Royal Society of London A* 310: 347–63.

Davies, P. C. W. 1996. "The Transfer of Viable Micro-Organisms between Planets." In *Evolution of Hydrothermal Ecosystems on Earth (and Mars?)*, ed. Gregory Bock and Jamie Goode. Proceedings of the CIBA Foundation Symposium 20. New York: Wiley.

Davies, P. C. W. 2010a. *The Eerie Silence*. London: Penguin.
Davies, P. C. W. 2010b. *The Origin of Life* London: Penguin.
Davies, P. C. W., and C. H. Lineweaver. 2005. "Searching for a Second Sample of Life on Earth." *Astrobiology* 5: 154–72.
de Duve, C. 1995. *Vital Dust: Life as a Cosmic Imperative*. New York: Basic Books, New York.
Haldane, J. B. S. 1929. "The Origin of Life." *The Rationalist Annual* 148: 3–10.
Maher, Kevin, and David Stephenson. 1988. "Impact Frustration of the Origin of Life." *Nature* 331: 612–14.
Miller, S. L. 1953. "A Production of Amino Acids under Possible Primitive Earth Conditions." *Science* 117: 528–29.
Monod, Jacques. 1972. *Chance and Necessity*. Translated by A. Wainhouse. London: Collins.
Oparin, A. I. 1924. "Proiskhozhdenie zhizny" ["The Origin of Life"]. Translated by Ann Synge in *The Origin of Life*, ed. J. D. Bernal, 199–234. London: Weidenfeld and Nicolson, London, 1967.
Pikuta, E. V., Richard B. Hoover, Brig Klyce, Paul C. W. Davies, and Pauline Davies. 2006. "Bacterial Utilization of L-Sugars and D-Amino Acids." *Proceedings of SPIE's 47th Annual Meeting*, San Diego, p. 63090A (August).
Sagan, Carl. 1995. "The Abundance of Life-Bearing Planets." *Bioastronomy News* 7(4): 1–4.
Sleep, Norman, Kevin J. Zahnle, James F. Kasting, and Harold J. Morowitz. 1989. "Annihilation of Ecosystems by Large Asteroid Impacts on the Early Earth." *Nature* 342: 139–42.
Walker, S. J., and P. C. W. Davies. 2012. "The Algorithmic Origins of Life." *Journal of the Royal Society Interface* 10, doi: 10.1098/rsif.2012.08

Index